EARTH AND MAN
A SYSTEMATIC GEOGRAPHY

EARTH AND MAN
A SYSTEMATIC GEOGRAPHY

Donald Steila Douglas C. Wilms Edward P. Leahy
East Carolina University

JOHN WILEY & SONS
New York • Chichester • Brisbane • Toronto • Singapore

Photo Research by Teri Leigh Stratford.
Photo Editor, Stella Kupferberg.

Library of Congress Cataloging in Publication Data:

Steila, Donald, 1939-
  Earth and man, a systematic geography.

  Includes bibliographies and index
  1. Geography—Text-books—1945- I. Wilms,
Douglas C., joint author. II. Leahy, Edward P.,
joint author. III. Title.
G128.S73 1981   910   80-19689
ISBN 0-471-04221-8

Printed in the United States of America

10 9 8 7

PREFACE

GEOGRAPHY IS AN EXCITING FIELD OF STUDY. ITS ORIGINS are at least as old as the ancient Greeks to whom we owe its name, and its current applications aid scientists and others in the resolution of modern social and environmental problems. As a discipline, geography is concerned with the earth and man. It is concerned with the patterns of interaction between humankind and the environments it inhabits. This interaction gives rise to spatial relationships—systematic arrangements of human and physical elements—that give character to area. This text focuses on these systematic arrangements.

Earth and Man: A Systematic Geography is written in a traditional format using contemporary illustrations. The major theme of the book is man-land relationships. This theme, established in the first chapter, is followed throughout. The book is adaptable to a variety of teaching strategies. It can be used for a one-semester or one-quarter introductory course, or for a two-semester sequence emphasizing physical landscape in one course and human landscape in the other. Additional readings can supplement the text for a two-semester sequence course.

Chapter 1 is designed as an introductory view of evolutionary forces, both physical and cultural, that give character to an area. Chapters 2 to 7 deal with the processes and elements of the physical environment—the atmosphere, global climates, soils and vegetation, and landforms—and their distribution over the earth. Chapters 8 to 16 deal with the human environment—population, culture traits, agriculture, extractive industries, energy, manufacturing, trade and transportation, and urban and political geography. Chapter 17, the final chapter, evaluates the role of people as occupants of the earth, examining cases of environmental deterioration and mankind's living in disharmony with nature.

Each chapter is followed by a list of key terms and concepts. These also appear in the glossary at the end of the text. Each chapter contains a set of study questions designed to help the student to review material presented in the chapter discussion. These are followed by a list of references for further study. Visual aids are widely used. Maps, graphs, tables, and photographs illustrate the text and amplify ideas presented throughout the book, while inserts are presented that elaborate on ideas, concepts, and personalities. In this regard we thank the many organizations that permitted the use of copyrighted material.

The purpose of this text is to acquaint the beginner with the principal themes of geography and to encourage the student to develop an appreciation and understanding of the discipline. In many instances, the material within each chapter is representative of material covered more fully in upper level college and university courses in geography departments throughout the country. It is our hope that students wishing to pursue an interest in these topics will avail themselves of these in-depth courses.

The metric system has been used throughout, with English units given in parentheses. This, we believe, is the most practical method of presenting data until readers are more familiar with the metric system. A number of conversion factors are given below.

(The Peoples Republic of China) is referred to as China, and Nationalist China is called Taiwan.

The word "man" as it appears in the title and text of this book is used in its generic sense, referring to the entire human family of men, women, and children. No sexist meaning is intended.

METRIC CONVERSIONS

Length
1 millimeter = 0.03937 inches
1 centimeter = 0.3937 inches
1 meter = 39.37 inches (3.28 feet)
1 kilometer = 0.62137 miles

Area
1 square meter (m^2) = 1550 square inches
1 hectare = 10,000 m^2 = 2.471 acres
1 square kilometer (km^2) = 0.3861 square miles

To Convert	Multiply by
Square kilometers to square miles	0.3861
Square miles to square kilometers	2.589998
Kilometers to miles	0.6214
Miles to kilometers	1.60935

Weight (Mass)
1 gram = 0.03527 ounces
1 kilogram = 2.2046 pounds
1 metric ton = 2204.6 pounds

Capacity (Volume)
1 liter = 1.0567 liquid quarts

Temperature
°C = 5/9 (°F −32)
°F = 9/5 (°C) +32

In 1979 the U.S. Board on Geographic Names approved Pinyin spelling of Chinese names. Pinyin spellings are the official Romanized forms of the Chinese language and now supersede the Wade-Giles system. The following table shows those letters and letter combinations whose sound values in the Pinyin scheme are different from English sound values.

Pinyin	English
c	ts
q	ch
x	sh
z	dz
zh	j

The Pinyin system of Chinese place names is used throughout the book, followed by the Wade-Giles system in parentheses. Thus, a reference to the capital of China appears as Beijing (Peking). For clarity, however, Zhonghua Renmin Gongheguo

We are grateful to our students and colleagues whose questions throughout the years generated the ideas presented here. To these individuals a debt of gratitude is owed; the responsibility for error rests with the writers. We wish to thank the many geographers whose work we have drawn on and apologize if the constraints of space have caused us to oversimplify their ideas. We also thank the many geographers who reviewed the manuscript and offered advice, criticism, and suggestions. Special appreciation is due to the following individuals who gave freely of their time to review portions of the manuscript: Charles Gritzner, Philip Shea, Charles Ziehr, and to Peter Fricke who wrote the section on "Ownership of the Oceans."

Finally, we thank Irving Cooper, geography editor at Wiley, for guiding our work to completion.

DONALD STEILA
DOUGLAS C. WILMS
EDWARD P. LEAHY

CONTENTS

1

THE FOCUS OF GEOGRAPHY
Understanding the Character of Area

WHY THE EARTH'S SURFACE EXISTS IN ITS PRESENT state is the focus of the geographer's attention. Each landscape is unique, a product of billions of years of evolutionary history. Long before man arrived to imprint his cultural and economic peculiarities upon the surface, nature was shaping the landscape, producing various colored soils, and evolving vegetation forms.

Geography is involved with the interrelationships that give character or personality to area, and includes interactions occurring within both cultural and physical environments. It is probably true that some forms of geographic analysis have been practiced since *Homo sapiens* first appeared approximately one million years ago. Before the disciplines of biology, chemistry, geology, mathematics, and physics came into being, man was analyzing the components of his habitat, speculating as to why certain combinations of environmental elements existed, determining the area over which they extended, and putting this information into graphic form. In other words, he was *practicing geography*. Undoubtedly, early hunters established mental associations regarding their prey's feeding, sleeping, and territorial behavior patterns. In doing so, they identified distinct areas where selected game could be found. Subsequently, the primordial hunter organized his life style in relation to the availability of these food supplies and to other survival necessities such as water and shelter. Later, when man estab-lished communities, constructed shelters, and expanded the use of resources, a cultural dimension was added to the landscape—a factor fostering the evolution of a new environmental personality.

There is little question that geography is an ancient field of study. Yet, it is also very modern. Today's local, national, and international governmental planners deal with an "analysis of area" using an approach much different from that of the primitive hunter, but the analysis remains similar in its goals. The city planner is a practitioner of geography and is capable of analyzing urban growth in relation to both the potential and constraints of the physical and socioeconomic environments. Similarly, national governmental agencies attempt integrative planning programs to enhance regional and interregional development in order to maintain viable economic systems.

A concern for *area* is implicit in all geographic studies. An individual landscape may be unique, but can share common characteristics with other landscapes. This recurrence of landscape elements provides a unifying factor to areas of the earth and distinguishes them as separate entities. The common element may be one of numerous factors including: economic activity, such as dairying or wheat farming; cultural elements, such as language or religion; or physical traits—mountainous terrain or desert vegetation for example. Areas of the earth having such homogeneously distributed components are called *regions*.

THE REGIONAL CONCEPT

The world is complex and overwhelming in size. In subdividing a large area into a comprehensible size, the geographer creates a *region*: an area defined and delimited on the basis of specific criteria that provide it with homogeneity. Regions may be of any size and may vary from the simple to the complex. Each region possesses some degree of internal unity that distinguishes it from adjacent regions. Regions, however, do not exist in reality; they are mental constructs, or intellectual tools, designed to bring meaning and understanding to the myriad of phenomena on the earth's surface.

The most common and best known regions are *uniform*, or *formal* regions. These are areas characterized by one or more predetermined criteria—attributes that are uniform throughout a given region. A nation-state is a political region; the Piedmont is a physiographic region; the Corn Belt is an agricultural region; a metropolis is an urban region; and the tropical rainforest is a climatic-vegetative region.

The second category of regions includes those areas functionally organized on the basis of linkage and circulation. These are known as *functional* or *nodal* regions because they usually have a center or node to which the region is tied. A city is a node connected to its surrounding trade area by roads, newspaper circulation, and various modes of telecommunications. Goods, services, people, and ideas move in and out of the city over these routes of circulation. Nodal regions—more limited in size—include a church, which acts as a node for the parish it serves, and a school as a node for a local school district. Some functional regions do not necessarily focus upon a central place. A river basin is an example.

Geographic analysis normally proceeds through one of two methods. The regional method requires the assumption that some part of the earth contains homogeneous characteristics; it then proceeds with an analysis of the area's component factors to explain the interrelationships giving the region identity. The *systematic method* involves an analysis of the variation of individual landscape elements, whether cultural, economic, or physical. These elements are examined separately to determine which variables, if any, contribute toward areal cohesiveness. The systematic approach is employed in this book.

UNDERSTANDING THE CHARACTER OF AREA

"Rome was not built in a day," is an old adage frequently used to stress the importance of time in creating anything of splendor or complexity. Most of this book emphasizes processes and characteristics that are viable parts of today's landscapes. It is important to keep in mind, however, that what exists today is an evolutionary product whose origins may go back to the beginning of time. A knowledge of the past is important to understanding the present. For this reason, a brief historical sketch of our planet is provided in the remainder of this chapter. The sketch will provide a framework for understanding the surface of the earth as it is today.

Planet Earth

The earth, together with the other members of the *solar system*, is believed to have originated approximately 5 billion years ago (Fig. 1.1).* Although debate continues over the means by which planets are formed, a widely accepted hypothesis suggests that they were created by the compaction of clouds containing cosmic dust and gas. According to some theorists, incipient planets formed within these clouds as cosmic matter became concentrated at locations of pronounced turbulence. Perhaps millions of years later, the accumulated masses of matter experienced sufficient heating by both gravitational compression and radioactivity to cause them to become molten. The present analysis of the earth begins just prior to its cooling and solidification through the loss of heat to outer space.

During the earth's molten stage its internal structure was evolving. Heavy elements were settling into the core of the fluid mass, while lighter

---

* Earth's oldest known rocks, found in Greenland, have been dated at 3.8 billion years. Isotopes have been discovered, however, that push the date back to 5.1 billion years ago, a figure generally rounded off to 5 billion years.

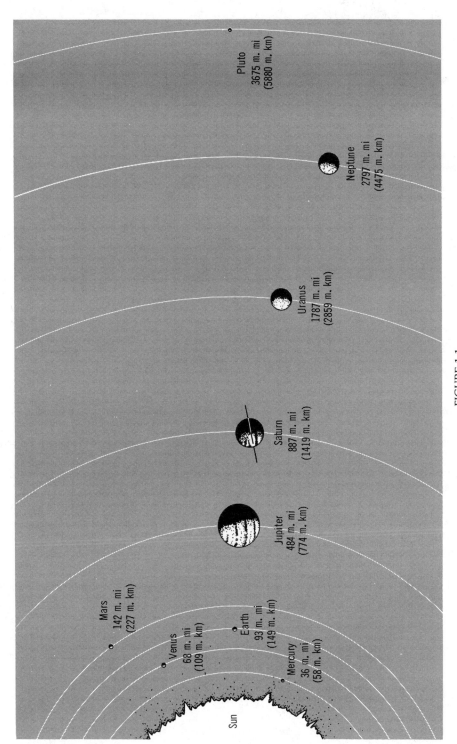

FIGURE 1.1

THE SOLAR SYSTEM. The earth is a member of the solar system, a celestial family composed of **a star (the sun)** and nine major planets plus smaller bodies. Most of the planet's orbits lie nearly in the same plane **(except for Pluto's orbit, which is highly eccentric)** and each planet revolves about the sun in the same direction.

UNDERSTANDING THE CHARACTER OF AREA

3

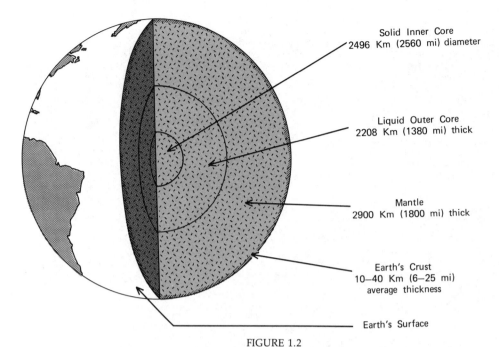

Solid Inner Core
2496 Km (2560 mi) diameter

Liquid Outer Core
2208 Km (1380 mi) thick

Mantle
2900 Km (1800 mi) thick

Earth's Crust
10—40 Km (6—25 mi)
average thickness

Earth's Surface

FIGURE 1.2

STRUCTURE OF THE EARTH. The *core*, composed of nickel and iron, is most likely solid in its innermost part; however, it is believed that the outer part is in a liquid state. The *mantle* is largely dense, magnesium-iron silicate rock. Surrounding the mantle are the lighter aluminum and magnesium silicate rocks of the earths *crust*.

matter was rising toward the surface. The resultant effect is that the earth is comprised of distinct layers, like the skins of an onion (Fig. 1.2). Each layer, however, has a unique composition. There are three such layers: (1) a very dense partly molten *core*, approximately 7000 km (4300 mi) in diameter, consisting mainly of metallic iron and nickel; (2) a *mantle* of dense rock matter, roughly 2900 km (1800 mi) thick, enclosing the core; and (3) a thin *crust* of lighter rock that varies in thickness (Fig. 1.3). The layers provide a gross diameter for the globe of about 13,000 km (8,000 mi) and a maximum circumference of approximately 40,000 km (24,860 mi). As shown in Figure 1.4, however, these average distances cannot be uniformly applied since the earth is not a perfect sphere.

The earth's crust has a very uneven surface. Large continents rise abruptly above the oceanic crust, and contrast with a succession of depressions (the ocean basins) that contain the planet's primary water bodies. These physical features are readily discernable to remote sensing satellites and astronauts orbiting the earth (Fig. 1.5).

So far as is known, the earth is the only planet in the solar system that contains large standing bodies of water. Our oceans and seas cover ap-

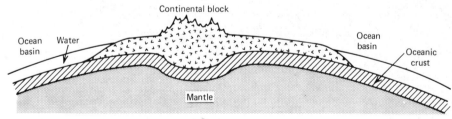

Continental block

Ocean basin
Water

Ocean basin

Oceanic crust

Mantle

FIGURE 1.3

THE EARTH'S CRUST. Continental land masses range from 20 to 60 km (12 to 36 mi) in thickness. This *continental crust* rests on denser earth materials that also extend across the earth, beneath large expanses of water, where it is known as *oceanic crust*. Oceanic crust is about 10 km (6 mi) thick.

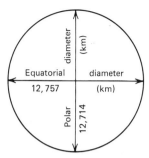

FIGURE 1.4

THE SHAPE OF THE EARTH. Because of greater centrifugal force at the equator and the fact that earth possesses plasticlike properties, the earth is not a perfect sphere. The equatorial diameter of 12,757 km (7927 mi) is greater than the polar diameter of 12,714 km (7900 mi) by about 43 km (27 mi).

proximately 71 percent of the globe. The Pacific Ocean alone spans more than one-third of the surface—an area large enough to contain all the land of the world and another continent the size of South America (Fig. 1.6). The other 29 percent of surface area is land. Our planet's large landmasses are referred to as *continents*; smaller bodies are called *islands*. The usage of both terms is dictated by custom rather than logic. Thus, although Africa, Asia, and Europe constitute one continuous landmass, we arbitrarily refer to them as separate entities.

FIGURE 1.5

SPACE VIEW. This shows India and Sri Lanka from a distance of 540 nautical miles above the earth's surface looking north with the Bay of Bengal to the right and the Arabian Sea to the left.

Earth's Changing Face

The face of the Earth has undergone dramatic alteration since its birth. Past thinking supported the view of a stable planet with fixed continents and ocean basins; current research suggests that continents have repeatedly broken apart and changed location. It is believed, furthermore, that the ocean floor adjusts to such changes by increasing in size in one area while decreasing in another. These views form the basis of what has been called plate tectonic theory.

Plate tectonic theory proposes that heat within the earth's mantle generates *convection* cells and that under appropriate conditions of heat and pressure the rock materials of the mantle become plastic and flow.* According to this theory the brittle, rocky material of the crust is dragged in large segments (or plates) along the upper limbs of convection cells—as if they were upon a conveyor belt.

Figure 1.7 is a simplified model of continental movement and seafloor development as proposed by the Plate Tectonic theory. Three large segments (or plates) of the crust are represented by A, B, and C. Plates A and C are being dragged westward by convection currents, while plate B is moving in an easterly direction. At point D the seafloor is being spread apart and new ocean floor is being created by upwelling of matter. Since the earth is not constantly being increased in size by this process, old seafloor must be reassimilated elsewhere. This occurs in many of the planet's oceanic trenches as illustrated at E. The Plate Tectonic theory holds that the ocean floor is being recycled while, at the same time, large crustal plates shift about and alter the global position of continents.

Based on the striking parallelism of the continental-shelf margins on the two sides of the Atlantic, fossil finds of identical animals and trees in South Africa and South America, geologic dating that shows the Atlantic Ocean floor to be younger than adjacent continents, and a host of other factors, the earth's large land masses are

* *Convection* is a process by which heat is transferred by moving matter. For example, convection develops in a pan of water heated by a burner. Over a heat source, heated water expands, becomes lighter, and rises to the surface as a current. At the same time, cooler water sinks to replace the rising mass. Consequently, circulation cells form as water rises over the heated spot, spreads out and cools, and descends on the cooler side.

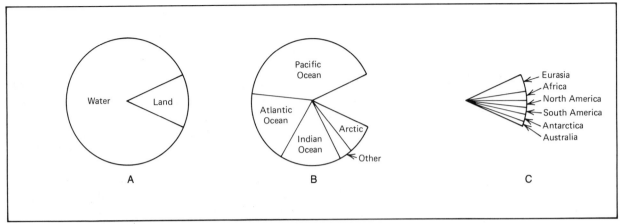

FIGURE 1.6
RELATIVE SIZES OF THE EARTH'S PRINCIPAL LAND AND WATER BODIES. (A) Surface areas. Total area equals 510,230,000 km² (197,000,000 mi²). (B) Water bodies. Total area equals 370,370,000 km² (143,000,000 mi²). (C) Landmasses. Total land area equals 134,860,000 km² (54,000,000 mi²).

thought to have formed a single supercontinent, called *Pangaea*, approximately 200 million years ago. The shape of Pangaea, as shown in Figure 1.8*A*, was determined by graphically fitting together the earth's landmasses along a line where the continental and oceanic crusts meet. (The overlap of continents is shown in black, gaps in white. These minor areas of disagreement have either been places that have been modified since Pangaea broke up, or else indicate meager data availability.) The Pangean supercontinent had the appearance of a great "V" that converged at the joint of southern Europe and northwest Africa. The ocean within the "V" is the ancient Tethys Sea. North America and Eurasia made up one arm of the "V"; Australia, Antarctica, and India together with South America and Africa comprised the other arm. These two arms of the "V" of Pan-

gaea have distinct names: the southern arm is called Gondwanaland; the northern arm is known as Laurasia.

Figure 1.8*B* illustrates the major crustal plates as they exist today, and arrows show their relative directions of motion. When one considers the vast distance between the Americas and the Afro-Eurasian landmass, the degree of continental movement appears incredible. However, because such movement occurred over a 200 million year time span, the distance traveled by a continent such as North America amounts to no more than a few centimeters per year.

If the present plate motions continue, the surface of the Earth will be quite different in the future. A map of the planet 50 million years in the future (predicated on present movement) is provided in Figure 1.8*C*. Notice that a large portion

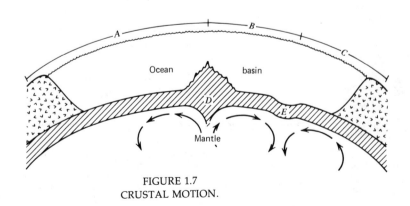

FIGURE 1.7
CRUSTAL MOTION.

FIGURE 1.8
THE CHANGING POSITIONS OF THE CONTINENTS. (A) Pangaea: the Supercontinent (200 million years ago). (B) present crustal plates and their direction of movement. (C) the future positions of the continents (50 million years in the future).

THE FOCUS OF GEOGRAPHY

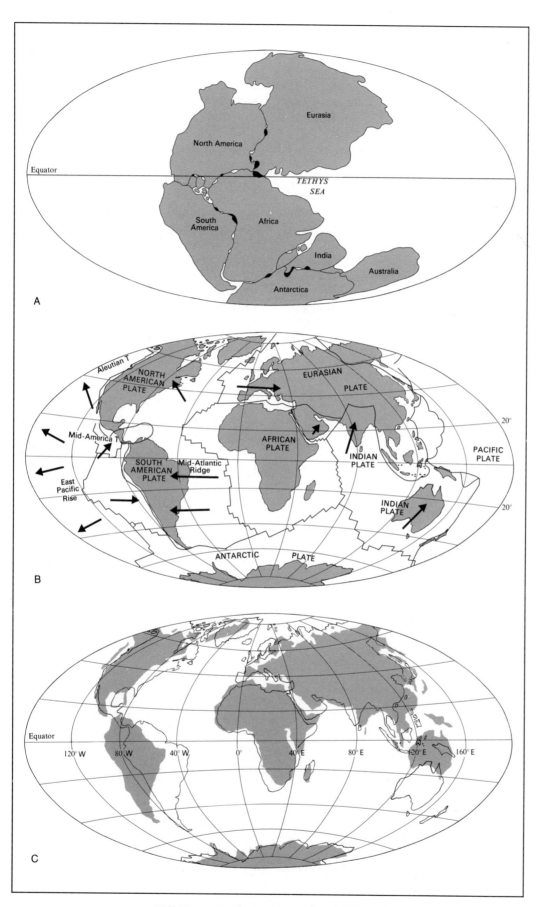

A

B

C

UNDERSTANDING THE CHARACTER OF AREA

of eastern Africa is separating from the continent. Australia and New Guinea will have moved considerably northward, and open waterways are found in the present locations of the Panama and Suez Canals.

"The only thing that is constant is change." This dictum has held true for the earth in the past and will undoubtedly typify its future. Mankind's biblical promise of "three-score and ten," however, is too short a span of time in which to assess (by direct observation) many changing aspects of our planet. One obviously cannot see continents moving; their motion is simply too slow. And, whereas there are many familiar landscapes that have exhibited little or no change over many years, this apparent stability is only a short-term illusion. Even the types of plants that could now be considered native to an area have not always occupied their present environment, nor has the atmosphere always been of the same composition (Fig. 1.9).

The earth's original atmosphere was not conducive to human habitation. It was hot, rich in ammonia and methane, and contained significant amounts of hydrogen and helium. Under the intense heat that characterized the atmosphere's birth, the lightest gases (hydrogen and helium)

soon escaped from the earth's gravitational field and were lost to outer space. Subsequently, ammonia and methane underwent decomposition by chemical reactions with sunlight and were broken down into their component elements. This activity released *nitrogen*, which now comprises 79.08 percent by volume of our atmosphere, and *carbon*. The atmosphere went through a lengthy transition. Mixtures of gases comparable to the present atmosphere did not appear until about 800 million years ago. And oxygen did not approach present levels for another 400 million years.

The development and maintenance of the atmosphere's oxygen supply resulted from the evolution of plant life within primordial seas. Approximately 3.3 billion years ago plants capable of releasing free oxygen into the atmosphere had evolved. Slowly (perhaps over hundreds of millions of years) plants contributed oxygen to the atmosphere; this paved the way for colonization of the landmasses by living organisms. It is oxygen, primarily, that filters out harmful radiation from space before it reaches the surface of the earth. In combination with other elements, oxygen provides precipitation and regulates heat exchanges.

Four hundred million years ago, familiar land-

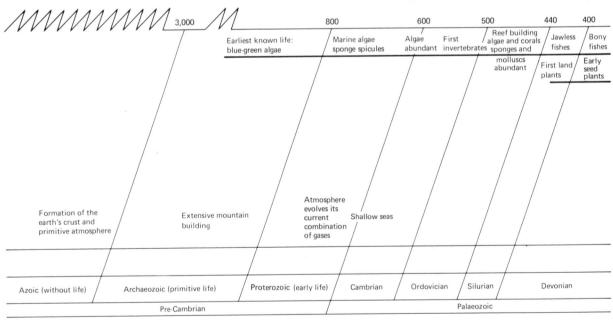

FIGURE 1.9
MAJOR EVENTS IN THE EARTH'S HISTORY.

scapes were barren, devoid of plants. Beginning with very simple forms of algae and fungi, the vegetation of the land gradually evolved. Although the evolution of species still continues, families of plants persist for lengthy periods of time. The most recent evolutionary plant forms that now dominate world vegetation—the conifer-cycad and the flowering tree families—have changed very little in the past 50 to 60 million years (Fig. 1.10).

While the earth's atmosphere and plant life were achieving their current composition, landscapes were being molded by processes operating on or near the crust. Streams carved valleys into the uplands and deposited sediments in the lowlands. The decay of the crust provided a basis for soil development. None of these processes operated in isolation. The atmosphere and gravity supplied most of the energy to shape the details of the surface. Rain, frost and thaw, running water, glacial activity, and the force of the wind all played a part. In turn, the relationship between vegetation formation and the atmosphere was critical. Without the protective screening-out of the sun's ultraviolet rays by oxygen, life as we know it would have been impossible. The character of vegetation depends upon the availability of atmospheric energy and of water. The particular fea-

tures of a soil depends upon the original composition of the earth's crust, the configuration of the surface, the influence of the atmosphere, and the impact of vegetation. Inherent soil characteristics enhanced or restricted the production of certain vegetation forms. The many elements of the physical environment interacted with processes that produced natural landscapes.

The formation of natural landscapes was accompanied by the evolution of *faunal* (animal) life forms; a development that was to have a profound effect upon landscape characteristics. The history of animals, which began with the appearance of single-celled organisms in the earth's primordial seas more than 600 million years ago, has been marked by the evolution of life forms of increasing complexity. Prior to about 350 million years ago, all fauna were marine organisms. Gradually, species evolved that were capable of surviving on land. Some of these early land animals (scorpions, cockroaches, and dragonflies) are still found on the earth. Testimony to the former existence of others, however, consists solely of remains preserved in rock.

Fossil evidence indicates that animal populations varied in composition throughout time. Using the dominant fossils in a geologic sequence of rocks, it is possible to visualize how and to guess

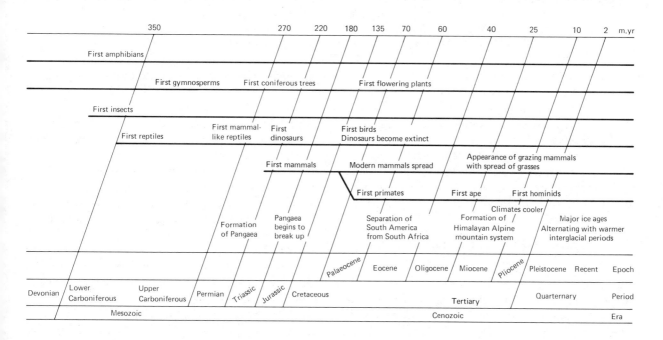

350 270 220 180 135 70 60 40 25 10 2 m.yr

First amphibians
First gymnosperms   First coniferous trees   First flowering plants
First insects
First reptiles   First mammal-like reptiles   First dinosaurs   First birds   Dinosaurs become extinct
First mammals   Modern mammals spread   Appearance of grazing mammals with spread of grasses
First primates   First ape   First hominids
Formation of Pangaea   Pangaea begins to break up   Separation of South America from South Africa   Formation of Himalayan Alpine mountain system   Climates cooler   Major ice ages Alternating with warmer interglacial periods

Palaeocene   Eocene   Oligocene   Miocene   Pliocene   Pleistocene   Recent   Epoch
Devonian   Lower Carboniferous   Upper Carboniferous   Permian   Triassic   Jurassic   Cretaceous   Tertiary   Quarternary   Period
Mesozoic   Cenozoic   Era

FIGURE 1.10
AN EXAMPLE OF MODERN CONIFERS:
A STAND OF PONDEROSA PINE IN THE DESCHUTES NATIONAL FOREST, OREGON.

approximately when species either appeared or disappeared from the earth. This record suggests that the first land animals were amphibians and insects who evolved about 400 million years ago. Thirty to forty million years later the age of reptiles began. This age was to last almost 250 million years, and was dominated by the dinosaur. Next came the age of mammals, a period beginning about 75 million years ago. Finally, man emerged.

Man and Man's Earth

Humans are related to the *primates* (mammals with five digits bearing flat nails on feet, or hands adapted for grasping). The fossil record of primates is scanty. Most primates were forest dwellers. Being generally intelligent, the primates were seldom trapped in bogs, tar pits, or other envi-

ronments conducive to quick burial and fossilization. Thus the origin of mankind remains conjectural, and a complete understanding of his evolution must await new discoveries and interpretations. Nonetheless, scraps of human bones and stone implements used by primordial man have been found. Analysis of this evidence suggests that the ancestral line leading to modern humans stretches back 5, perhaps even 6, million years. This may seem long to us, but is really a very short interval compared to the length of earth history (Fig. 1.11).

The progenitor having the potential from which the genus *Homo,* man, could evolve was called *Ramapithecus*—a small primate that had become different from the ancestors of present-day apes approximately 15 million years ago. From this stock evolved three hominid (manlike) types that were to live upon earth contemporaneously with each other. For reasons not yet understood, two of these species became extinct about a million years ago. The third species, *Homo habilus* (handy man), was a creature with a large brain and teeth that were human in pattern, and who had the ability to walk upright. This species was not strikingly different in appearance from the hominids with whom he coexisted. More significant than physical appearance, however, is the fact that he had developed newly emergent behavioral patterns atypical of life forms preceding him.

Homo habilus could walk upright, and could therefore use his hands for other tasks. He fashioned and used stone tools, and probably wooden implements as well. He learned to communicate through vocalization, and learned the use of fire. He was a food gatherer like his fellow primates, but differed from them in the manner in which he obtained and partook of his meals. Primates, in general, gather and consume food as opportunities arise. *Habilus,* however, was a systematic gatherer, first collecting food and then returning to a home base wherein his social group shared their daily finds. This communal sharing within an organized social structure was original in earth history and indicative of the element of humanness within the evolving genus *Homo*. Yet, the evolutionary and social processes leading to true man remained incomplete.

The next species to emerge (at least 1½ million years ago) was *Homo erectus* (upright man). Like modern man, *erectus* had essentially all the outward physical characteristics typical of modern man, except that he was shorter. Only the head with its flattened skull, prominent brow ridges, and protruding jaw were reminiscent of *habilus* ancestors. Also, whereas the previous *Homo* line was concentrated in Africa, *erectus* began spreading over the rest of the planet. At least a million years ago he made his way across the narrow strip of land joining the continents of Africa and Eurasia and was in the vanguard of mankind's ultimate domination of the earth.

In the past million years of human history, evolutionary change, cultural growth, and technology developed at an increasing pace. The last phases of human evolution involved raising the cranium to a domelike shape to accomodate a larger brain, and reducing the size of molar teeth and protrusion of the jaw. Accompanying these changes was an equally important internal biologic reorganization that gave rise to better nerve networks and brain centers. The step from *erectus* to *Homo sapiens* (wise man) is estimated to have occurred about 1 million years ago, and the refinement to *Homo sapiens sapiens* (modern man) approximately 50,000 years ago. More than likely, these transitions did not occur just once, but probably many times and in many places. Not all of

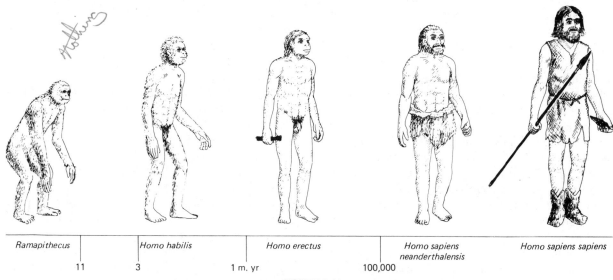

Ramapithecus		Homo habilis		Homo erectus		Homo sapiens neanderthalensis		Homo sapiens sapiens
	11		3		1 m. yr		100,000	

FIGURE 1.11
REPRESENTATIVES OF HUMAN PREDECESSORS.

the newly developed *sapiens* populations (such as the ice-adapted Neanderthales of western Europe) survived their various evolutionary routes. In summary, our ancestors, who had slowly migrated into parts of Europe, Asia, and Africa, represented a pool of human genes which, through mixing, were often reassembled into better and better combinations—much as hybridization under controlled conditions takes place today.

In addition to sharing genes, neighboring populations of ancestral man were also exchanging beliefs, attitudes, behavioralisms, and technological developments. Some of these were assimilated into the lives of the peoples affected, while others were rejected; the end product of the societal processes, however, was the diversification of world cultures. Evolution of culture and society accompanied biologic changes. These events were just as important toward developing the eventual character of modern man and the area he inhabits as his physical development.

The first *Homo habilus* was a simple food gatherer, eating edible vegetable matter and scavanged meat. His impact upon the environment was no more than that of other primates then alive or now extant. Gradually, his descendents learned to make sophisticated tools, hunt game, use fire, domesticate animals, cultivate crops, and finally harness power. These behavioral developments enabled humankind to advance as earth's dominating organism, and to become an important agent in transforming the character of area. Detailed analysis of landscape modification through cultural processes will be reserved for later chapters. At this point only brief examples of landscapes that illustrate man's role in affecting the character of area will be examined.

Most of our planet's land area has been occupied by people at one time or another. How people transform the earth's surface to produce "cultural landscapes" depends largely on their societal attitudes, objectives, and accumulated technology. Consider, for example, the land-use patterns shown in Figure 1.12*A* and *B*. Both landscapes have certain common physical traits. Each lies upon a river floodplain; climates are relatively similar; and forest dominates the native vegetation. Within the two landscapes the primary economic activity is agricultural. But each landscape has distinctive characteristics.

Figure 1.12*A* illustrates the impact of French

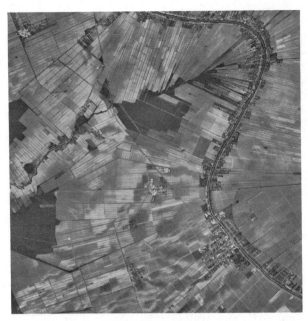

FIGURE 1.12A
LAND-USE PATTERNS ALONG THE MISSISSIPPI RIVER, BAYOU LAFOURCHE, ASSUMPTION PARISH, LOUISIANA.

colonists on the geography of southern Louisiana. When the French established their Louisiana settlements along the Mississippi River in the seventeenth century, they developed distinctive land-use patterns. Evidence of their manner of land settlement and subdivision is still present today. French settlers normally used the long-lot system in dividing the land along streams. Houses and roads were built on better drained land of natural levees adjacent to rivers. Homesteads were closely spaced and formed line villages along both sides of the river, with narrow lots extending back from them. In this way, all settlers had access to land and water transportation. Since the Mississippi is not straight in this area, the configuration of lots had to be adjusted to the curvature of the river. Inside the bends of the river, lots converged toward the apex, while outside the bends the lots frequently fanned out.

Figure 1.12*B* is a Chinese landscape located near Beijing (Peking), where virtually every square meter of arable land is in some form of crop production. The intensive use of land results from population pressures and the historical orientation of the Chinese toward an agrarian based culture. Hand tillage techniques and a tradition of dividing farms among sons encouraged the formation of small, fragmented fields in this area.

Land use for animal husbandry is relatively absent.

The above examples illustrate land-use patterns in two diverse cultures. An investigation of these two locations would reveal, moreover, that the structures each culture group has built, the crops cultivated, and the manner of living provide additional elements that foster uniqueness (personality) for each location. The social patterns of a group and their level of technology are not static, but vary with time, and in attempting to explain the dynamic forces that provide character to an area, one must also consider the impact of previous inhabitants and their contribution to the evolution of a cultural landscape. When a cultural attribute of a group changes, the change is frequently reflected in the physical environment. And, where different cultures occupy an area at different times, each group may leave its identifying signature upon the area inhabited. In North America, the changing landscape of the Great Plains offers a classic study in these concepts.

When the first Europeans arrived in the Great Plains, they found aboriginal cultures that were agriculturally oriented. The daily routine of the Indians centered on the subsistence cultivation of corn, beans, and squash. About once a year, the tribes went on a major bison hunt to supplement their vegetable diet and to obtain hides, sinew, bone, and other raw materials. Hunting did not occupy much of their time. This way of life, however, was drastically changed by the introduction of the horse. The Indians obtained the first horses after the Spaniards settled New Mexico in 1598. By 1800, the use of the horse had spread throughout the tribes of the Great Plains. The mobility

FIGURE 1.12B
LAND-USE PATTERNS ALONG THE CHANG JIANG (YANGSTZE RIVER), CHINA.

UNDERSTANDING THE CHARACTER OF AREA

FIGURE 1.13
CHEYENNE INDIAN CAMP NEAR FORT LARAMIE, WYOMING ABOUT 1880.

offered by the horse resulted in the convergence of diverse aboriginal groups onto the plains to take advantage of the material wealth afforded by hunting the great bison herds. Within 100 years of the introduction of the horse, the Indian population of the plains had tripled to an estimated 150,000. A new way of life was established, one dependent upon following bison herds. Thus, subsistence agriculture was largely abandoned and the aboriginals of the plains became nomadic hunters; gardening simply was not as profitable as hunting. Social life and the nomadic routine of the Great Plains people became closely tied to the migrational behavior of the buffalo. During most of the year a number of families lived together as migratory bands following scattered, small herds. In the summer, however, the small buffalo herds would reunite into massive herds that blackened the plains. The Indian bands would unite at a summer encampment for tribal ceremonies and a communal hunt. Figure 1.13 represents a large Cheyenne encampment near Fort Laramie in Wyoming, and provides an indication of what the character of the region was like in the late 1800s.

Due to the constant shifting of encampments and the use of transportable shelters, little physical evidence remains of a way of life that dominated the Great Plains throughout most of the nineteenth century.

A major change in the character of the Great Plains region occurred when President Lincoln endorsed the Homestead Act of 1862, giving settlers 160 acres of land if they worked it for five years. Thousands of cattlemen and "sodbusters" thronged to the area. The new occupants took a different view of land resources than the aboriginal populations. The newcomers built permanent shelters; where wood was not available, they built them of sod (Fig. 1.14). Houses, barns, graneries, and a host of other buildings designed to be permanent replaced the Indian teepee dwelling. Fields were laid out and crops seeded. In a few decades the concept of individual land ownership, imposed upon the area by a new culture, changed the economic and societal structure of the region.

A look at today's Great Plains reveals that the area has yet another personality. In part, its cultural traits are rooted in its richness of past peoples—each of whom contributed their attitudes, objectives, and technologies to mastering the use of the region's resources. Yet, the major aspects of the present landscape are largely a result of evolving technology. Where sod houses once stood, modern homes are found; giant grain silos have replaced the small graneries of the farmstead; bustling cities occupy sites of what were once small hamlets; and massive farm machinery now cultivates mile-long fields that previously required countless hours of human and animal labor (Fig. 1.15).

From this overview of the habitation of the Great Plains we can gain insight into the physical manifestations of the region's sequent occupance by culture groups. Equally important to the area's changing character, however, have been human traits such as: political affiliations, religious views, languages, and ethnic heritages. The geography of the Great Plains, as with any region, represents an intermeshing of forces—both physical and cultural, giving rise to unique combinations of elements that provide for regional identity.

This chapter has presented several concepts regarding the evolution of an area's character (or personality). Clearly, the study of geography is fascinating, but complex. It deals with numerous phenomena, both physical and cultural, and strives to explain the interrelated processes that are active in producing the landscape. To comprehend the geography of an area, one must proceed with an orderly analysis of environmental elements and processes and their interaction with other components of the earth system. After our brief excursion into the historical background of the earth and the development of humankind, we are now prepared to undertake our study of geography. The chapters that follow begin with an analysis of earth's physical factors and conclude with a discussion of man and his use of planetary resources.

FIGURE 1.14
SOD HOUSE AND FRONTIER FAMILY NEAR CALLAWAY, NEBRASKA ABOUT 1892.

FIGURE 1.15

A GREAT PLAINS HOMESTEAD NEAR CLIFFORD, NORTH DAKOTA. It is well protected from wind and snow by a wind break of trees and shrubs.

KEY TERMS AND CONCEPTS

geography	oceanic crust
region	polar diameter
regional method	equatorial diameter
systematic method	plate tectonic theory
solar system	convection
earth structure	drifting continents
core	origin of the
mantle	atmosphere
crust	origin of plant life
continental crust	origin of man

Ramapithecus   *Homo sapiens*
hominid   *Homo sapiens sapiens*
Homo habilus   cultural landscape
Homo erectus

DISCUSSION QUESTIONS

1. Define geography. What can be considered the primary concern of the geographer?
2. Why can geography be considered one of the most ancient of mankind's accumulated bodies of knowledge?

3. Describe the difference between the regional and systematic approaches to the study of geography. What do both have in common as an ultimate goal?

4. How do uniform and nodal regions differ? Give two examples of each.

5. Name the earth's nearest star. How does this star differ in appearance, to the unaided eye, from other stars of the universe? Why?

6. Describe the layered structure of the earth's interior.

7. Explain the logical reason(s) for the difference in length between the equatorial and polar diameters of the earth.

8. Explain the theory that accounts for the changing positions of the continents.

9. What relationship exists between the atmosphere's oxygen supply and plant life.

10. Describe the evolutionary developments leading to the emergence of modern man.

11. What characteristics of mankind makes him unique among the primates?

12. List a variety of means by which humans affect the character of area.

13. How does the concept of sequent occupance apply in the description of the geography of an area?

REFERENCES FOR FURTHER STUDY

Broek, Jan O. M., and John W. Webb, *A Geography of Mankind,* McGraw-Hill Book Co., New York, 1978.

James, Preston E., *American Geography: Inventory and Prospects,* University of Syracuse Press, Syracuse, New York, 1964.

Landsberg, Helmut E., "The Origin of the Atmosphere," *Scientific American,* 189 (August): 82–86, 1953.

Leakey, Richard E., and Roger Lewin, *Origins,* E. P. Dutton, New York, 1977.

McAlester, A. L., *The History of Life,* Prentice-Hall, Inc., Englewood Cliffs, N.J. 1967.

Urey, Harold C., "The Origin of the Earth," *Scientific American,* 187 (October): 53–60, 1952.

Wilson, J. T., ed., *Continents Adrift,* W. H. Freeman and Co., San Francisco, 1970.

2

THE ATMOSPHERE
Its Structure and Dynamics

I T IS RAINING OUTSIDE AND WE REACH FOR AN UMBRELLA. Newspapers state that drought stricken ranchers in the West are culling starving cattle from their herds. Beef prices soon rise. An increase in atmospheric pollutants brings ominous warnings of an impending ice age. The manner in which men clothe themselves, build their homes, and raise crops, and the types of activities in which they indulge, are all affected by the behavior of the atmosphere (Fig. 2.1). Even the landscape—its soils, vegetation, and landforms—is inextricably part of the atmospheric environment and a visible expression of its available energy.

COMPOSITION AND STRUCTURE OF THE ATMOSPHERE

The atmosphere is made up of a mixture of gases (mainly nitrogen and oxygen), particulate matter, and water (Fig. 2.2). This mixture is held to the earth by gravity and has an outer limit of approximately 1000 km (600 mi). The density of the atmosphere is greatest near the earth's surface and decreases rapidly with altitude. Indeed, one-half of the mass of the atmosphere lies within 5 km (3 mi) of the earth's surface, and a total of 99 percent is found below 100 km (60 mi). This is illustrated in Figure 2.3.

Having evolved at the very base of this "ocean of air," mankind's existence is dependent upon a dense atmosphere—one in which air molecules are packed closely together. When ascending into the rarified gases above the earth, one must be enclosed by artificial earth-surface environments. Thus, the cabins of jet-liners are pressurized for passenger comfort, high-altitude mountain climbers carry supplemental supplies of oxygen, and space-age astronauts walking upon the moon wear spacesuits that simulate earth-like atmospheres.

Pressure decreases with increasing height above the Earth's surface, and temperatures within the lower atmosphere also normally diminish with an increase in altitude. Familiar summer retreats to the mountains have long been justified as providing an escape from the oppressive heat of the lowlands. As illustrated in Figure 2.4, however, the temperature patterns aloft do not continually decrease relative to distance from the earth's surface. Upper atmospheric observations often reveal signs of temperature stagnation—meaning no change at all—and even of temperature reversals. Only in recent decades have these apparently perplexing thermal characteristics become explainable in terms of processes that operate within our atmosphere.

Thermal Structure of the Atmosphere

Because temperature varies with altitude, it is possible to identify distinct layers, or *strata*, of air within the atmosphere (Fig. 2.5). The lowest layer

FIGURE 2.1
WEATHER IN THE NEWS. The weather affects human activities and is reported by the news media daily. Occasionally weather is sufficiently dramatic to provide "headline" news.

of the atmosphere is called the *troposphere*. Elliptical in shape, the troposphere is deepest at the equator (approximately 20 km or 12 mi) and shallowest at the poles (about 10 km or 6 mi). It contains practically all of the atmosphere's water vapor and particulate matter, and is the region in which our weather is generated. The troposphere

is the most dynamic portion of the atmosphere.

Throughout the troposphere average temperatures are warmest near the surface of the earth and normally decrease as altitude increases, the result of the atmosphere being primarily heated by the earth's surface. The last statement may seem to be a contradiction since it is the sun that heats and

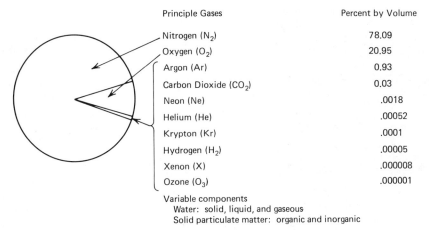

Principle Gases	Percent by Volume
Nitrogen (N_2)	78.09
Oxygen (O_2)	20.95
Argon (Ar)	0.93
Carbon Dioxide (CO_2)	0.03
Neon (Ne)	.0018
Helium (He)	.00052
Krypton (Kr)	.0001
Hydrogen (H_2)	.00005
Xenon (X)	.000008
Ozone (O_3)	.000001

Variable components
Water: solid, liquid, and gaseous
Solid particulate matter: organic and inorganic

FIGURE 2.2
COMPOSITION OF THE ATMOSPHERE.

animates the atmosphere. Actually, both of these statements are true. It is the sun that is the dominant energy supplier to the atmosphere. However, solar energy that is emitted at a very high temperature, arrives at the atmosphere as *short-wave radiation*—a form of energy that the earth's envelop of gases has little capacity to absorb.* This energy travels through the atmosphere as though the air were essentially transparent. The surface of the earth, being opaque, absorbs energy. Thus the surface experiences a temperature increase and, in turn, is capable of radiating energy. Being at a much lower temperature than the sun's surface (average Earth temperature is 12° C (54° F) the sun's is 6,000° C (10,500° F)), the earth emits *long-wave radiation*—a form of energy the atmosphere can absorb. In other words, the form of energy is converted at the earth's surface. The re-emitted terrestrial radiation now passes back through the atmosphere, warming the air from the surface upward.

The sequence of energy absorption by our atmosphere has often been likened to the processes that heat greenhouses, even though the analogy is not perfect (Figure 2.6). Shortwave radiation from the sun may pass through the glass of a greenhouse as if it were transparent. It then heats inter-surfaces (soil, boxes, wooden shelves, etc.) which in turn become long-wave radiators of en-

ergy. It is not easy for long-wave radiation to escape through the structure, except through conduction via the glass panes; thus, energy is absorbed by the air within the greenhouse and temperature increases. A parallel process is in operation when one opens a car door on a sunny, but cool, day. The interior of the car is found to be considerably warmer than conditions outside the car.

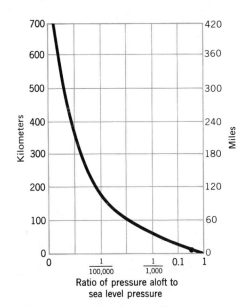

Ratio of pressure aloft to
sea level pressure

FIGURE 2.3
AIR DENSITY ALOFT. This graph shows the decrease of atmospheric pressure with elevation above sea level. The dot indicates the point on the curve where pressure is reduced to one-half the value at sea level.

---

* *Short-wave radiation* is referred to herein as the sun's radiation, which includes energy wave lengths in the range from 0.2 to 3 microns. *Long-wave radiation* emitted by the earth is largely in the range from 3 to 50 microns.

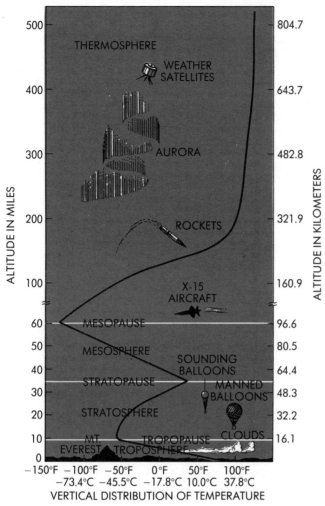

FIGURE 2.4
TEMPERATURE VARIATION WITH ALTITUDE.

Just as more heat energy is felt when one stands directly in front of a fireplace, so the lower atmosphere is warmest nearest the radiating surface. Although the rate of temperature change with altitude varies from place to place, the long-term global average amounts to about 6.5C°/1000 m (3.5F°/1000 ft). The average figure quoted above is commonly referred to as the *normal temperature lapse rate* (Fig. 2.7). This normal drop in temperature does not continue throughout the entire atmosphere, but is interrupted by a layer in which warming occurs. The level at which temperatures first cease dropping is called the tropopause—the upper limit of the troposphere. At this level average temperatures are about −60° C (−70° F).

Beyond the troposphere is the *stratosphere*, a region of air featuring a low-lying *isothermal layer* (where temperature remains constant with increasing height). Beyond the isothermal layer, to a distance of approximately 50 km (30 mi) from the surface, temperatures rise to about 0° C (32° F) at the stratospheric boundary (the *stratopause*). The reversal of the thermal pattern from that of the troposphere's is explained by the presence of *ozone*. The gas ozone (O_3) is known to form within the stratosphere through the bombardment of molecular oxygen (O_2) by ultraviolet rays from the sun. In this process ultraviolet rays are largely absorbed. The ozone prevents harmful radiation from reaching the earth's surface. Most ozone is concentrated between about 15 km (9 mi) to 55 km (35 mi) from the earth's surface, a region known as the *ozone layer*. A discussion of this layer's crit-

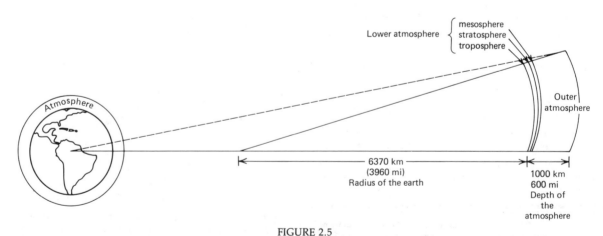

FIGURE 2.5
CROSS-SECTIONAL VIEW OF EARTH AND ITS ATMOSPHERE. The lower atmosphere is subdivided into layers based on thermal characteristics. Although many people conceive of the earth's gaseous envelope as continuing for great distances into space, its effective limits should be viewed as comprising a very thin layer relative to the size of the earth.

THE ATMOSPHERE

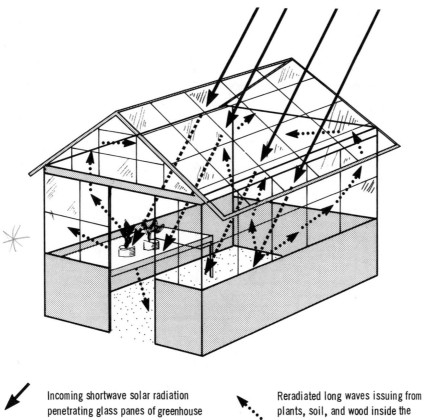

| | Incoming shortwave solar radiation penetrating glass panes of greenhouse | | Reradiated long waves issuing from plants, soil, and wood inside the greenhouse |

FIGURE 2.6
THE GREENHOUSE EFFECT.

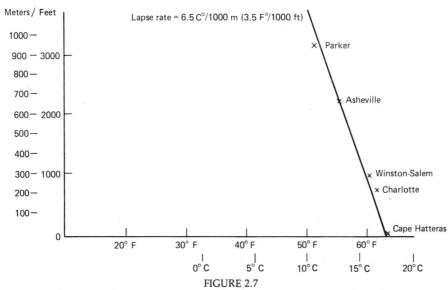

FIGURE 2.7

THE NORMAL LAPSE RATE. Mean annual temperatures of selected weather stations in North Carolina and the predictable normal lapse rate are plotted above. Notice the considerable degree of agreement between them.

COMPOSITION AND STRUCTURE OF THE ATMOSPHERE

ical significance and the possible impact of alter-ation of its characteristics is provided in the following inset.

THE DEATH OF THE OZONE LAYER?

Consumer needs are met in various ways by engineers and chemists. Occasionally, however, industrial products that make our lives easier can also damage the environment. In fact, they can even harm human life. A classic example of this is the present concern over the potential destruction of the ozone layer by something as commonplace as household spray (aerosol) cans.

In practically every household is a variety of spray cans, ranging from mosquito repellents to deodorants. These aerosol cans have been traditionally pressurized with chemical compounds known as halomethanes, of which the most widely used is chlorofluoromethane. Chlorofluoromethane, a combination of chlorine (Cl_2), fluorine (F_2), and carbon (C), is essentially inert (will not readily react with other chemicals) at surface temperatures and pressure. This property was once considered an advantage. Since these chemicals do not react with gases in the atmosphere or with chemicals in surface waters, they appeared to be harmless and simply accumulated in the environment. In 1974, however, researchers at the University of California proved that, under ultraviolet radiation, chlorofluoromethane dissociated (split up into atoms). They felt that this discovery could have serious implications for the survival of the ozone layer.

If halomethanes are released to the lower atmosphere in abundance, some scientists postulate that they will also be carried to higher levels, including the ozone layer, by air circulation systems. At the ozone level, where ultraviolet radiation is intense, the freed halomethane atoms, expecially the chlorine atom (Cl), are available for a series of chemical reactions with ozone (O_3) to produce molecular oxygen. The reaction sequence theoretically decreases both the amount of ozone present and also the number of free oxygen atoms readily available to form more ozone. Relating the incidence of skin cancer to ultraviolet radiation exposure, it was proposed that a mere 5 percent reduction in stratospheric ozone would result in 20,000 to 60,000 new cases of skin cancer annually. Further decreases could cause organic mutations and threaten the existence of all unprotected organisms.

Industrial scientists argue that tests performed with halomethanes at surface temperatures are unrealistic when applied to the much colder upper atmosphere, and that the dangers perceived in the University of California studies are nonexistent. Other arguments point out that emissions from supersonic aircraft and refrigerants pose a more serious threat to the ozone layer than halomethanes released by aerosol cans. Whether the spray-can danger to the stratosphere is real or imagined, the controversy over the threat to the ozone layer has sharpened public awareness of ozone's importance to surface life and has resulted in governmental programs aimed at determining any possible damage to the Earth's vital ultraviolet radiation screen.

The *mesosphere* is the next highest region. In many respects it parallels the troposphere in temperature characteristics. Extending outward to approximately 80 km (48 mi), temperature in the mesosphere decreases to its minimum, approximately −83° C (−120° F), at the *mesopause*. Beyond the mesosphere lies a realm of extremely rarified air, the *thermosphere*.

ATMOSPHERIC ENERGY AND TEMPERATURE RESPONSE

Streaming outward in all directions from the sun is a relatively steady flow of energy. Our planet receives a minute amount (approximately one two-billionth) of the total energy emitted (Fig. 2.8), although this supplies virtually all the energy needed to heat and animate our atmosphere. Solar energy intercepted by the earth is called *insolation*, a term that distinguishes it from total sun energy. Although insolation is crucial to atmospheric processes, the earth plays a key role in how that energy becomes functional (Fig. 2.9A).

Of the energy received at the top of the troposphere, only 15 percent is absorbed by the atmosphere and has a direct effect upon air temperature. Of the remaining energy, 42 percent is lost to outer space by reflection and scattering from

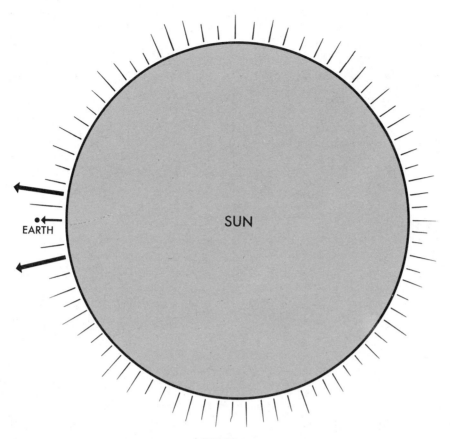

FIGURE 2.8
INSOLATION. If the orbit of the earth, with a diameter of about 298 million km (179 million m), is considered to represent a radiation sphere around the sun, it is logical that earth, having only a 12.7 thousand km (7.6 thousand m) diameter, intercepts only a minute amount of the total energy the sun produces. This amount is distinguished from the total by calling it insolation.

clouds, from the earth's surface, and from particulate matter in the air; and 43 percent passes through the atmosphere to be absorbed at the earth's surface.

The conversion of solar energy to long-wave radiation at the earth's surface is the principle mechanism by which our atmosphere is heated. As shown in Figure 2.9B, 24 percent of the energy absorbed at the surface is reradiated from the earth as *ground* or *terrestrial radiation*, 16 percent is absorbed by the atmosphere, and 8 percent is lost to outer space. Another 23 percent is temporarily tied up as *latent heat* in *evaporation processes;* that is, in changing the state of water from liquid to gas. This energy is later released to heat the atmosphere when water vapor condenses, as in cloud formation.

Finally, to retain a balance between energy absorbed and energy reradiated, the atmosphere itself exchanges energy with both the earth's surface and outer space. As shown in Figure 2.9C, approximately 4 percent is returned to the earth's surface while 50 percent is lost to outer space.

The example used above assumes that a balance exists for the entire globe between energy receipts and losses over long time periods. Considering the last several centuries, during which earth climates have remained relatively stable, the concept is realistic. But large daily and seasonal imbalances in energy absorbed versus energy reradiated are common. And, since atmospheric temperature is an indirect measure of the energy contained in a unit of air, considerable variation is expected in this weather element.

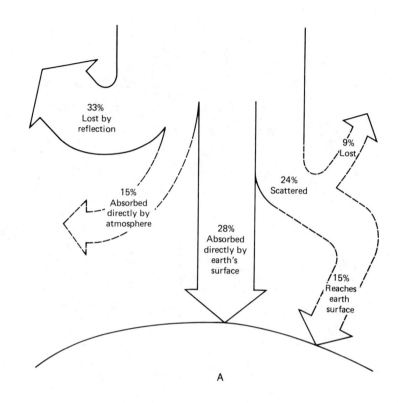

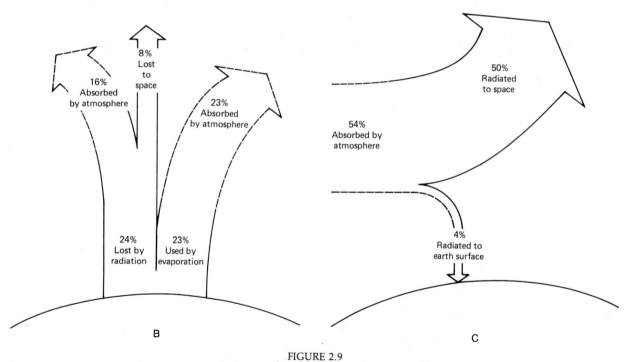

FIGURE 2.9
ENERGY EXCHANGES IN THE EARTH-ATMOSPHERE SYSTEM. A. Distribution of solar radiation. B. Longwave radiation exchanges between the earth and atmosphere. C. Radiation losses of the atmosphere.

THE ATMOSPHERE

Daily Temperature Variation

Any substance able to absorb energy is also able to radiate energy. The intensity of the energy radiated by a body depends upon its temperature. Figure 2.10 is a graph of air temperature in Columbia, Missouri, on an equinox date. During the daylight period when insolation is high, relatively large amounts of terrestrial radiation permit temperatures to climb. At night, temperatures drop as energy is radiated from the surface. Just as a frying pan continues to give up heat after the burner is turned off, so a heated earth continues to release its stored energy at night.

A curious aspect of the daily temperature curve is that it is not perfectly synchronized with incoming solar radiation (Fig. 2.11). Returning to the analogy of the frying pan, we are all aware that the pan does not immediately become hot when we place it upon the burner. Rather, there is a time lag during which the pan must absorb heat prior to reradiating it. The earth has a similar response to energy inputs. Daily temperatures begin to rise shortly after sunrise and normally reach their peak at mid afternoon—not at noon when insolation is at its maximum.

Seasonal Temperature Variation

At an average earth-distance from the sun, the flow of solar energy to our atmosphere equals 2

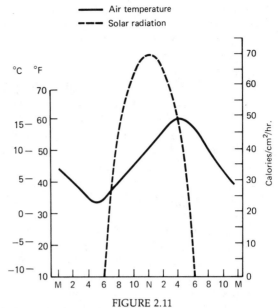

FIGURE 2.11

TEMPERATURE AND INSOLATION AT COLUMBIA, MISSOURI (March 21, 1978).

calories/cm²/minute.* Because this energy-supply rate varies little, it is known as the *solar constant*. However, no matter how constant the supply, energy is not uniformly distributed over the globe. This holds true both latitudinally—from equator to pole—and for a location's seasonal variation in energy.

Only a small fraction of the earth receives energy at the solar constant rate. As seen in Figure 2.9, a considerable portion of incoming radiation is subject to depletion in the atmosphere. In addition, orbital distance, earth shape, and the orientation of the planet's axis all contribute to the energy budget, and to consequent thermal patterns of earth locations.

The earth's distance from the sun varies from an average 149.5 million km (92.9 million mi) by 2.4 million km (1.5 million mi). This degree of departure from a perfect circular orbit only amounts to slightly more than 1 percent. Thus, the effect of varying orbital distance upon global heating and seasonal temperature change is small, but can increase or reduce a location's receipt of energy by as much as 3.5 percent. The earth is farthest from the sun about July 3rd, approaching the

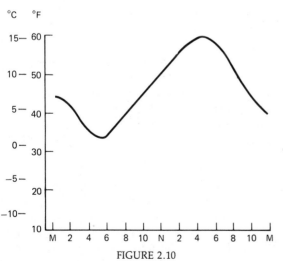

FIGURE 2.10

HOURLY TEMPERATURE IN COLUMBIA, MISSOURI (March 21, 1978).

* A *calorie* is the heat required to raise the temperature of one gram of pure water from 14.5° C to 15.5° C.

peak of the summer season in the Northern Hemisphere. Much more critical to our planet's energy distribution patterns than distance from the sun are the next topics to be discussed—earth shape and axial orientation.

The small portion of total sun radiation intercepted by the earth arrives essentially as bundles of energy traveling in paths parallel to one another. Since the earth is spherical in shape only one such stream of energy can arrive at an angle that is perpendicular to the surface. The remainder forms angles of less than 90° with the planet (Fig. 2.12). The significance of this fact is that energy intensity—heating power—is directly related to the angle at which radiation is received. Figure 2.13 illustrates this principle. Note that the band of energy intercepted at an angle of 90° heats and illuminates a surface area essentially equivalent to the cross-sectional area of the energy stream. Due to the earth's curvature, an equivalent amount of energy arriving at a low angle is distributed over a much larger surface area. Thus, as sun angles decrease there is less intense radiation per unit of surface area.

Finally, if the earth's axis were perpendicular to its orbital plane there would still be no seasons. The sun would always be overhead at the equator. As we move toward the poles, the sun would appear lower in the sky because its altitude would decrease one degree for each degree of latitudinal distance from the equator (Fig. 2.14). If this were

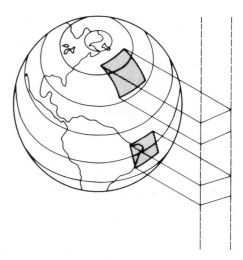

FIGURE 2.13
THE RELATIONSHIP OF INSOLATION INTENSITY TO ANGULAR RECEIPT OF SOLAR RADIATION.

the case energy distribution over the earth would graphically look much as in Figure 2.15. Near the equator temperature would be considerably higher than at present, and people would avoid living there. The poles, on the other hand, would be warmer than they are now, and higher latitudes would be somewhat more attractive for settlement. However, since the axis is tilted 23½° away from the perpendicular, the position of the sun's vertical rays can likewise vary by the same number of latitudinal degress from the equator. Thus, energy concentration by vertical (and near-vertical) sun rays shifts to and from the Northern and Southern Hemispheres, a result of the axis always maintaining the same orientation in space (Fig. 2.16). The consequence is that the intensity of solar radiation varies not only by latitude, but also, for any given location, by time of year.

An additional aspect of energy receipt (as affected by axial tilt) relates to the length of illumination period. When the North Pole leans toward the sun, the Northern Hemisphere's daylight period exceeds 12 hours. The effect is that considerably more than one half of the Northern Hemisphere is illuminated, resulting in locations experiencing a shortened journey through the darkened side of the globe that we call night (Fig. 2.17). At the same time the Southern Hemisphere is less well lighted; with less than 50 percent of its area illuminated, daylight is less than 12 hours in length.

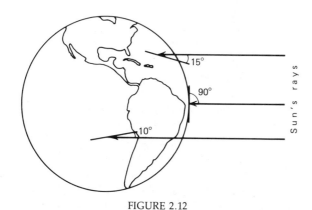

FIGURE 2.12
THE ANGULAR RECEIPT OF SOLAR ENERGY. Because the earth is spherical and intercepts sun rays as bands of energy traveling parallel to one another, there is only one location on earth where the sun's rays are perpendicular (at 90°) to the surface. All other rays of energy reach the surface at acute angles.

THE ATMOSPHERE

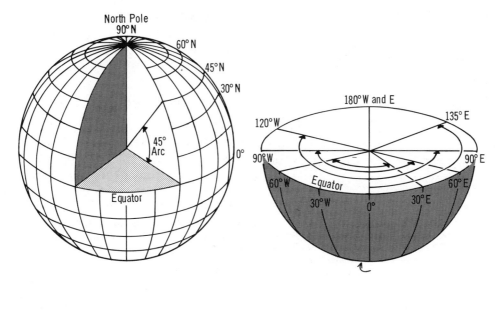

A

B

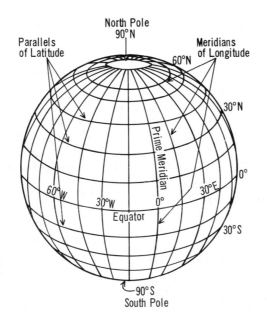

C

FIGURE 2.14

LATITUDE AND LONGITUDE. *Latitude* is angular distance (A) between the equatorial plane and location of a point (p) on the earth's surface. A line connecting all points at the same latitude forms a circular plane known as a *parallel of latitude*. *Longitude* is the angular distance (B) of a location in an east-west direction. As shown in (C), it is measured from the *prime meridian* (0° Longitude).

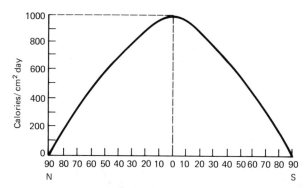

FIGURE 2.15
APPROXIMATE INSOLATION FOR AN EARTH WITH A VERTICAL AXIS. If the earth had an axis that was perpendicular to the plane of the ecliptic, each day would experience a maximum of solar radiation receipt at the equator and a minimum at the poles, very much as illustrated above. (The data in the above graph represents an average for the equinox dates.)

The uniform rhythm of the seasons is a direct response to the cyclic variation of solar energy concentration. In the Northern Hemisphere, for example, summer begins about June 21st, the day of the *summer solstice* (Fig. 2.16).* On this day, the sun's vertical rays are striking the earth at 23½° N. latitude. The sun is not only high in the northern sky, providing high potential energy, but is also above the horizon longer than at any other time of the year. The surface gains more heat than it loses, and temperatures rise. At the other extreme is the *winter solstice* that occurs around December 22nd. The sun's vertical rays are now found at 23½° S. latitude. Northern Hemisphere residents view the sun low in the sky and days are short, more energy is lost than received, and temperatures decline. Intermediate between the two extremes are the *spring* and *autumnal equinoxes* (approximately March 21 and September 23). These are the only days of the year in which the earth is illuminated from pole to pole, with the sun's vertical rays being found on the equator. All points on the planet (except the poles) experience equivalent alternating periods of night and day. In the Southern Hemisphere the seasons are reversed. Summer begins about December 22nd, fall

on approximately March 21st, winter starts around June 21st, and spring begins around September 23rd.

Surface Temperature Variation

Points along any given parallel of latitude have the potential for receiving exactly the same amount of solar energy. Yet an examination of the lower atmosphere's temperature patterns reveals marked contrasts for locations lying in close proximity. This is not surprising, since the atmosphere's dominant energy supply emanates from the surface, and the surface is by no means uniform.

Substances absorb and reradiate energy at different rates. On a global scale, the most important substances affecting thermal patterns are land and water. Mineral matter tends to absorb and reradiate heat readily. So land masses heat quickly and also cool quickly. The *temperature range* (difference between maximum and minimum temperatures) is large over landmasses, a characteristic of climate that is referred to as *continentality*. The major water bodies, on the other hand, heat and cool slowly. Compared to landmasses, water bodies get neither extremely hot nor cold. As a result, temperature range is comparatively small in marine climates.

As illustrated in Figure 2.18, four basic factors are normally attributed to the different rates at which land and water bodies heat and cool:

1. *Land is opaque and water is transparent.* This means that solar radiation is concentrated in a thin surface layer on landmasses, but, sun rays penetrate very deep into water bodies—dispersing light energy throughout a larger volume of matter.

2. *Evaporation is greater over water bodies.* Evaporation ties up energy, making it unavailable for surface heating. Since less evaporation occurs over land areas, more energy is available to heat them.

3. *Water bodies are turbulent.* Whereas the landmasses are static and lose little energy to adjacent earth materials, water bodies are in constant motion, redistributing absorbed energy both latitudinally and vertically.

4. *It takes more energy to heat up water.* An equivalent temperature increase in compa-

---

* Because our year of 365 days is 0.25 days short of the time it takes the earth to revolve about the sun, the date provided for the initiation of each season is only approximate, and may vary by as much as two days in any year.

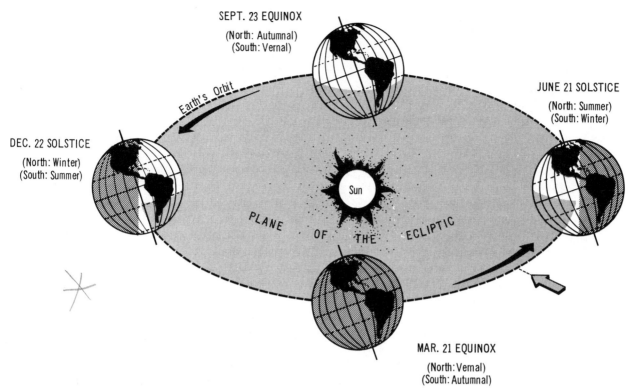

SEPT. 23 EQUINOX
(North: Autumnal)
(South: Vernal)

JUNE 21 SOLSTICE
(North: Summer)
(South: Winter)

DEC. 22 SOLSTICE
(North: Winter)
(South: Summer)

Earth's Orbit

PLANE OF THE ECLIPTIC

Sun

MAR. 21 EQUINOX
(North: Vernal)
(South: Autumnal)

FIGURE 2.16
ILLUMINATION AND HEATING OF THE EARTH. The intensity of solar heating and the amount of surface area illuminated increase in the hemisphere of the earth oriented toward the sun and decrease when the same hemisphere is directed away from the sun.

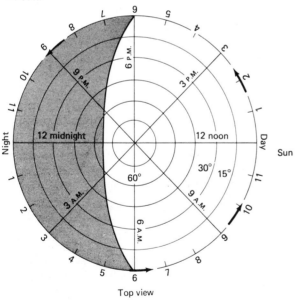

Top view

FIGURE 2.17
ILLUMINATION DURATION. On the diagram above is shown the number of hours of sunlight for any latitude in the Northern Hemisphere on June 21. The thick line separates night and day.

rable volumes of water and earth materials requires about three to five times more energy for the water.

The impact of the earth's two primary surface materials—land and water—upon thermal patterns is depicted on maps by means of *isotherms* (lines connecting points of equal temperature). Isotherms for January and July are shown on Figure 2.19. Note that isotherms in the Southern Hemisphere, over unbroken water surfaces, have a general east-west trend. If the earth's surface were of uniform composition all isotherms could be expected to follow this pattern; that is, to be closely aligned with the parallels of latitude. Where large land and water masses alternate (as in the Northern Hemisphere) the isotherms become convoluted. The seasonal shift in the position of isotherms is relatively moderate over the oceans. Over the large landmasses isotherms shift great distances, bending toward the equator in winter and toward the poles in summer.

Isotherms that form closed loops identify centers of seasonally high and low temperature.

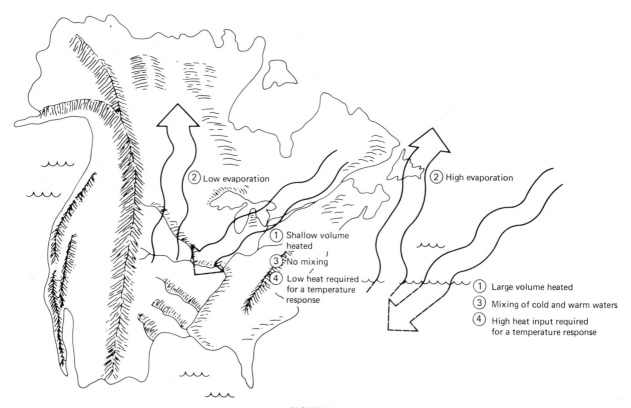

FIGURE 2.18
FACTORS AFFECTING THE COOLING AND HEATING OF THE EARTH'S SURFACE.

These are geographically significant hot and cold spots—areas that frequently affect large scale weather systems and provide character to the earth's regional climates. The January map reveals two such cold spots over the planet's largest land masses. Over Siberia a center of low temperature forms in January with average temperatures below −50° C (−58° F). Similarly, in North America there is a region where monthly temperatures fall below −34° C (−30° F). Both of these areas serve as sources for very dense, cold, polar air masses that effect winter climates far to the south. Because of the smaller size of North America and its indented northern coastline, the degree of continental cooling is not quite as great as in Asia. In July, high temperature centers dominate the map. North Africa, southwestern Asia, and the southwestern United States are all areas of intense surface heating, with mean monthly temperatures exceeding 30° C (86° F), and in some cases 35° C (95° F). Continental hot and cold spots are seasonal in na-

ture. Only two temperature centers on the earth are considered to be permanent. Both are cold—Greenland and Antarctica.

ATMOSPHERIC CIRCULATION

About three times more solar energy arrives at the equator than at the poles. But radiation from the earth's surface is much more evenly distributed by latitude (Fig. 2.20). Consequently, tropical latitudes receive more energy than they reradiate, and polar latitudes radiate more than they receive. Since polar regions are not getting colder or the tropics warmer, some mechanism must be operating to transport heat from low to high latitudes. That mechanism is the earth's atmospheric and oceanic circulation systems.

Horizontal movement of air is known as *wind*. Although wind is a commonly recognized weather element, the processes by which it gains momen-

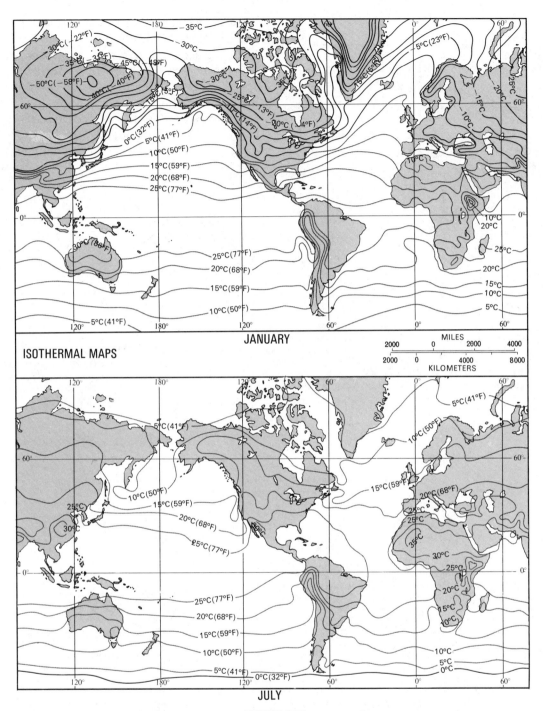

FIGURE 2.19
WORLD MEAN SEA-LEVEL TEMPERATURES IN °C (°F): (A) JANUARY, (B) JULY.

From Strahler *Modern Physical Geography* © 1978, John Wiley & Sons, New York.

ATMOSPHERIC CIRCULATION

33

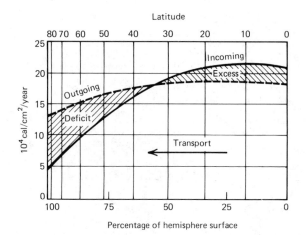

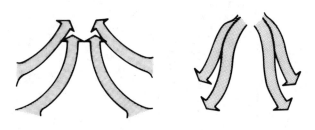

distance constitutes *pressure gradient force*, the primary factor determining the wind's speed and intitial direction of movement. When pressure differences are great, as in a hurricane, wind speed is high; when the variation is minor, there is little air motion.

The pressure gradient force concept is illustrated in Figure 2.22. Just as isotherms were previously used to analyze temperature patterns, this graphic example uses *isobars* (lines connecting

FIGURE 2.20

LATITUDINAL RADIATION BALANCE. *Source:* D. Riley and L. Spolton, *World Weather and Climate* (London: Cambridge University Press, 1974), p. 7. Reprinted by permission of the publisher.

FIGURE 2.21

LOW AND HIGH PRESSURE SYSTEMS. Pressure systems are characterized by definite airflow patterns. In low-pressure systems, air is converging and ascending; whereas, in high-pressure systems, air is diverging and descending.

tum and direction are less well understood by the general public. Air is set in motion by differences in air pressure and always moves from regions of high pressure to regions of low pressure (Fig. 2.21). The difference in air pressure over a given

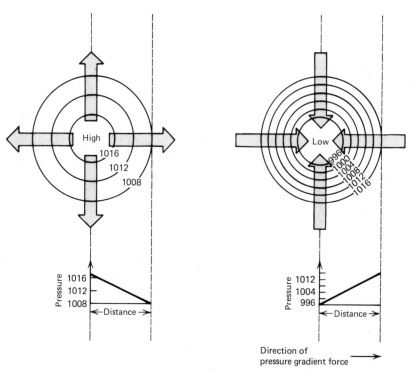

FIGURE 2.22
PRESSURE GRADIENT FORCE.

THE ATMOSPHERE

points of equal barometric pressure) to identify pressure patterns. The basic principle of pressure gradient force is that pressure-fields slope (have a gradient) in the atmosphere. This is illustrated by graphs that are drawn from the center to the outer limits of two pressure cells, high and low. Wind speed is greatest where there is the largest change in pressure per unit distance. In these illustrations, the low-pressure cell will foster higher velocity winds than will the high-pressure cell. The initial motion of air in both cases is at right angles to the isobars, always moving toward lower pressure. Isobaric pressure is expressed in millibars. A *millibar* is a unit of atmospheric pressure equal to one-thousandth of a bar. A bar is a force of one million dynes per square centimeter.

Although pressure gradient force dominates in determining initial air motion and direction, there are other factors that affect ultimate wind direction—in particular, the *Coriolis effect* and *frictional drag*. The Coriolis effect results from the rotation of the earth and appears to deflect all freely moving bodies from their intended paths. This "apparent force" acts to the right of the direction of motion in the Northern Hemisphere and to the left in the Southern Hemisphere. The strength of the Coriolis effect is related to both wind velocity and latitude. As air speed increases, so does the Coriolis deflection. Similarly, as latitude increases the Coriolis effect increases in magnitude. Its influence on moving air is essentially nonexistent near the equator and reaches a maximum at the poles. In a frictionless state the apparent Coriolis force and pressure gradient force are often found to be in a state of balance. The result is that winds are deflected from their intended path by as much

as 90° and follow a track that parallels the isobars rather than crossing them (Fig. 2.23). Frictionless air movements of this nature are known as *geostrophic winds,* an ideal state that is most characteristically found in the upper atmosphere (above 1000 meters/3000'). Near the earth's surface the force of friction of air with the ground counteracts the Coriolis effect and prevents air from turning parallel to the isobars. Rather, it flows at acute angles of 20° to 45° with the isobars (Fig. 2.24). Thus, if a high pressure cell were located directly to the north of a location in the Northern Hemisphere, the pressure gradient force would act in a direction from north-to-south and earth rotation would favor deflecting the moving air from east to west. But because of frictional drag, the resultant wind would be directed from the northeast to the southwest.

Global-Scale Pressure and Wind Systems

Since wind depends upon pressure variation and since air motion can, in turn, affect pressure patterns, it is appropriate to examine the global characteristics of these two weather elements together. As an aide in this discussion, frequent reference should be made to the schematic model shown in Figure 2.25 and to the isobaric maps of Figure 2.26.

On both of the isobaric maps a zone of low pressure is found closely aligned with the equator. This pressure feature, called the *equatorial low pressure trough (doldrums,* in wind terminology) is a direct result of energy surpluses experienced at the equator. Hence, doldrums are thermally in-

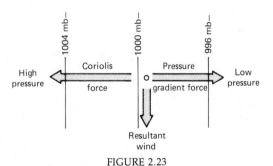

FIGURE 2.23
GEOSTROPHIC WIND. Under frictionless conditions where pressure gradient force and Coriolis effect are balanced, the wind blows parallel to the isobars.

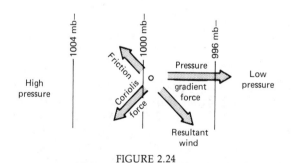

FIGURE 2.24
THE EFFECT OF FRICTION. Friction counteracts the Coriolis effect to deflect wind so that there is a drift of air across the isobars toward regions of lower pressure.

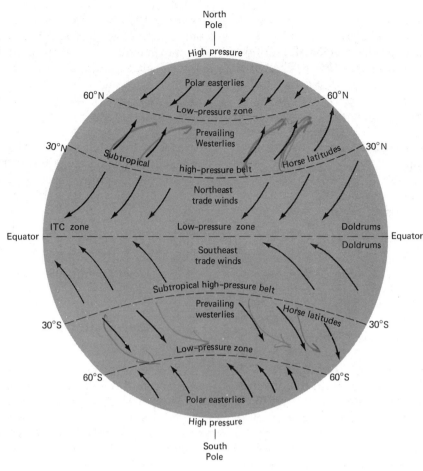

FIGURE 2.25
A SCHEMATIC DIAGRAM OF GLOBAL PRESSURE AND WIND SYSTEMS.

duced. Air becomes light and buoyant when heated, and has a tendency to rise from the surface much like a hot-air balloon. There is no persistent wind direction operating in such an area. The pressure gradient is low, calms are common, and light surface winds come from all directions of the compass. Sailors once dreaded crossing the doldrums and the possibility of resting becalmed for long periods.

Air that ascends in the doldrum zone carries surplus heat to upper altitudes. As shown in Figure 2.25, a portion of this upper air is believed to migrate slowly to the poles. Most of it, however, descends between 20 to 30° latitude to form the *subtropical high pressure systems (horse latitudes).* Horse latitudes are not continuous about the earth. Rather they are found as semipermanent

cells located over the oceans. They are dynamic systems from which air descends and flows toward the equator as the *north-east* and *south-east trades* and toward the poles as the *prevailing westerlies* of the midlatitudes. Calms are not uncommon in the horse latitudes, occurring as much as 25 percent of the time. Horse traders, caught at sea for lengthy periods of time, are thought to have given these pressure systems their name. With freshwater running short, many horses were sacrificed by being thrown overboard. During the remainder of the year winds are variable. The descending dry air normally provides for fair weather, clear skies, and slight precipitation. Some of the largest desert areas of the world are found in the horse latitudes.

As already mentioned, the equatorward flank of

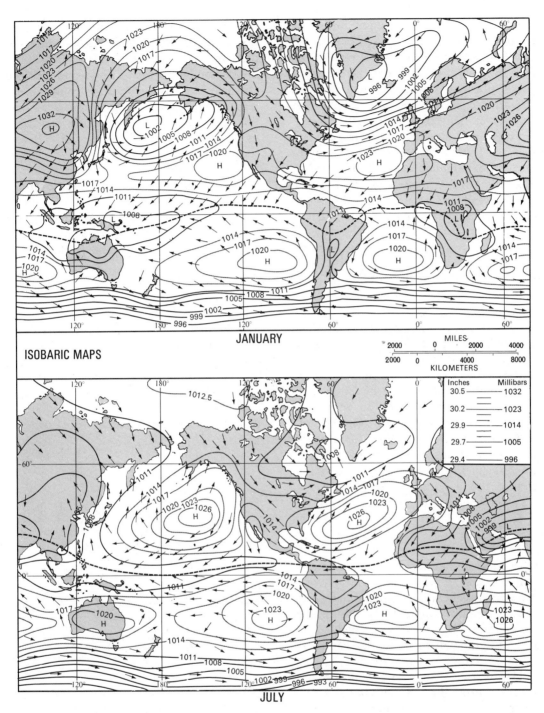

JANUARY

	MILES			
2000	0	2000	4000	
2000	0	4000	8000	
	KILOMETERS			

Inches		Millibars
30.5		1032
30.2		1023
29.9		1014
29.7		1005
29.4		996

JULY

FIGURE 2.26
WORLD DISTRIBUTION OF SEA-LEVEL PRESSURE IN MILLIBARS. (A) January; (B) July.

From Strahler *Modern Physical Geography* © 1978, John Wiley & Sons, New York.

ATMOSPHERIC CIRCULATION

the subtropical high pressure systems forms the *trade-wind* belts. These occur approximately between 5 and 30° N and S. The trades result from the pressure gradient that exists between the subtropical high pressure cells and the equatorial trough. As air moves equatorward it is further subject to Coriolis deflection, causing the resultant flow to be from the northeast in the Northern Hemisphere *(northeast trades)* and from the southeast in the Southern Hemisphere *(southeast trades)*. In each case the wind is named for its direction of origin. (All winds are identified in this way.) The trades are the most regular and persistent of all planetary wind systems. During the days of sailing vessels, they were favored regions of travel.

On the poleward side of the subtropical high pressure cells, between latitudes 35 and 60° N and S, lies the belt of *prevailing westerlies*. Although winds with a westerly component predominate in this region, frequent storms and outpourings of polar air provide for highly variable airflow patterns. In the Southern Hemisphere, where the ocean is essentially uninterrupted by landmasses and where the westerlies blow, the winds are more persistent than their northern counterparts. They also have considerable strength and a reputation for their wild and stormy character.

Radiational heat losses in the vicinity of the poles produce thermal high pressure cells that generate air movements (the *polar easterlies*) toward the equator. Although it was once believed that the easterly flow of air was persistent, its exact characteristics are yet to be determined.

The polar easterlies and prevailing westerlies converge at the *polar front*, a boundary between air that is relatively cold and dense and that which is warmer and more buoyant. Along this contact surface are regions of low pressure, *subpolar lows*. Large storm systems that travel in an easterly direction (resulting from diverse air masses converging at low pressure centers) are frequent along the polar front. During the summer the polar front is positioned in the upper mid-latitudes. In winter, it pushes toward the equator to bring variable winds and precipitation to subtropical regions.

It must be borne in mind that the foregoing discussion is based on some very gross generalizations about the distribution of surface pressure and wind systems. The influence of land and water bodies on air circulation was largely ignored, and regional and local variations in wind direction were not considered. In addition, the discussion began with a cause-and-effect relationship; that is, the equatorial trough was said to result from surplus energy at the equator. In fact, the area of most intense heating varies according as the vertical rays of the sun shift in their latitudinal position. Moreover, the wind systems themselves shift in positon, although not as much as the vertical rays of the sun. Latitudinal migration of wind belts is as much as 10 to 15° over the oceans and somewhat greater over landmasses.

Above the surface winds (aloft of about 1000 meters or 3300 feet), the general air flow displays yet another set of patterns. Lacking the frictional drag associated with surface winds, the Coriolis effect is here more pronounced. These upper level winds are schematically shown in Figure 2.27. Air equatorward of the subtropical highs moves essentially westward. Poleward of the highs the circulation takes the form of a great spiral (counter-clockwise in the Northern Hemisphere) that converges toward the poles. These west winds, called the *upper-air westerlies*, may develop convoluted waves that often effect the circulation of surface winds.

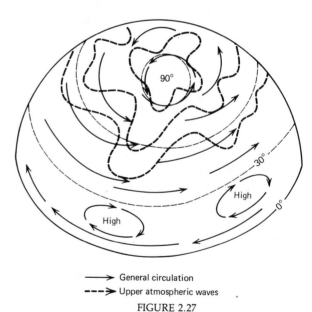

→ General circulation
----→ Upper atmospheric waves

FIGURE 2.27
CIRCULATION OF THE UPPER ATMOSPHERE. The general flow of the upper atmosphere is illustrated. Also shown are possible wave patterns that could occur in the upper westerlies.

Regional Wind Systems

The controls that affect air motion on a global scale also operate at regional levels. One of the best known regional wind systems is the *monsoon*. Monsoons are winds that essentially reverse their direction of flow between summer and winter. The best developed monsoon circulation pattern is found in south and east Asia. This area will be used as an example.

During the summer, intense heating of the Asian continent produces a thermally induced low-pressure system in south Asia. This causes a lower-atmospheric movement of air with a south-westerly flow from the Indian Ocean and the western Pacific toward the continent (Fig. 2.28). This *summer monsoon* is made up of warm and humid air that produces heavy rainfall as it is forced to rise over mountains and hills.

Radiational cooling of the Asian continent during the winter leads to the formation of a strongly developed high-pressure cell in Siberia. The winds blowing out of this area have a northeasterly component. This *winter monsoon* is dry (originating in

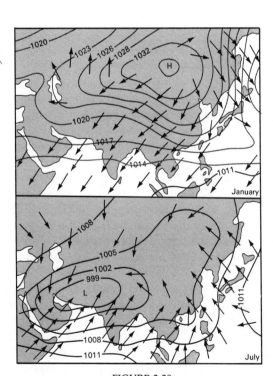

FIGURE 2.28
WIND DIRECTIONS DURING THE ASIAN MONSOONS.
(A) January; (B) July.

From Strahler *Modern Physical Geography* © 1978, John Wiley & Sons, New York.

the interior of the continent, it has little opportunity to absorb moisture) and brings clear weather to most of the monsoon lands. The term monsoon, meaning "reversing", was introduced by Arab traders who would sail from East Africa for India with the onset of the summer winds. In winter, reverse winds carried them home.

Local Wind Systems

It is possible to apply the principles associated with global pressure and wind patterns to local conditions. Two examples follow. Daily *land* and *sea breezes* are typical of tropical coasts, and are of some importance in the middle latitudes during the summer. These breezes arise from a daily reversal of wind patterns occurring in a narrow strip (a few kilometers) of land near the coast. During the day, the *sea breeze* blows from the cooler water body to the hotter land. At night, because of ground radiational heat losses, the *land breeze* flows from the cooler land to the warmer water surface.

Valley and mountain breezes have an airflow pattern similar to that of land and sea breezes. Air in an enclosed mountain valley or on slopes exposed to the sun heats rapidly during the day. As the warmed air expands, it ascends along mountain slopes, producing a *valley breeze*, At night, the upper slopes cool rapidly. Cooler, denser air then slides, as a *mountain breeze*, down the mountainsides by gravitational drainage to the valley below.

ATMOSPHERIC MOISTURE

It has been pointed out that the atmosphere's thermal patterns and wind systems are intimately related to the manner in which energy is distributed over the globe. In a similar fashion, the moisture content of the air is also energy dependent. Movement of water into the atmosphere, the characteristics of its behavior on its journey, and its subsequent return to the surface are all conditioned by the energy available.

The complex set of processes comprising evaporation, rainfall, and runoff is called the *hydrologic cycle* (Figure 2.29). Based on long-term averages for the United States, it is estimated that approximately 76 cm (30 in) of precipitation is inter-

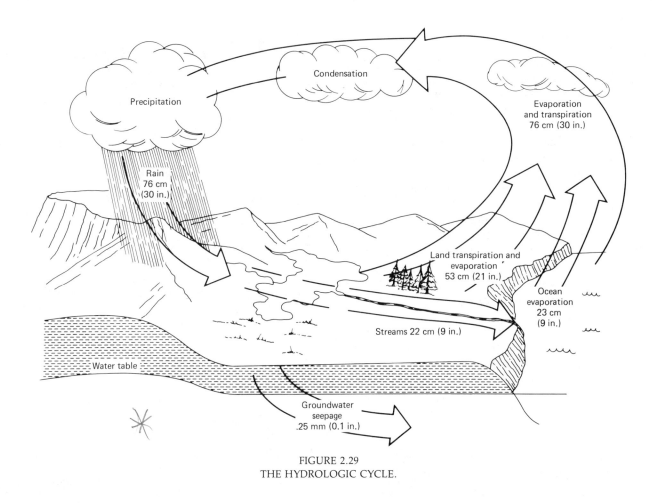

FIGURE 2.29
THE HYDROLOGIC CYCLE.

Labels in figure:
Condensation
Precipitation
Evaporation and transpiration 76 cm (30 in.)
Rain 76 cm (30 in.)
Land transpiration and evaporation 53 cm (21 in.)
Ocean evaporation 23 cm (9 in.)
Streams 22 cm (9 in.)
Water table
Groundwater seepage .25 mm (0.1 in.)

cepted at the surface each year. About 70 percent (53 cm) of this moisture is directly returned to the atmosphere through evaporation and transpiration by plants. The remaining 30 percent (22 cm) provides for surface freshwater supplies and for recharging groundwater resources. This is the same water that eventually returns to the oceans via streams, as surface runoff. The oceans through evaporation, in turn, resupply the continental atmosphere with an amount of moisture sufficient to maintain a long-term water-balance. This exchange of moisture between sea, air, and land is an endless cycle. A glass of water drawn from the tap in Oklahoma City, for example, may contain water molecules that were transpired from a Rocky Mountain pine tree or a stalk of Kansas corn, or some that were evaporated from the Pacific Ocean. Others may have come from a glacier, or from stored groundwater centuries old. All water molecules have participated in countless phase changes in the hydrologic cycle, on land, sea, in the air, and in a variety of geographic locations.

Water makes up a small percentage (less than 5 percent) of the atmosphere's total volume. But water is one of the most important weather variables. And, although water can be found in all three of its known states—solid, liquid, and gaseous—the vapor is the most common state.

Water vapor enters the atmosphere by evaporation from land and open water surfaces and by transpiration from plants—processes collectively referred to as *evapotranspiration*. The rate at which these processes proceed depends largely upon the availability of solar energy. The higher the temperature of the air, the more moisture the air can hold (Fig. 2.30). Thus, it can be expected that evaporation and transpiration are at a maximum in warm and moist climates. Favored locations are the tropics, subtropics, and warm oceanic drifts.

Water vapor absorbed into the atmosphere at

ground level is normally distributed to the air above the surface by atmospheric turbulence, a mixing action that takes place in unsettled air. In stagnant air, evaporation and transpiration cease as soon as a thin layer of air adjacent to the water supply becomes saturated with moisture. But stagnant air is a relatively uncommon phenomon.

The water vapor content of air determines its *humidity* status, a weather element that has been represented by several mathematical indices. One such index is *relative humidity*: the ratio of water vapor present in the air to the amount the air could hold if saturated at the same temperature.

$$\frac{\% \text{ Relative}}{\text{humidity}} = \frac{\text{water vapor present}}{\text{water vapor holding capacity}} \times 100$$

Other humidity indices include: *absolute humidity* (grams of moisture per cubic meter of air) and *specific humidity* (grams of moisture per kilogram of air).

Transfer of moisture to the atmosphere as water vapor requires an expenditure of energy. The energy that transforms water from liquid into gas may be thought of as temporarily locked up (or stored) in the water vapor. Unavailable for the immediate warming of either the surface or the atmosphere, it is called the *latent heat of vaporization,* and represents a huge potential energy reserve within the atmosphere. Latent heat of vaporization is about 585 calories per gram of water at 20° C (68° F). Note that latent heat differs from sensible heat which can be measured with a thermometer.

Water vapor diffused into the atmosphere over an evaporating or transpiring surface does not normally stay in its original locale. Rather, varying atmospheric pressure patterns cause winds to shift and redistribute their moisture load. In this process, the stage is set for mechanisms whereby water will, once again, change its form and simultaneously release latent heat. This occurs when the air is saturated and no longer capable of holding the water vapor it has absorbed.

Practically all of the air within the troposphere contains some moisture. Normally, most surface air contains less moisture than it is capable of holding. There are times, however, when the surrounding air is saturated; that is, there is 100 percent relative humidity. When such conditions occur, the air is said to have reached its *dew-point;* that is, a temperature at which it is fully saturated and below which *condensation* will occur. Condensation is evaporation in reverse. It represents the metamorphosis of water from gas to liquid droplets, and occurs in conjunction with a decrease in air temperature. Figure 2.31 illustrates an analogy between a sponge and an atmosphere's ability to hold moisure. In part (*A*) of the figure the sponge has taken up some water, but can absorb more. Similarly, the parcel of air contains water vapor, but is not saturated. Part (*B*) shows both air and the sponge holding their capacity amounts—saturated. In (*C*) the temperature has dropped and has reduced the atmosphere's ability to hold moisture, some of which must be released from the vapor state. The consequence is the condensation of liquid droplets, incipient cloud formation. Our parallel with the sponge shows it being squeezed, reducing its holding capacity, and releasing water. If the dew-point happens to be below freezing, *sublimation* rather than condensation may occur. Sublimation is a process in which water vapor changes from a gas into a solid

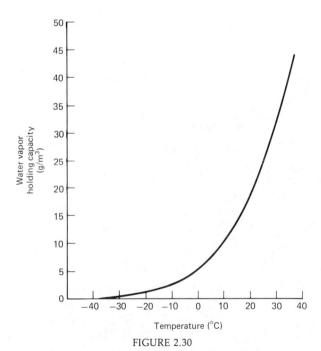

FIGURE 2.30
WATER VAPOR HOLDING CAPACITY OF AIR. As temperature rises, air has an increased capacity to hold moisture. The increase is logarithmic; that is, the capacity increases at an increasing rate for equivalent temperature increments. (This graph permits calculation of the amount of water vapor that will condense from a volume of saturated air when cooled.)

ATMOSPHERIC MOISTURE

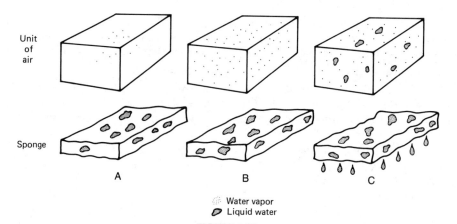

FIGURE 2.31
ANALOGY OF THE SATURATION AND DEW POINT TEMPERATURE OF AIR AND THE
WATER HOLDING CAPACITY OF A SPONGE. (See text for explanation.)

(or from a solid into a gas) and is the principal mechanism by which snow crystals are formed. The visible expression of atmospheric condensation (or sublimation) is the formation of fog at the earth's surface and of clouds aloft.

Fog is essentially a cloud whose base is in con-

tact with the earth's surface. It forms when saturated air experiences a temperature drop. Assume that a location has a mid-afternoon temperature of 10° C (50° F) and 80 percent relative humidity. Referring to Figure 2.30, we can see that the air is capable of holding 10 g/m³ of moisture. Since the

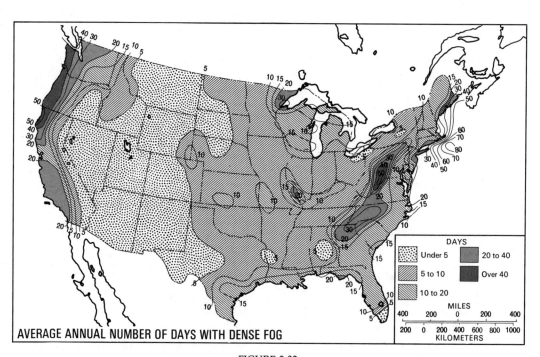

AVERAGE ANNUAL NUMBER OF DAYS WITH DENSE FOG

DAYS

Under 5   20 to 40
5 to 10   Over 40
10 to 20

MILES
400 200 0 200 400
200 0 200 400 600 800 1000
KILOMETERS

FIGURE 2.32
AVERAGE NUMBER OF DAYS WITH FOG IN THE UNITED STATES. The most fog-prone areas in the United States are along the New England and Pacific Coasts, where moist air is cooled by cold ocean waters. Other areas of frequent occurrence are in mountainous regions. Fog seldom occurs in the warm, dry air of the deserts.

relative humidity is only 80 percent, the amount of water-vapor actually held by the air is 8 g/m^3. If the night-time temperature, because of radiational heat loss, drops to 0° C (32° F)—where the air can only hold 5 g/m^3 of moisture in vapor form—a condensation of 3 g/m^3 of moisture will occur and fog will result. This is an example of *radiation fog;* that is, fog produced by temperature decreasing because of energy losses through ground radiation. Another way in which fog can form is when warm, moist air is chilled below its dew-point by coming in contact with a cold surface. Because this normally occurs in association with the horizontal movement of air, it is referred to as *advection fog* (Fig. 2.32).

Clouds are patches of air removed from the earth's surface that contain water in liquid and/or solid form. The controlling factor in the development of clouds is the lifting of air. Air that ascends experiences environments of continually decreasing atmospheric pressure. This affects rising air by permitting it to expand and cool as it lifts. The rate of this temperature decrease with distance from the surface is predictable and known as the *adiabatic temperature lapse rate* (Fig. 2.33). After such a parcel of rising air has cooled below its dew point condensation takes place, and several hundred water droplets may form within a single cubic centimeter of air. Their average size ranges from 10 to 20 microns (.0004 to .0008 in), small enough to be kept aloft by the bouyancy of the atmosphere.

Clouds can be classified into family groupings acording to their height and shape (Fig. 2.34). Since the source of moisture for cloud development originates at the surface, clouds with low bases normally contain the greatest density of water; high clouds hold the least. This criterion provides for three altitudinal classes: (1) the *cirrus group*—or high family (found above 6000 m or 20,000 ft)—is composed of thin, wispy clouds, made up largely of ice-crystals; (2) the *alto group*—or intermediate level family (lying between 2000 and 6000 m)—contains denser and deeper clouds; and (3) the *strato group* is the low-level family (below 2000 m or 6500 ft) and contains the deepest and densest clouds.

Cloud shape is primarily determined by the processes that cause the air to lift. Broad-scale ascension, as with warm fronts, normally leads to

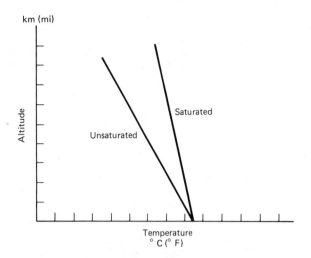

FIGURE 2.33
ADIABATIC HEATING AND COOLING. Lifting air experiences cooling; subsidence of air induces warming. Unsaturated air changes temperature at the rate of 10 C°/1000 m (5.5 F°/1000 ft). Saturated air, when lifted, decreases at about 6 C°/1000 m (3.2 F°/1000 ft). This lower rate results from the release of latent heat.

thick, blanketlike forms. These horizontally structured cloud shapes are referred to as *stratiform.* Rapid lifting of air promotes the vertical development of clouds, or a *cumuliform* shape. The height and shape categories, when combined, provide a cloud's classification. For example, a cloud in the *cirrus family* that is vertically developed is a *cirrocumulus.*

An average sized raindrop has an approximate volume equal to 1 million cloud droplets. In the warm clouds (those with temperatures above freezing) of the tropics, raindrops are believed to form through coalescence of droplets. This involves oversized cloud droplets (larger than 20 microns) colliding with smaller drops and absorbing them. As a droplet grows, its rate of fall increases, leading to more frequent collisions and additional growth. Turbulence and overturning within the cloud maintain this growth process until the mass of the droplet becomes sufficiently heavy to fall to the ground.

In the middle latitudes, where clouds extend into altitudes where temperatures are below freezing, raindrop formation is believed to proceed somewhat differently. These cool clouds contain liquid water droplets at temperatures well below freezing (*supercooled water*) along with frozen

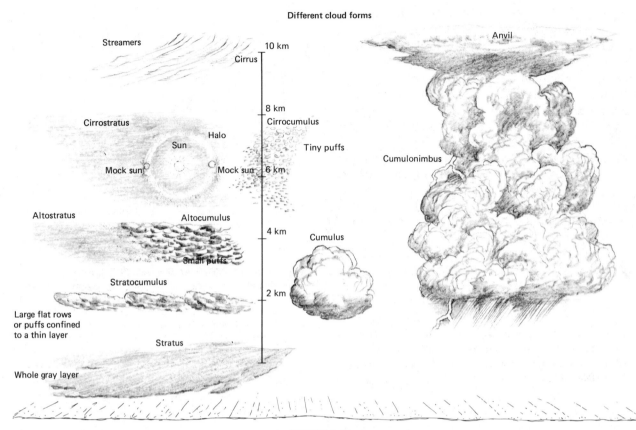

FIGURE 2.34
CLOUD FAMILIES.

water droplets. The process involves water vapor crystallizing on the ice and simultaneously reducing atmospheric humidity. The latter situation permits the supercooled water to evaporate into the air and become available to repeat the cycle.

Theoretically, all liquid water would eventually disappear and the ice crystals would become large enough to fall. If the crystals fall through warm air, they melt and arrive as raindrops; if they fall through cold air, they reach the ground as snow.

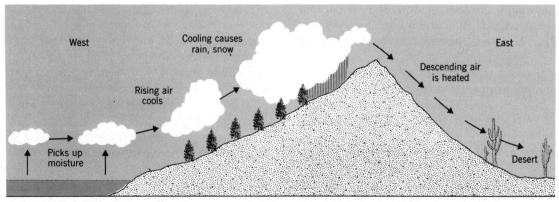

FIGURE 2.35
THE OROGRAPHIC EFFECT. In areas where the prevailing wind direction is perpendicular to a mountain range, the influence of topography on local climate can be dramatic. Forced ascent on the windward side of mountains may provide copious precipitation. The leeward side, experiencing descending air, is often moisture deficient.

THE ATMOSPHERE

The most common mechanisms by which air rises to form clouds (and precipitation) is *orographic, convective,* and *frontal-induced condensation. Orographic clouds* develop when large masses of moist air are forced to ascend the slopes of a mountain range, as shown in Figure 2.35. In areas where mountain ranges lie in opposition to the prevailing wind direction, windward slopes (where adiabatic cooling takes place with ascent) are usually cloudy and wet; leeward slopes (where adiabatic heating takes place with air descent) tend to be dry with clear skies.

Convective clouds result from the lifting of warm air because it is lighter than surrounding air. Convective clouds are most commonly found in tropical climates and in the middle latitudes during the summer. Solar heating warms certain surfaces more than others. The overlying heated air becomes bouyant and lifts from the surface, much like an updraft in a chimney. Air that spontaneously rises because it is warmer than surrounding air is said to be *unstable.* To determine whether air will lift or be *stable* (resist lifting) it is necessary to compare the cooling rate of lifting air to that of the surrounding still air. Figure 2.36 *A* illustrates a situation in which air is forced to rise in stable air. Note that as the air parcel rises

it becomes cooler than the surrounding atmosphere. When no longer forced upward, this air will sink back to the surface. Convective clouds cannot form under stable conditions. Figure 2.36 *B* represents convective lifting of unstable air. As the parcel rises from the surface it cools at a lower rate than its surrounding air. Thus, it is lighter, more buoyant, and continues to rise. When saturation and condensation occur, the parcel's rate of cooling is even lower because of released latent heat. This results in an even greater difference between the parcel's and the still air's temperatures, increased atmospheric instability, and a hastening of the parcel's upward ascent. The resulting cloud form (due to instability and adiabatic cooling) is puffy cumulus, a few kilometers in diameter. The air removed from the surface is replaced by downdrafts of air between clouds, creating a patchwork of clouds and open sky. The variation in motion between drafts (*thermals*) and downdrafts is often the cause of a bumpy airplane trip.

When convectional activity is pronounced thunderstorms are apt to form (Fig. 2.37). Generated by intense updrafts, thunderstorms are tall and dense cumulus rain clouds, called *cumulonimbus.* Air rises within the cloud system in what appears

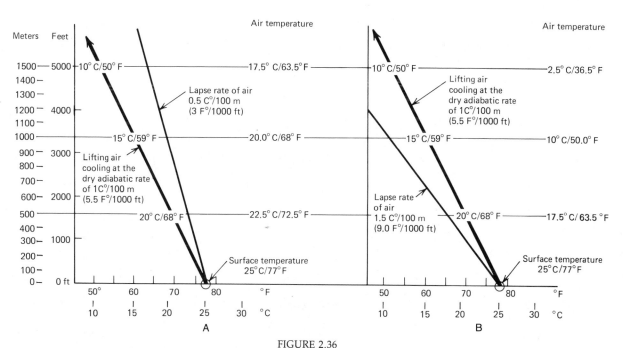

FIGURE 2.36
TEMPERATURE RELATIONSHIPS OF LIFTING AIR UNDER STABLE AND UNSTABLE ATMOSPHERIC CONDITIONS.

ATMOSPHERIC MOISTURE

FIGURE 2.37
AN APPROACHING THUNDERSTORM. A fully developed cumulonimbus cloud (thunderhead) photographed in Oklahoma at an altitude of 9000 m (30,000 ft). The top of the thunderhead is more than 13,500 m (45,000 ft) high.

to be a series of bubbles and has an accompanying downdraft associated with falling precipitation. Figure 2.38 illustrates the characteristics of a mature thunderstorm. The cumulonimbus cloud may contain rain, snow, and ice crystals, even in midsummer. The snow and ice, however, usually melt before reaching the surface. Winds are violent within the storm, creating highly charged electrical fields that produce lightning and thunder. Vertical wind speeds are vigorous, attaining speeds up to 900 m per minute (3000 ft per minute) at low and intermediate levels of the storm. After the storm extends to higher altitudes (between 6 to 12 km or 4 to 8 mi), vertical velocity diminishes and the cloud's top is often dragged downwind with upper atmospheric flow to form an *anvil head*. As the storm continues to develop, a normal sequence of events is that the downdraft keeps expanding until it cuts off the inflow of warm surface air. Without a fresh supply of warm air the thunderstorm loses its energy and slowly dissipates.

Frontal clouds are formed when relatively warm, moist air is forced to rise over denser and cooler air masses (Fig. 2.39). The resultant lifting produces adiabatic cooling and condensation in a manner parallel to air being forced to rise up the side of a mountain. These processes are operative in the middle latitudes and high latitudes where

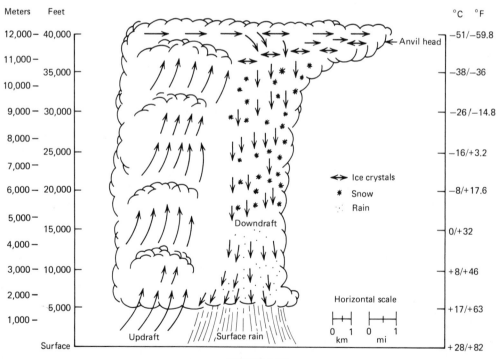

FIGURE 2.38
STRUCTURE OF A THUNDERSTORM.

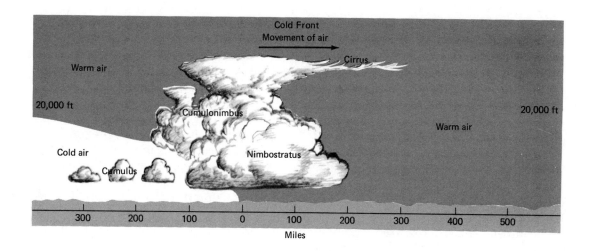

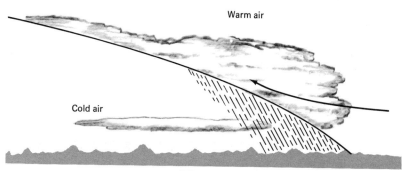

FIGURE 2.39
GENERAL CHARACTERISTICS OF FRONTAL CLOUDS.

they often represent the primary moisture supplying mechanism for an area.

Warm front clouds tend to horizontally structure. Cold front clouds are usually well developed in the vertical. Precipitation from warm front clouds is generally of low intensity (drizzle or light rain), but of long duration. Moistening the soil very gradually, this form of rainfall is usually beneficial to plants and to the recharging of soil water reserves. Cold front precipitation is normally intense and of brief duration. The old expression "it's raining cats and dogs," referring to cloudbursts discharging copious amounts of water is applicable to the cold front. Large amounts of water reaching the surface over very short intervals of time permit little infiltration of rainfall into the soil. The result is that soil erosion and localized flooding often occur. The origin of frontal systems and their behavior will be discussed in greater detail in the last section of this chapter.

Clouds are important weather variables, supplying the moisture that provides precipitation to the earth's landmasses. It is from this ultimate source that rivers are nourished, plants are sustained, and fresh water is made available to man. The several forms of precipitation, resulting from clouds are now listed and explained.

1. *Rain* is the most common and ubiquitous form of precipitation. Rain is comprised of water droplets larger than 0.5 mm (0.02 in.).

2. *Drizzle* occurs when cloud droplets have not attained much growth. The average droplet size is less than 0.5 mm.

3. *Sleet* results when raindrops form in warm, upper air and fall through colder, lower lying air with temperatures below freezing.

4. *Snow* is formed by ice-crystal growth in lifted air that has temperatures below freezing.

5. *Hail* is frozen water droplets. Hail forms when cloud droplets are carried in updrafts to parts of the cloud where temperatures are below freezing. If overturning within the cloud is active, repetition of the process freezes the droplet; induces collision with other droplets for growth and; refreezes the added water to produce hailstones.

Global Precipitation

When the distribution of precipitation over the earth's surface is mapped, patterns or zones of marked contrast can be recognized. Amounts of precipitation vary latitudinally, seasonally, and in reliability as well as in overall quantity. If all of the earth's annual precipitation could be collected at the surface at one time, the depth of water would average about one meter. However, this figure hides more than it reveals, because: (1) the mean depth over the continents, where man lives, is only two-thirds of the average global value; and (2) variation in rainfall over landmasses can range from less than 10 cm (4 in.) to over 1,000 cm (400 in.) per year.

The specific precipitation characteristics of a location are determined by two primary factors: air motion and the atmosphere's initial moisture status. In regions where air is lifting, adiabatic cooling normally results in cloud formation and precipitation. Recalling our previous discussion of atmospheric circulation, we may expect the tropical convergence zone and the polar front regions of the middle latitudes to be favored locations for such action (Fig. 2.40). Where air subsides, as in the subtropic highs, adiabatic warming is unfavorable for cloud development and precipitation is relatively low.

The atmosphere's moisture status is dependent also upon the initial surface temperature of the air and its region of origin. If surface temperatures are high, the atmosphere can absorb relatively large quantities of water vapor. When temperature is low, the air's moisture-storing capacity is low. Thus, we can predict high potential water-vapor storage near the equator and minimal stor-

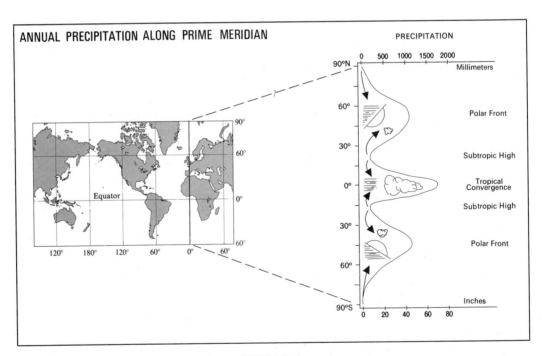

FIGURE 2.40

MEAN ANNUAL PRECIPITATION AMOUNTS ALONG THE PRIME MERIDIAN. As discussed in this chapter, precipitation totals are relatively high in regions of air convergence, but low in areas of divergence. Temperature's role in determining moisture availability is also important. The tropics tend to be high, whereas polar areas are low.

age at the poles. However, a mass of air may have the ability to store large quantities of water, but moisture cannot be absorbed if the region of origin of the air is a dry, continental location.

Figure 2.41 is a map of mean annual precipitation for the world. Patterns are identified by *isohyets*—lines connecting points of equal precipitation. The equatorial region (with warm, moist, converging air) is seen to have some of the world's wettest climates. Indeed, in some tropical areas where orographic effects are significant, rainfall may exceed 1200 cm (480 in.) per year. Polar areas (with cool, dry, subsiding air) are low in total moisture receipts. Wet, west coasts are found in middle latitudes; these areas receive the impact of oceanic, westerly air masses. Equatorward of these locations are the dry, subtropical west coasts that are subject to the descending air of the subtropical high pressure systems. Throughout the remainder of the middle latitudes, precipitation is modest, except for interior continental locations (distant from moisture sources) that tend to be moisture deficient. The dominant controls over moisture distribution patterns in this part of the globe are: (1) the varying position of the polar front and its frequent storm systems; and (2) intense surface heating during the summer that causes convective showers, especially in subtropical realms.

Mean annual precipitation is an important aspect of a location's climate. Yet, mean annual precipitation provides information on only one aspect of the complex moisture picture. Equally significant are: (1) the seasonal distribution of rainfall (Does it arrive in the winter, the growing season, or is it evenly distributed throughout the year?); (2) the intensity of rainfall (Does it fall in torrential downpours that cause flash floods and rapid runoff, or does it fall in the form of a light drizzle that can soak into the soil?) and; (3) the reliability of precipitation amounts (Can the same amount be predicted from year to year, or is it undependable?). These questions will be discussed in detail in the following chapter.

ATMOSPHERIC DISTURBANCES

Migrating cyclonic systems are part of the large-scale motion of the atmosphere. These systems alter our daily weather and constitute a basic part of an area's climate. Cyclonic systems are identified by the inspiraling and lifting of air; that is, a low-pressure circulation pattern. When these systems are weak, cloudiness and light rain normally occur; when intense, their winds may be violent and precipitation heavy. Cyclonic systems can be classified into three general categories on the basis of size. From largest to smallest, they are the *extratropical cyclone*, the *tropical cyclone*, and the *tornado*.

The *extratropical cyclone* is a common weather disturbance throughout the middle and high latitudes. It forms along the boundary between air masses containing different properties. An *air mass* is defined as a large body of air within which the vertical gradients of temperature and moisture are relatively uniform. Temperature and moisture characteristics of air are largely determined by the region in which the air originates. For example, in the interior of northern Canada and Siberia the atmosphere is cold and normally low in moisture content. Air masses forming over the Gulf of Mexico are relatively warm and tend to have greater amounts of moisture.

Although an air mass's physical characteristics can be modified as it moves into a region where surface temperatures and/or water availability differ from those of its source region, its region of origin is still important to its classification. Figure 2.42 provides a generalized map of air mass source regions with a code to simplify air mass identification. The capital letters (A, AA, P, T, E) relate to thermal characteristics of the air mass and can roughly be equated to their latitudinal position of origin. In their respective order the capital letter A means *Arctic*, AA is *Antarctic*, P is *Polar*, T is *Tropical*, and E is *Equatorial*. The lower case letters (c,m) apply to the nature of the surface, whether it be land (*c, continental*) or water (*m, marine*). Thus, a cP air mass would be *polar continental* (cold and relatively dry), and mT would be *tropical marine* (warm and moist). Arrows indicate the general direction of air mass movements. When two air masses having different temperature and humidity properties converge, they seldom merge to form a smooth gradient. Rather, they maintain a distinct boundary surface and adjust to convergence by having the warm, light air lift in advance of or over the cooler, denser air mass. Figure 2.43 illustrates this process taking

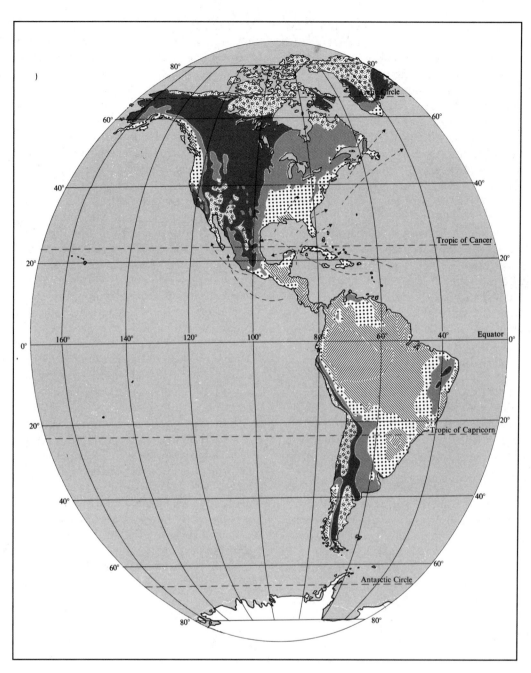

FIGURE 2.41
MEAN ANNUAL PRECIPITATION OF THE CONTINENTS.

place in the origin, development, and decay of an extratropical cyclone as it might appear on a sequence of daily weather maps. In *A*, two air masses are in contact with one another and form a *front*. On a weather map, such as is seen in newspapers or on the television screen, a front is identified as a line where two air masses meet at the earth's surface. In reality, however, the contact surface between the two air masses is three dimensional. In the example provided, cold air is pushing equatorward (Figures *B* and *C*) along the ground around and under a tongue of warm air

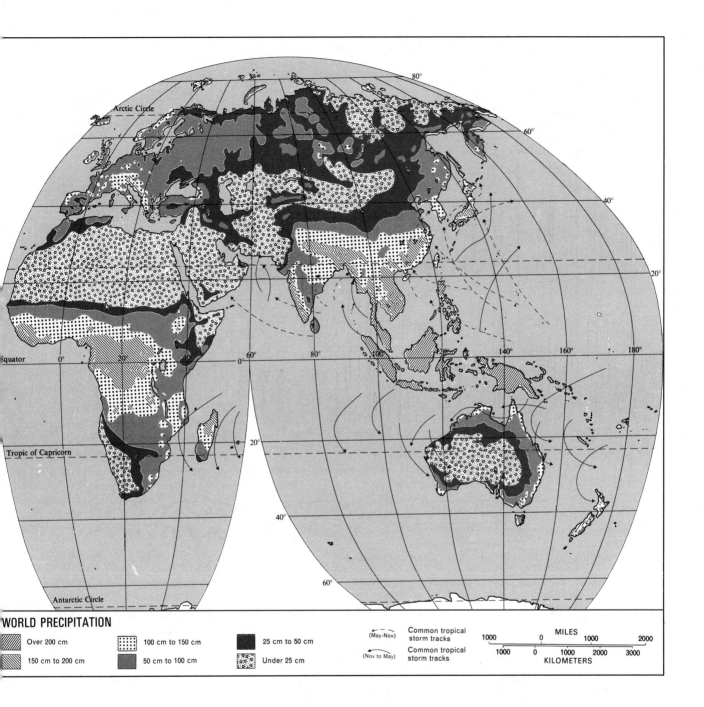

WORLD PRECIPITATION

Over 200 cm	100 cm to 150 cm	25 cm to 50 cm
150 cm to 200 cm	50 cm to 100 cm	Under 25 cm

(May-Nov) Common tropical storm tracks

(Nov to May) Common tropical storm tracks

MILES
1000   0   1000   2000

KILOMETERS
1000   0   1000   2000   3000

that is moving poleward. The latter is known as the *warm sector* of the storm system. The resultant lifting and convergence of air forms a core of low pressure at the peak of the indentation, or wave, between the two air masses. Air movement toward the low, coupled with Coriolis deflection, causes a counterclockwise flow (in the Northern Hemisphere) of air around the low. In the Southern Hemisphere similar conditions produce a clockwise air flow pattern. The leading edge of the storm system—where warm air is replacing cold air—is called the *warm front*. The trailing edge of

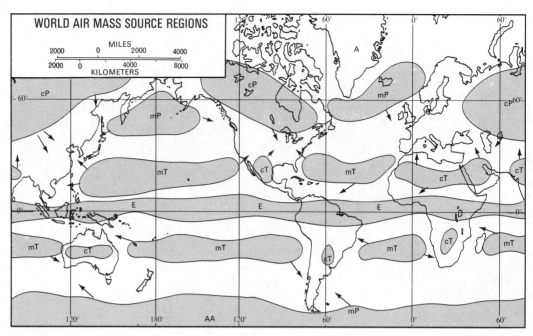

FIGURE 2.42
AIR MASSES AND THEIR SOURCE REGIONS.

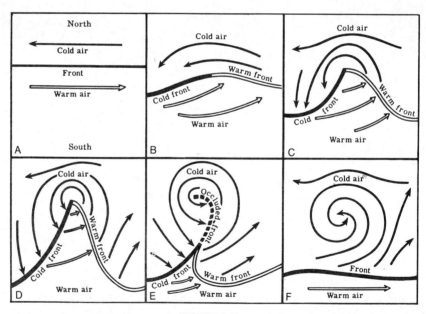

FIGURE 2.43
STAGES IN THE DEVELOPMENT AND OCCLUSION OF AN EXTRATROPICAL CY-
CLONE. (A) Initial stage. (B) Incipient cyclonic circulation. (C) Warm sector between
fronts becomes well defined. (D) Cold front closing in on warm front. (E) Occlusion.
(F) Dissipation stage.

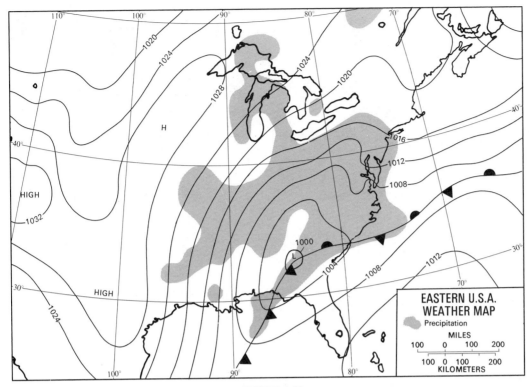

FIGURE 2.44
SURFACE WEATHER MAP DEPICTING AN EXTRATROPICAL CYCLONIC STORM.

the storm—where cold air is forcing warm air aloft—is the *cold front*. Along both of these fronts appreciable cloudiness and precipitation frequently occur (Fig. 2.44). Eventually the cold front will squeeze out the less dense air of the warm sector and merge with the cool air in advance of the warm front (Fig. 2.43E). When this occurs, the system is said to be *occluded,* and the storm system is in a dissipating stage.

Extratropical cyclones are the largest (ranging from 200 to over 3,000 km in diameter) and most frequently occurring of any cyclonic system. Once formed, they travel a general eastward path, called a *storm track,* moving at an average speed of 30 to 50 km (18 to 30 mi) per hour (Fig. 2.45).

Tropical cyclones, also known as *hurricanes,* are much smaller weather disturbances. Although they only range between 150 to 500 km (100 to 300 mi) in diameter, they attain wind speeds of 120 to 200 km (75 to 125 mi) per hour and more. A tropical cyclone is one of the most powerful and dev-

astating storm systems in the atmosphere (Fig. 2.46).

Tropical cyclones form over warm ocean waters (temperatures of 27° C/80° F, or more) between latitudes 8 to 15°N and 8 to 15°S. These warm, moist surfaces produce unstable air with a high water-vapor content. Under the right conditions—when upper atmospheric air is diverging—the lifting and condensation of oceanic air releases huge quantities of latent heat. The storm system then intensifies and atmospheric pressure drops rapidly. After forming, the system moves westward and northwestward with the trades and often curves to the northeast at about 30 to 35° latitude (Fig. 2.47). The storm is born in an oceanic environment. If it moves inland, its source of energy—latent heat of condensation—is cut-off and the system degenerates.

Tornadoes are the smallest and the most violent of all cyclonic storms. Tornadoes normally form along squall lines that travel in advance of a cold

ATMOSPHERIC DISTURBANCES

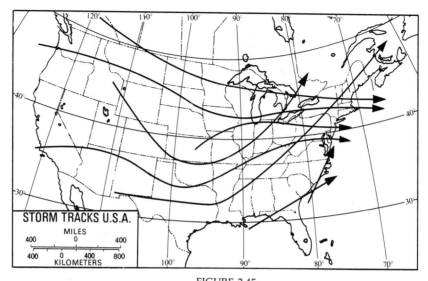

FIGURE 2.45
COMMON TRACKS OF EXTRATROPICAL STORMS ACROSS THE CONTIGUOUS UNITED STATES. The most northerly storm tracks are those most frequently followed during midsummer. As winter approaches, the frontal action of storms tend to pass along the more southerly routes.

front. They occur in every month of the year but are most common in spring and summer. As with a hurricane, a tornado requires the release of latent heat for its development. Such conditions are ideally met when rapidly moving continental polar air masses force moist tropical air to lift and form towering cumulus clouds. Tornado clouds develop a rotational motion and form huge dark funnels with air pressure considerably lower than the surrounding atmosphere. These storms have been known to dip down to the surface with wind speeds exceeding 800 km (500 mi) per hour. Destruction of surface features is most severe under the funnel, an area with a diameter of 90 to 460

meters (300 to 1500 ft). This is illustrated in Figure 2.48. Although very high wind speeds are responsible for a significant part of the damage wrought by tornadoes, the low pressure of the storm system is equally destructive. Strong pressure differentials can build between the exterior and interior walls and roofs of buildings. As the pressure suddenly and dramatically drops outside of a house, for example, the walls and roof can literally be blown out from the foundation. The suction within the funnel can be so great that water from streams and lakes can be drawn up into the funnel's cloud. The strange tale in Ripley's *Believe it or Not* about a rainstorm in which fish and frogs fell from the

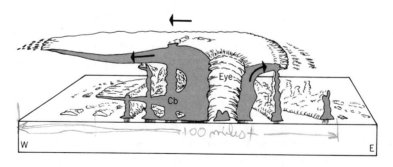

FIGURE 2.46
THE VERTICAL STRUCTURE OF A HURRICANE.
Cb = Cumulonimbus clouds.

THE ATMOSPHERE
54

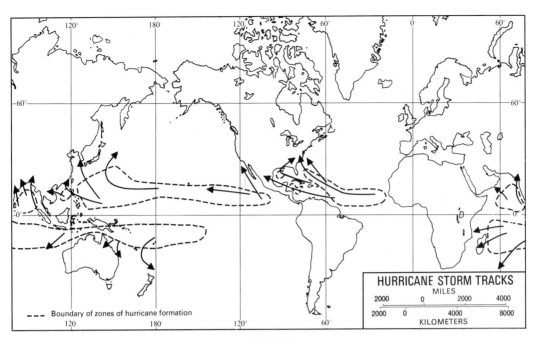

FIGURE 2.47
TRACKS OF MAJOR HURRICANES FOR THE YEARS 1954 TO 1965.

sky was probably the result of a tornado sucking up a pond and its contents that were later dropped as the storm system decreased in strength. As can be seen in Figure 2.49, the frequency of tornado occurrence is not uniformly distributed throughout the United States. It is greatest in the Great Plains, midwest and southeastern states, and relatively low in the west and northeast.

This chapter has dealt with the dynamics of the atmosphere—the active elements that provide for the complex daily pattern of weather from place to place over the earth. It might seem that the

FIGURE 2.48
A TORNADO FUNNEL AT ENID, OKLAHOMA.

ATMOSPHERIC DISTURBANCES

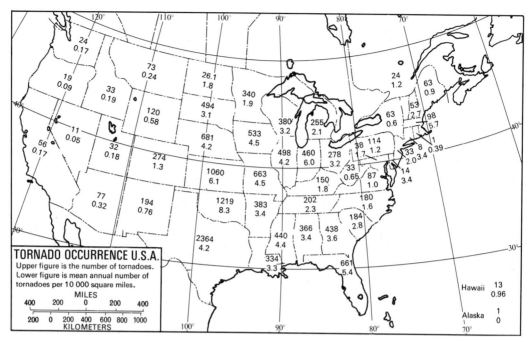

FIGURE 2.49
FREQUENCY OF TORNADOES IN THE UNITED STATES, 1956 TO 1971.

number of parameters that have been considered, coupled with their variable nature, would make atmospheric characteristics of the earth difficult to comprehend for more than a single location for a short period of time. However, certain parameters reoccur in predictable combinations over broad enough areas to permit the identification of large-scale climatic regions. These parameters will be considered in the following chapter.

KEY TERMS AND CONCEPTS

composition of the atmosphere
density of the atmosphere
thermal structure of the atmosphere
troposphere
short-wave radiation

long-wave radiation
normal temperature lapse rate
greenhouse effect
stratosphere
ozone layer
mesosphere
insolation

ground (terrestrial) radiation
latent heat
daily temperature variation
seasonal temperature variation
summer solstice
winter solstice
autumnal equinox
vernal equinox
temperature range
continentality
isothermal maps
atmospheric pressure
pressure gradient force
isobars
Coriolis effect
Geostrophic wind
global-scale pressure patterns
global-scale wind systems

equatorial low pressure trough
doldrums
subtropical high pressure systems
horse latitudes
trade winds
prevailing westerly winds
polar easterly winds
Upper-air westerlies
monsoon winds
land and sea breezes
valley and mountain breezes
hydrologic cycle
evapotranspiration
humidity
relative humidity
latent heat of vaporization
dew point
condensation

fog
clouds
unsaturated adiabatic lapse rate
saturated adiabatic lapse rate
orographic clouds
convective clouds
unstable air
stable air
thunderstorms

frontal clouds
global precipitation patterns
isohyet maps
extratropical cyclone *midlatitude cyclone*
air mass
weather fronts
tropical cyclones (hurricanes)
tornadoes

DISCUSSION QUESTIONS

1. Devise a general statement that can be applied to the expected changes in temperature and atmospheric pressure as one ascends in altitude within the troposphere. Which changes most rapidly?

2. Why is short-wave energy less effective than long-wave energy in heating the atmosphere?

3. Since the earth is farthest from the sun in July, and is then receiving less energy than at any other time of the year, why is July the hottest month for most Northern Hemisphere land stations?

4. Explain how a position is located on the earth's surface by using latitude and longitude.

5. How would monthly temperature graphs differ for locations: (1) along the northwest coast of the United States, and (2) in central Nebraska? What explanation can be offered for the differences in the two graphs?

6. What is the relationship between pressure gradient force and wind velocity?

7. What are isotherms? How can isothermal maps be useful?

8. What effect do large landmasses have on isotherm patterns in January and July?

9. Describe the locations of the globe's principal pressure systems, giving their approximate latitude. Why do these systems shift in latitude during the year?

10. Explain the Coriolis effect. What causes this effect?

11. Describe the patterns of surface winds and explain how they are related to global pressure systems.

12. What are the sources of atmospheric moisture?

13. Define humidity.

14. What is the dry adiabatic lapse rate? How does this differ from the normal temperature lapse rate? Why does the wet adiabatic lapse rate differ from the dry adiabatic lapse rate? Explain how adiabatic cooling is related to condensation and precipitation producing processes.

15. Define air mass.

16. Explain how an extratropical cyclonic storm forms along the polar front. Describe the structure of the storm's cold and warm fronts. How do extratropical cyclones influence weather in the middle and high latitudes?

17. Describe latitudinal variation in precipitation. Identify the atmospheric controls that contribute to areas of high and low precipitation.

REFERENCES FOR FURTHER STUDY

Byers, Horace R., *General Meteorology*, Fourth Edition, McGraw-Hill Book Co., New York, 1974.

Chang, Jen-Hu., *Climate and Agriculture: An Ecological Survey*, Aldine Publishing Co., Chicago, 1968.

Miller, David H., *A Survey Course: The Energy and Mass Budget at the Surface of the Earth*, Commission on College Geography, Publication No. 7, Association of American Geographers, Washington, D.C., 1968.

Petterssen, Sverre, *Introduction to Meteorology*, McGraw-Hill Book Co., New York, 1969.

Sellers, William D., *Physical Climatology*, University of Chicago Press, Chicago, 1965.

Strahler, Arthur N., *The Earth Sciences*, 2nd ed., Harper and Row, New York, 1971.

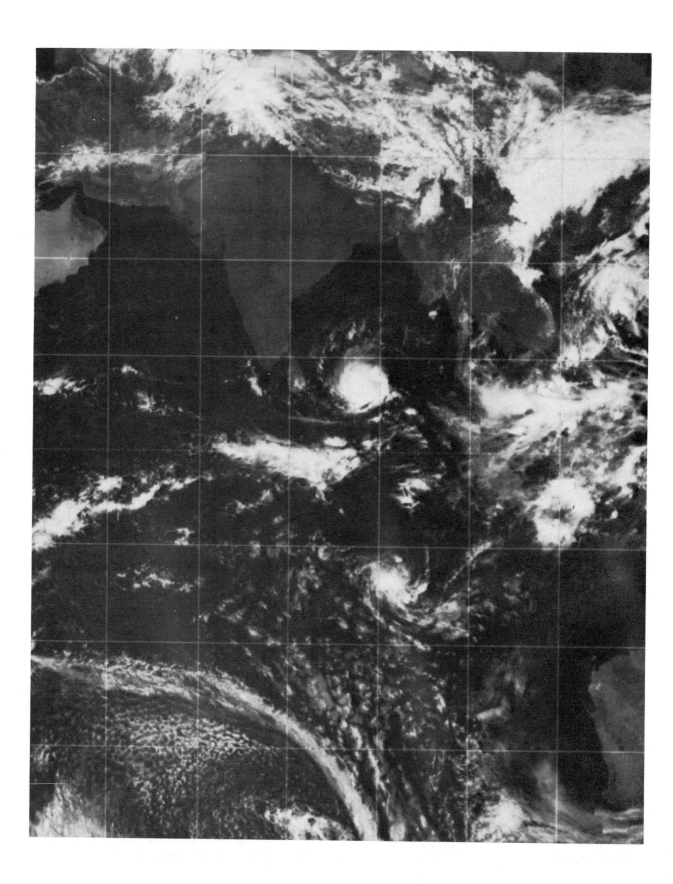

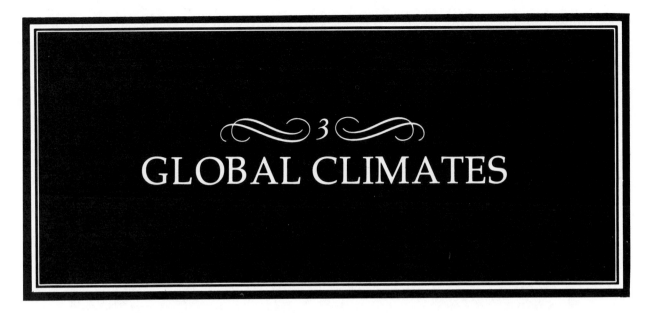

3
GLOBAL CLIMATES

CHAPTER 2 DEALT WITH INDIVIDUAL ELEMENTS OF THE atmosphere. Considered together, these elements constitute weather. Weather may change from day to day, from day to night, and occasionally from hour to hour. Properly used, the term weather should only be applied to describe the state of the atmosphere during short intervals of time. *Climate,* on the other hand, is the generalized description of atmospheric conditions over an extended time scale. Decisions about whether to play tennis, put on a swim suit, or carry an umbrella, are based upon the short-term state of the atmosphere. Planning for irrigation scheduling, storage of heating fuel for the winter months, and the type of vegetables to be grown in the garden, however, require a consideration of climate.

A location's climate is determined by interacting atmospheric processes. Inasmuch as there are a large number of different combinations of atmospheric processes, there are also a large number of climatic types. To provide for an orderly comprehension of these many diverse climates, a classification system is normally employed. This procedure involves grouping together, at different levels of generalization, locations sharing common climatic characteristics. This procedure minimizes the number of variables to be examined and understood at one time.

Numerous systems for classifying climates have been devised. One particularly suited for identi-

fying large-scale regional climates was designed by a German botanist, Wladimir Köppen (1846–1940). The underlying assumption of the Köppen system is that specific forms of natural vegetation are found associated with unique thermal and moisture regimes. As a result, plant associations* are reflections of *climatic regions*—areas in which homogeneous sets of climatic conditions are found. The temperature and moisture conditions that exist at the recognized boundaries of plant associations, in turn, identify the limits (or basic requirements) for each climatic type. Although Köppen's approach is not without its shortcomings, it uses easily obtained climatic data, is readily comprehensible, and offers a useful global perspective for the study of climatic types.

Köppen's classification system is structured about an alphabetical code—using both capital and lower-case letters—that is used for generalizing climatic data. At the most general level, five climatic realms are recognized. Proceeding from the equator to the poles and identified by capital letters, they are: the A climates, or winterless realm of the humid tropics; the B climates which are moisture deficient; the C, or humid and mild winter climates of the lower middle latitudes; the D, or humid and severe winter realm found in

---

* An association is an unique assemblage of plant forms comprised of mixed species that inhabit an area.

upper middle latitudes of the Northern Hemisphere; and the E, polar or summerless climatic types of high latitudes (Fig. 3.1).

A CLIMATES
(THE HUMID TROPICAL REALM)

The world's A climates are humid and winterless. The word "humid" signifies that a location receives enough precipitation to produce surplus moisture at the surface during at least one season of the year. The word "winterless," according to the Köppen system, means that no mean monthly temperature falls below 18° C (64.4° F). This temperature requirement was chosen because of its rather close correlation with the poleward boundary of tropical vegetation associations.

The A climates are dominated by equatorial and tropical air masses and generally occupy an area

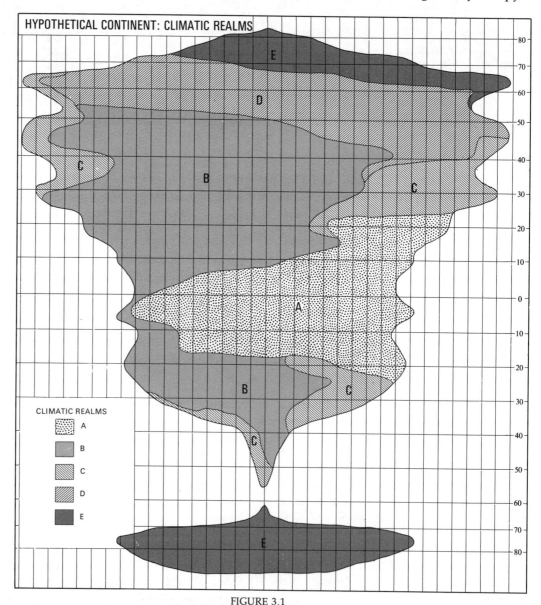

FIGURE 3.1
SCHEMATIC CONTINENT WITH CLIMATIC REALMS. This diagram represents all of the earth's landmasses squeezed together at their respective latitudes. The distribution of the five climatic realms are shown in their appropriate latitudinal position and are proportionally representative of the areas they occupy.

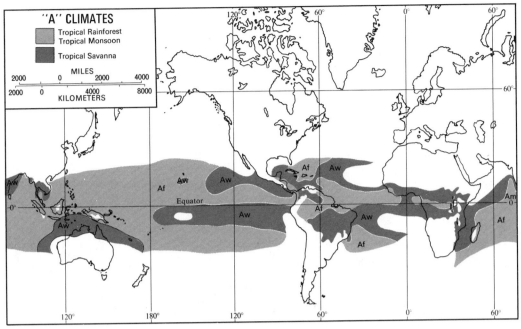

FIGURE 3.2
TROPICAL CLIMATES OF THE EARTH.

extending from 23½° S to 23½° N latitude (Fig. 3.2). Throughout this area the sun is directly overhead during part of the year, annual insolation rates are high, warm temperatures prevail, and normal air flow is equatorward with a westerly component.

Tropical Rainforest Climate (Af)

The Af climate* is dominant in lowland areas that lie within 10° latitude of the equator and also along more poleward and windward tropical coasts. These conditions are met best in the Amazon Basin of South America, eastern coasts of Central America, parts of the Congo Basin of Africa, eastern Madagascar, Indonesia, and in the Philippines.

Because of proximity to the equator, Af climates are among the world's leading recipients of insolation. High energy availability coupled with an essentially uniform daylight period throughout the year provides for both high air temperatures and little seasonal temperature variation. Each

month of the year averages close to 27° C (80° F), and their annual range can be as small as 2 to 3C°. Indeed, the *diurnal range* (difference between day and night-time temperatures) of 8 to 10C° is considerably larger than the monthly variation (Fig. 3.3). Contrary to the once popular belief that this area represented the "torrid zone" of unbearable heat, maximum temperature readings in the tropical rainforest are much lower than for many middle-latitude locations, and rarely surpass 38° C (100° F). The primary thermal characteristics for the region consist of an absence of seasons and a repetitive daily temperature cycle.

Pressure and wind patterns of the tropical rainforest region are largely determined by the seasonal shifts of the equatorial trough. This low pressure area moves north and south across the equator, and is a region toward which the trade winds converge and lift. The trade winds pass over great distances of tropical ocean and absorb large quantities of moisture along the way. Visitors to the rainforest climates frequently complain about the high humidity of the air, claiming that its effects are debilitating. This high moisture content coupled with a surplus energy regime promotes development of unstable air and convectional activity. Short mid-afternoon thunderstorms

---

* In Köppen's alphabetic code, the lower-case letter *f* identifies climatic types with no distinct dry season. In the tropics this means that every month must receive at least 60 mm (2.4 in.) of rainfall.

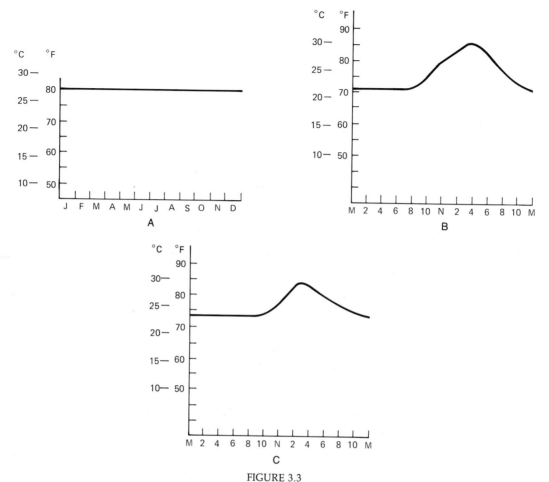

FIGURE 3.3
TEMPERATURE GRAPHS FOR BELEM, BRAZIL (1° 27′ S, 48° 29′ W). (A) Mean monthly temperature. (B) Hourly temperature (January 1944). (C) Hourly temperature (July 1944).

and copious amounts of rainfall are common. Heavy annual rainfall in excess of 2000 mm (80 in.) is characteristic of this climatic realm (Fig. 3.4). No month is truly dry, and all months receive at least 60 mm (2.4 in.) of rain. Although short-period droughts do occur, the tropical rainforest climatic region regularly receives an annual surplus water supply. As a consequence, the area is normally able to meet its evapotranspiration needs.

The Af region's warm climate and abundance of water has fostered an environment favorable to the evolution of one of the earth's most luxuriant plant associations. The growing season is unending, and many plants exhibit no signs of the dormancy so typical of middle-latitude plants. However, the lush vegetation that clothes the surface is but a mask over a soil that (in most of the rainforest climate area) is naturally infertile. These

very old and weathered soils have only a meager supply of nutrients to meet the crop needs of the indigenous populations. Thus, as the map of world population reveals, extensive parts of this region support very low population densities.

Tropical Savanna Climate (Aw)

Poleward of the wet tropical regions, between latitudes 5 to 25° N and S, are the *Tropical Savanna* (Aw) climates.* Figure 3.5 reveals that no large temperature contrasts are found in this area; yet, poleward through the region, the first hints of seasonal temperature variation (typical of the middle

---

* The code letter *w* identifies climatic types with a dry season during the winter. In the winterless realm of the tropics, *w* is related to the time of year when the sun is lowest in the sky.

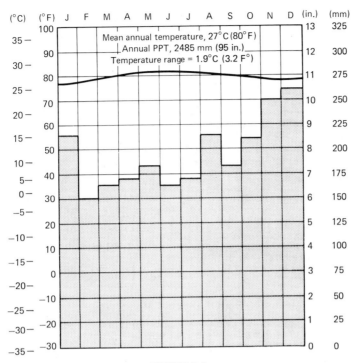

FIGURE 3.4
CLIMOGRAPH FOR SINGAPORE (TROPICAL RAINFOREST).

latitudes) appear. The most distinguishing feature of these winterless climates is a well defined dry period that coincides with the time of year when the sun is lowest in the sky. The length of the dry period varies from about two months near the equator to six months in locations farthest from the rainforest (Fig. 3.6). Tropical savanna climates were so named because their vegetation forms differ markedly from those of adjacent rainforest areas. Consisting mainly of tall grasses, 2 to 4 meters (6 to 12 feet) high, the structural character of the vegetation is directly related to seasonal water deficiencies.*

Rainfall in the savanna is determined by the seasonal migration of the global wind systems. During the solstices the equatorial trough is displaced poleward into the hemisphere receiving most solar energy. Thus, in June the area of convergence moves northward and in December toward the south. With the convergence of humid, unstable air, comes the rainy season and a supply of water that often exceeds 1000 mm (40 in.). This is the season in which the savanna vegetation becomes lush and green, streams are swollen, large expanses of lowland are flooded, and temperatures are somewhat moderated by the extensive cloud cover and cooling effect of the rain (Fig. 3.7).

The rains end for the Aw regions when the shifting of the wind systems result in the retreat towards the equator of the convergence zone and its replacement by the descending air of the subtropical high pressure systems. This relatively stable, drier air has minimal opportunity for ascent or producing rainfall. As a result, soil moisture is depleted; grasses become dry and brittle; trees lose their foliage; and streamflow diminishes to a mere trickle or may be absent. Indeed, the climate of the dry period is frequently akin to that of the deserts and may be considered a case of "seasonal aridity."*

---

* Controversy exists over whether the savanna grasslands are natural vegetation associations or culturally induced. This topic is discussed in Chapter 4.

* An observant student of climatology will find that this book uses the term "seasonal aridity" instead of "seasonal drought." The latter term, widely used in the professional literature for years, is erroneous. Drought implies conditions drier than normally expected. Seasonal dryness in the Aw climates, however, is expected. Thus, it is a normal situation of moisture deficiency and needs a terminology that expresses conditions as they commonly occur.

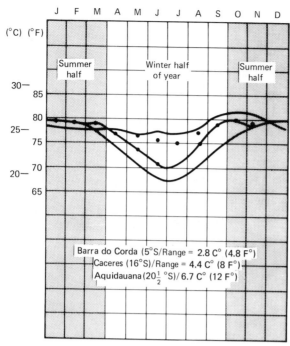

FIGURE 3.5

TEMPERATURE IN THE TROPICAL SAVANNA CLIMATE. The three temperature curves are for locations in Brazil. At 5° S latitude the range is 2.8 C° (5 F°), at 16° S it is 4.4 C° (8 F°), and at 20½° S it reaches 6.7 C° (12 F°). Since these examples represent southern hemisphere locations, the summer half of the year occurs between September 23 and March 21.

Such a pattern of contrasting water availability from season to season has an important influence on economic activities. Although the temperature regime may be suitable to crop growth throughout the entire year, moisture requirements of plants restrict agricultural activity largely to the rainy season. Nomadic herding, still practiced in parts of Africa, follows migration paths that coincide with the greening of forage in relation to the shifting latitudinal position of the rains.

Tropical Monsoon Climate (Am)*

The monsoon tropics have a climatic type sharing the characteristics of both the rainforest and the savanna. Monsoon tropics are found along the west coasts of India and Burma, in the Indochina Peninsula, the Philippines, the western Guinea

---

*The lowercase letter *m* is used to identify the monsoon climates of the tropics.

Coast of Africa, northeastern South America, and along windward coasts of the Caribbean. The term "monsoon" is not restricted to regions affected solely by a monsoonal wind circulation pattern. Rather, the term identifies forested tropical climates with distinct dry periods.

The mean monthly temperatures of the monsoon climates are very similar to those of the rainforest, although the more poleward location of some monsoon regions provides for a greater annual range. One of the peculiar aspects of the temperature pattern, however, is that the maximum is reached in late spring rather than in summer when insolation is at its maximum (Fig. 3.8). Lower summer temperatures result from solar energy being checked in its path to the surface by heavy and persistent rainclouds of the high-sun period.

Precipitation in the monsoon region is heavy when warm, unstable winds are blowing onshore. When a mountain barrier backs a coastline, these rain-making processes can be further accentuated, providing for rain totals that may easily exceed 2000 mm (80 in.) in a matter of a few months. The summertime rainfall pattern, accompanied by dense cloud cover, is followed by a dry season with a tendency for sunny weather and infrequent showers.

The regular seasonal variation in airflow and precipitation patterns is basic to the agricultural systems of monsoon lands. This is particularly true of monsoon India, where the summer season provides about 90 percent of the annual water supply—a period critical for crops such as rice. Each year, the time of arrival and the character of the wet monsoon season is awaited with anxiety. If it is late and accompanied by meager rains, planting time and irrigation schedules are disrupted, threatening famine and mass starvation. When it arrives early and with intense rainfall, flooding, drowned crops, and massive landslides occur.

B CLIMATES (THE MOISTURE DEFICIENT REALM)

The B climates comprise the climatic realm in which there is an annual moisture deficiency and where precipitation is unpredictable. B climates

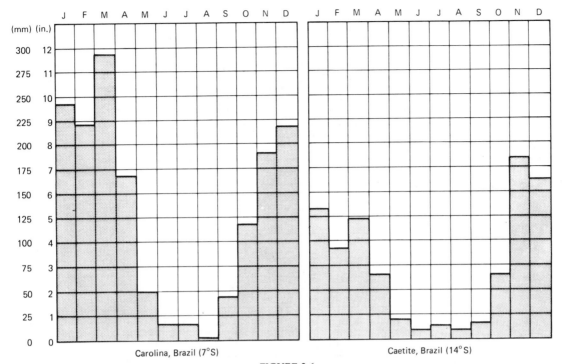

FIGURE 3.6

EXAMPLES OF RAINFALL PATTERNS IN THE TROPICAL SAVANNA CLIMATE. Carolina, Brazil (7° S), near the rainforest boundary, has only three months with rainfall less than 25 mm (1 in.), its annual total equals 1575 mm (62 in.). Caetite, Brazil (14° S), has five months with less than 25 mm of rainfall and annual totals amounting to 889 mm (35 in.).

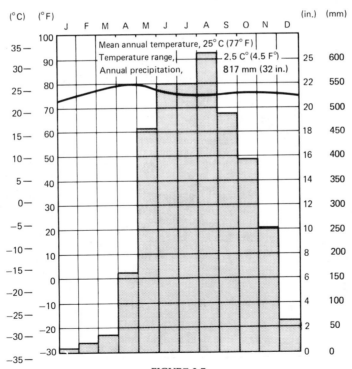

FIGURE 3.7

CLIMOGRAPH FOR MARACAY, VENEZUELA (TROPICAL SAVANNA).

B CLIMATES (THE MOISTURE DEFICIENT REALM)

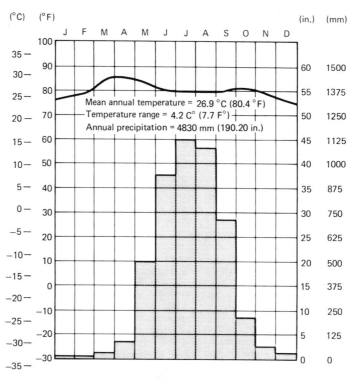

FIGURE 3.8
CLIMOGRAPH FOR MOULMEIN, BURMA (TROPICAL MONSOON).

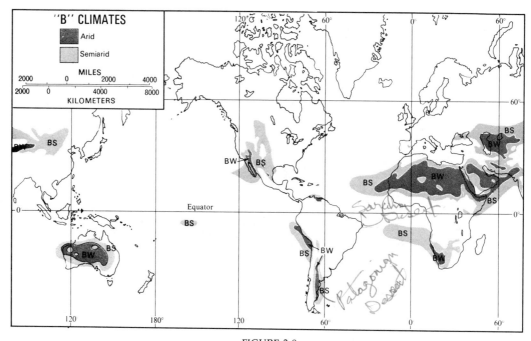

FIGURE 3.9
DRY CLIMATES OF THE EARTH.

range in latitude from the equator to about 50°N and S and are classified as either BW (arid) or BS (semiarid) types. Tropical forms of the B climate (identified by Köppen with a lower case "h") are found in the Sonorran region of the southwestern United States and northwest Mexico, along the coast of Peru and Chile, in central Australia, in southwestern Asia extending from Saudi Arabia through Pakistan, and in both North and South Africa (Fig. 3.9). These dry regions are primarily the product of subsidence of air in the subtropic highs. As the descending air is adiabatically heated, its relative humidity is lowered and its capacity to hold moisture increases. Thus, conditions are unfavorable for clouds to form and for rain to develop. In the middle latitudes, dry climates (coded with a lower case "k") are found in the intermontane basins and Great Plains of North America, in the Soviet Union and northern China, and in southern and western Argentina. The climatic controls creating B climates in these regions are great distances from moist air sources, and mountain barriers that create a rainshadow effect.

Arid Climates (BW)*

Arid climatic regimes occupy approximately 20 percent of the world's land area. A high incidence of clear skies permits maximum solar radiation on land surfaces. Thus, the desert surface heats very rapidly after sunrise, raising ground temperatures and creating very steep, near-surface atmospheric lapse rates. Maximum air temperatures in the tropical deserts (BWh) are higher than for any climatic region—readings above 50° C (122° F) being common. The world's official high temperature of 58° C (136° F) was recorded in Azizia, Libya, on September 13, 1922. After sunset, the pattern reverses. The clear desert skies permit rapid loss of energy through terrestrial radiation, and temperatures drop quickly. As a consequence, temperature patterns exhibit high ranges and maxima, both daily and seasonally (Fig. 3.10). The primary exception to this generalization is found in coastal deserts where cool ocean currents may modify temperature extremes considerably. With temperature maxima usually above the norm for their lat-

itudinal location, the atmosphere of desert regimes possesses high energy availability. The relationship between all this energy available for evapotranspiring moisture and the environmental water supply is a critical factor in determining an area's degree of aridity.

Moisture supply in arid regions is both meager and erratic, and mean precipitation figures provide scant information on expected amounts of available water during any given period. In a 10-year span at Las Vegas, Nevada, for example, annual precipitation ranged from 14.2 to 141 mm (0.56 to 5.53 in.), as shown in Table 3.1. This is a variation from the annual precipitation norm, 99 mm (3.9 in), of 14 and 142 percent respectively. Individual months experience even greater departures. August had a percentage range of from less than 2 to over 539 percent for the same time period. In general, most desert regions receive less than 250 mm (10 in) of precipitation annually.

The moisture that does reach the desert surface is not only unreliable in pattern and low in total amount, but also generally has minimal opportunity to infiltrate the soil. Desert landscapes baked by the sun are not often receptive to water infiltration. Moreover, much of the rainfall arrives as intense showers of short duration. Surface water either becomes rapidly channeled and infiltrates into the more porous stream beds to recharge groundwater storage, or is lost through intense evaporation on the hot desert surfaces.

Scarcity of water has always represented a threat to life in arid climatic regimes. Prior to the Industrial Revolution most desert settlements were found near natural occurring oases or permanently flowing streams such as the Nile in Africa (Fig. 3.11). Today, with the ability to store water and to pump groundwater, both urbanization and agriculture have flourished within the deserts. Nevertheless, the problem persists. Reservoirs have a relatively short period of usefulness due to sediments that accumulate in their standing water.* In numerous areas withdrawal of groundwater has exceeded the rate of recharge, and wells must be sunk deeper. A few decades ago, irrigation and groundwater reclamation were heralded as a panacea that would permit unrestricted development and occupance of the deserts. Experience has

---

* BW designations are provided for climates in which precipitation is less than 50 percent of the area's moisture requirements.

* 100 years is the average life span of a dam and reservoir.

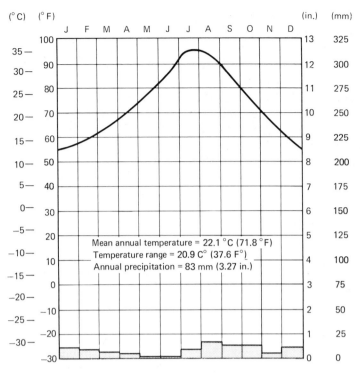

FIGURE 3.10
CLIMOGRAPH FOR YUMA, ARIZONA (DESERT).

TABLE 3.1 MONTHLY PRECIPITATION FOR LAS VEGAS, NEVADA IN MILLIMETERS (INCHES)

	January	February	March	April	May	June
1951	4.8(0.19)	0.5(0.02)	0.7(0.03)	1.0(0.04)	2.5(0.10)	0.(0.00)
1952	27.2(1.07)	T	38.1(1.50)	14.5(0.57)	T	0.2(0.01)
1953	.2(0.01)	0.00	0.7(0.03)	T	0.7(0.03)	T
1954	23.1(0.91)	0.5(0.02)	20.6(0.81)	T	0.(0.00)	0.2(0.01)
1955	35.6(1.40)	3.3(1.3)	T	2.79(0.11)	0.7(0.03)	9.9(0.39)
1956	5.8(0.23)	0.(0.00)	0(0.00)	2.03(0.08)	T	0(0.00)
1957	6.6(0.26)	4.16(0.16)	2.5(0.10)	13.9(0.55)	3.8(0.15)	T
1958	10.9(0.43)	18.5(0.73)	13.4(0.53)	16.3(0.64)	4.5(0.18)	0.00
1959	2.79(0.11)	18.2(0.72)	T	T	T	T
1960	14.2(0.56)	10.6(0.42)	1.3(0.05)	2.5(0.10)	T	T
Period	13.2(0.52)	5.5(0.22)	7.8(0.31)	5.3(0.21)	1.2(.05)	1.0(0.04)
Years	10	10	10	10	10	10
Normal	13.46(0.53)	11.17(0.44)	8.89(0.35)	5.84(0.23)	2.03(0.08)	1.02(0.04)

T = trace

proven, however, that future uses of desert lands must be planned in accordance with their meager water resources.

Semiarid Climates (BS)*

Semiarid climates are transitional between desert and humid types. Their temperatures are similar to those of the arid regimes, except that maxima are not quite as high (Fig. 3.12). Extremes of temperature are large, however, and can range from below $-40°$ C ($-40°$ F) in winter to above $40°$ C ($104°$ F) in summer. These conditions occur in both the northern Great Plains of North America and in the interior of Asia.

As in arid climates, semiarid regions are deficient in precipitation. Moisture supply supports grasses but is inadequate to maintain a forest cover. Like the arid climates, precipitation is very unpredictable. In certain years rainfall may exceed evapotranspiration demands, making the region (or parts of it) temporarily humid. In other years, high temperatures and below normal water supplies result in desert-like atmospheric conditions.

---

* BS designations are provided for climates in which precipitation meets more than 50 percent of the area's moisture requirements in at least one season, but less than 100 percent in both seasons.

The wary farmer of the semiarid region does not institute land-use practices geared to years of above normal rainfall, but has to plan his economic activities according to the more regularly occurring dry years. He continually faces the threat of drought and must also cope with thunderstorms, hail, and tornadoes.

C CLIMATES
(THE HUMID, MILD-WINTER REALM)

The middle latitudes are characterized by marked seasonal temperature contrasts. Unlike the tropics, the middle latitudes are areas where polar and tropical air masses meet. Thus, in addition to the cyclical pattern of heating and cooling, the frontal conditions common throughout much of this realm provide for unsettled weather and surplus precipitation in at least one season of the year. These mild winter representatives of the humid middle latitude climates are those in which temperatures are usually not low enough to support a cold season snow cover. Although snow may occasionally fall, it is normally of short duration and is not retained on the ground for any

(continued)

July	August	September	October	November	December	Annual
13.7(0.54)	4.1(0.16)	24.9(0.16)	1.3(0.05)	11.2(0.44)	5.8(0.23)	70.6(2.78)
11.9(0.47)	12.4(0.49)	22.1(0.87)	0.(0.00)	4.0(0.16)	10.4(0.41)	141.0(5.53)
5.84(0.23)	2.5(0.10)	5.0(0.10)	3.0(0.12)	0.5(0.02)	T	14.2(0.56)
40.9(1.61)	7.1(0.28)	11.4(0.45)	0.50(0.02)	9.1(0.36)	6.0(0.24)	119.6(4.71)
39.4(1.55)	44.2(1.74)	0.00	T	0.7(0.03)	0.5(0.02)	137.2(5.40)
41.6(1.64)	0(0.00)	0(0.00)	2.2(0.09)	0.00	0.00	51.8(2.04)
10.4(0.41)	65.8(2.59)	T	12.4(0.49)	6.6(0.26)	2.5(0.01)	126.5(4.98)
0.2(0.01)	4.5(0.18)	5.3(0.21)	16.0(0.63)	24.4(0.96)	0.00	114.8(4.52)
0.50(0.02)	8.3(0.33)	.25(0.01)	7.9(0.31)	27.7(1.09)	35.0(1.38)	105.9(4.17)
10.4(0.41)	T	5.8(0.23)	12.4(0.49)	47.7(1.88)	6.6(0.26)	111.8(4.40)
17.5(0.69)	14.9(0.59)	7.1(0.28)	6.0(0.24)	13.2(0.52)	6.6(0.26)	99.8(3.93)
10	10	10	10	10	10	10
12.7(0.50)	12.19(0.48)	8.63(0.34)	5.08(0.20)	7.87(0.31)	10.16(0.40)	99.0(3.90)

Source. National Weather Service.

FIGURE 3.11

OASIS IN THE SOUF, ALGERIA. Date palms are planted in excavated depressions to enable their roots to reach groundwater.

significant length of time. There are four primary subdivisions of the C climates: *humid subtropical, marine west coast, Mediterranean,* and *subtropical monsoon* (Fig. 3.13).

Humid Subtropical Climate (Cfa)*

The humid subtropical climatic type represents that part of the mild winter realm in which no dry season occurs and in which summers are long and hot. These areas are located on the east side of continents, roughly between latitudes 25 and 35°

---

* In C climates, one or more months must exceed 10° C (50° F) and no month be below 0° C (32° F); the small letter *f* in the middle latitudes means that all months have at least 30 mm (1.2 in) of precipitation; and *a* identifies a hot summer in which at least four months exceed 10° C (50° F), with the warmest month averaging 22° C (72° F) or more.

N and S. and occur in: southeastern United States, southeastern Europe, eastern China, Taiwan, southern Japan, northeastern Argentina, Uruguay, the Natal coast of South Africa, and eastern Australia.

The temperature range for a typical humid subtropical station is moderately large, the results of seasonal contrasts in energy received from the sun, as well as with the effect of continental location (Fig. 3.14). The range at Savannah, Georgia, for example, is about 17 C° (30 F°).

Summer in these regions are climatically similar to that of the wet tropical climate types, except that temperature maxima are greater. High temperatures coupled with the season's high humidity provide for hot and sultry summer weather. The humid air acts as an insulating blanket in the

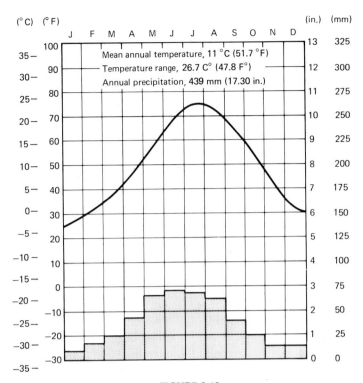

FIGURE 3.12
CLIMOGRAPH FOR GOODLAND, KANSAS (SEMIARID).

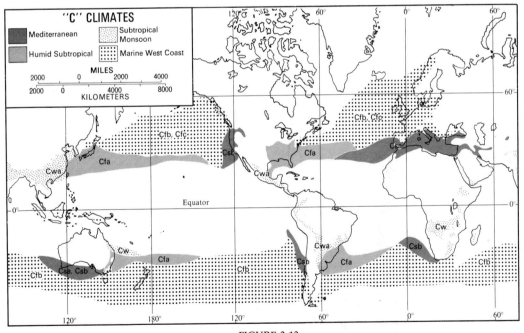

FIGURE 3.13
THE HUMID, MILD WINTER CLIMATES OF THE EARTH.

C CLIMATES (THE HUMID, MILD-WINTER REALM)

atmosphere, permitting little loss of heat by ground radiation. Thus, the daily range in summer temperature is normally small. The warmest month usually averages about 27° C (80° F), with daily peaks that may exceed 38° C (100° F). Summer is also long. The period of time between the last frost of spring and the first frost of autumn—the *growing season*—is seven months or more. The region's similarity to tropical climates vanishes with the approach of winter, when polar air moves into the subtropical latitudes more frequently. Constant fluctuations between tropical and polar air masses along the polar front provide for considerable variation in day-to-day temperatures during winter months, while monthly temperatures average between 5 and 12° C (41 to 54° F) for most stations.

The humid subtropics receive ample precipitation, averaging from 750 mm (30 in.) along the region's semiarid borders to over 1500 mm (60 in.) along the tropical margins. The primary determinants of this moisture supply are summertime convection and wintertime polar-front activity. It has already been suggested that summers are like the tropics with respect to temperature. The same holds true for rainfall. During summer the region is dominated by unstable tropical air that emanates from the western flanks of the subtropic highs. Variation in surface heating and weak atmospheric disturbances lead to air convergence and the formation of cumulus clouds from which frequent mid-afternoon and evening thunderstorms are generated. As can be seen in Figure 3.14, these warm air masses contain large quantities of moisture. As the cool season approaches, convectional activity diminishes dramatically. Precipitation at this time is provided by the clashing of air masses in middle-latitude cyclonic storms. Overcast skies commonly foster gray, rainy days. But these are interspersed with pleasant winter days when coats are thrown aside and replaced by sweaters. Snow infrequently falls when vigorous polar outbursts push equatorward. Seldom does it stay on the ground for more than a few days. In addition to these two primary controls on precipitation, some humid subtropical stations (as in central and southern Florida) also experience noticeable late summer or early fall rainfall maxima as a result of tropical cyclonic

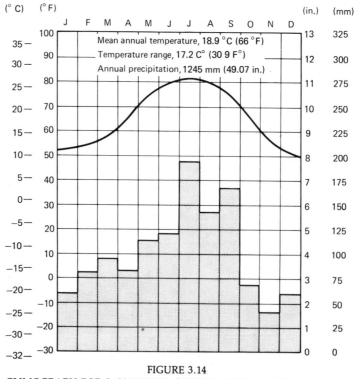

FIGURE 3.14
CLIMOGRAPH FOR SAVANNAH, GEORGIA (HUMID SUBTROPICAL).

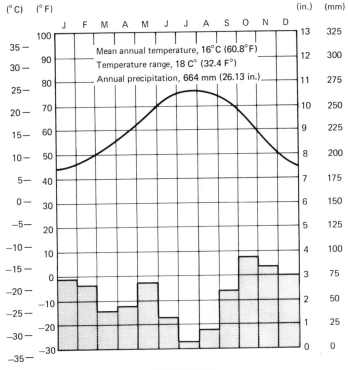

FIGURE 3.15

CLIMOGRAPH FOR ROME, ITALY (MEDITERRANEAN, HOT SUMMER).

storm activity. The humid subtropics are some of the most naturally productive regions of the middle latitudes. This climatic type supports a lush growth of vegetation that permits economic diversity. In the United States, for example, the long frost free period is conducive to the rapid growth of trees that support thriving pulp, plywood, and lumber industries; to the grazing of livestock; and to the growing of specialty crops such as cotton, tobacco, and citrus fruits.

Mediterranean Climates (Cs)*

Mediterranean climates are subtropical regions with dry summers and mild, rainy winters. They are best developed on the west side of continents, between latitudes 30 and 45° N and S. The largest expanse of this climate type is found along the margins of the Mediterranean Sea, from whence its name is derived. Other areas with a Mediterranean climate include central and coastal Califor-

nia, central Chile, the tip of South Africa, and southern Australia (Fig. 3.13).

The primary climatic controls in the dry-summer subtropics are associated with the shifting of the global wind systems. In summer these regions are under the influence of the stable subsiding air of the subtropic high-pressure cells. During the winter, when the subtropic highs shift equatorward, the unsettled westerlies with their cyclonic storm systems are dominant.

Temperature characteristics are similar to those of the humid subtropics, except in summer. Because of lower atmospheric humidity during the warm season, greater radiational heat losses result in lower night-time temperatures and larger diurnal ranges. Monthly temperatures during the warm season seldom exceed 27° C (80° F), even though the daily maxima may be well above 38° C (100° F). The Mediterranean climate is subdivided into two types based upon summer temperatures. The hot summer varient (Csa) is found at inland locations and along the borders of the warm Mediterranean Sea. As illustrated in Figure 3.15, their warmest month exceeds 22° C (72° F), and at least

* The lower case letter s identifies a dry summer season.

C CLIMATES (THE HUMID, MILD-WINTER REALM)

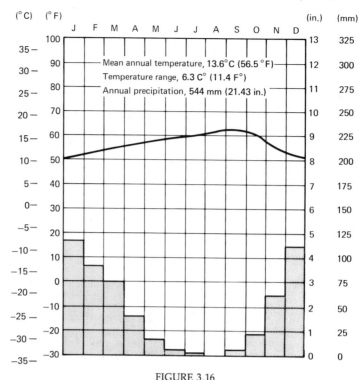

FIGURE 3.16
CLIMOGRAPH FOR SAN FRANCISCO, CALIFORNIA (MEDITERRA-
NEAN, WARM SUMMER).

four months have average temperatures above 10°
C (50° F). The second type (Csb*) is located along
coasts where the marine influence and cool ocean
currents moderate summer extremes and maximum temperatures are much lower. San Francisco, California, with a warm month maxima of
18° C (64° F), is an excellent example of this type
(Fig. 3.16).

Precipitation in Mediterranean climatic regions
is concentrated in the winter half of the year, and
averages from 350 to 650 mm (14-26 in) per year.
During the cool season, maritime air masses—
both tropical and polar—converge; this leads to
overcast and rainy skies at least 40 percent of the
time. Most of the precipitation falls as rain, with
snow being confined to highland locations. Infrequently, however, the polar margins of this climatic type will experience snowfall of short duration even at low elevations. Although total
rainfall is low, it does occur in the cool season

when loss of moisture by evapotranspiration is
minimized. Summer stands in sharp contrast to
the wet winter as a period of "seasonal aridity."
Under the influence of the subtropic highs, descending air has little potential for producing rain.
Low relative humidity and an offshore airflow
pattern result in a summer moisture regime similar to that of the desert. This distinctive rainfall
regime has had a pronounced effect upon the region's land-use patterns. To accommodate the
growth of crops, large-scale irrigation works have
been constructed for high-value plants, and rapidly maturing grains—such as barley and wheat—
have been planted in nonirrigated areas. This is
a climatic region popular for both vacations and
settlement.

Marine West Coast Climates (Cfb,c)*

Lying poleward of the Mediterranean region on
the west side of continents (between latitudes 40

---

* The lower case letter *b* identifies "warm summer" climates in
which at least four months exceed 10° C (50° F), yet the hottest month
is less than 22° C (72° F).

* The letter *c* identifies "short summer" types, where three months
or less have average temperatures greater than 10° C (50° F).

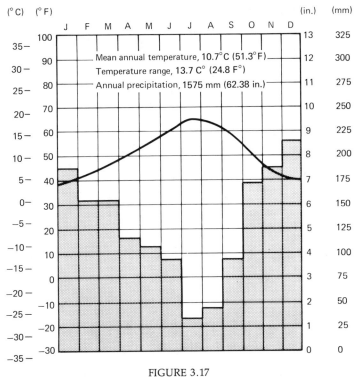

FIGURE 3.17
CLIMOGRAPH FOR STARTUP, WASHINGTON (MARINE WEST COAST).

and 60° N and S) are the marine west coast climates (Fig. 3.13).* The principal areas in which they occur are: the west coast of North America from California to Alaska; the British Isles and northwest Europe; southern Chile; southeastern Australia and New Zealand.

The chief distinguishing characteristic of this climate type is that temperatures are very moderate for their latitudinal location. Whereas inland climates at the same latitude have bitter winters and hot summers, marine climates are oceanic in nature with mild winters and summers. The temperature of the warmest month is normally between 15 and 20° C (59 to 68° F), with daily maxima only infrequently exceeding 25° C (77° F). The coldest months range no more than 17 C° (30 F°) below the warmest month and are between 5 to 15° C (9 to 27° F) warmer than the latitudinal norm. In addition to the small range in mean monthly temperature, the diurnal range is also small.

The marine west coast climate is humid. No

month is considered to be truly dry. But there is a distinct seasonal concentration of rainfall, being the wettest half of the year (Fig. 3.17). The principal sources of moisture are maritime air masses that originate to the west of these climates. In their predominant flow pattern toward the east they experience frontal convergence that results in extensive cloud development and profuse rainfall. These areas rank as some of the cloudiest and most fog-prone on the globe. Cloudiness during the cool season may be present more than 70 percent of the time. Gray leaden skies have a tendency to release persistent light rain or drizzle for days at a time. Stations recording 150 rain days a year are not uncommon. Bahia Felix, Chile, near the Straights of Magellan, represents an extreme case, averaging 325 rain days per year. But most locations have a summer that is both sunny and pleasant. This is a result of the subtropic highs shifting to their most poleward position and serving to block frontal storms and steer them toward higher latitudes. Nevertheless, warm-season storms do occur. Rainfall totals are very reliable but vary widely from one marine climate to another. In

* Although most of the earth's marine west coast climates are found as described above, a few cases of east coast locations exist.

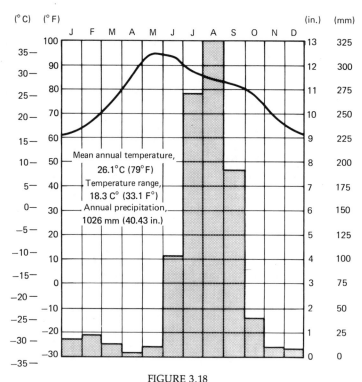

FIGURE 3.18
CLIMOGRAPH FOR ALLAHABAD, INDIA (SUBTROPICAL MONSOON).

lowland areas they may be as little as 500 mm (20 in), while in some coastal regions backed by mountains they are in excess of 2500 mm (100 in). The moist air masses of marine climates are also conducive to the formation of dense ground fog. Averages of 40 to 60 days of fog per year are normal. Snow occasionally falls, but, as in the subtropics, lasts for a few days at most. Only at higher altitudes can a winter snowpack be sustained.

The marine climate is essentially free of such violent storms as tornadoes and hurricanes. However, the westerly winds arriving at the coast after a long journey over ocean water can attain high velocities and be a hazard to shipping. Provincial wind names such as the "roaring forties" or the "screaming sixties" have been applied to the westerlies of the South Pacific and are terms indicative of their fierceness.

Subtropical Monsoon Climate (Cw)*

This subtropical climate has a seasonal rainfall pattern that is the reverse of the Mediterranean

climatic type. Precipitation is heavy during the warm season and very light during the winter. Subtropical climates are found on the poleward margins of the tropical monsoon and savanna regions and represent a subtropical extension of the latter types. Principal areas of occurrence include: northern India; southwestern China; a narrow zone along the northeast coast of Australia; parts of Mexico, Paraguay, and eastern Bolivia; and the plateaus of tropical and subtropical Africa.

The primary difference between the subtropical monsoon climate and its tropical counterparts is the seasonal temperature pattern. Increased distance from the equator coupled with an interior continental location provide for stronger seasonal contrasts in temperature and winter monthly temperatures that fall short of the tropical climatic requirements. Figure 3.18 illustrates the characteristics of this climate type. Allahabad, India, is located about 25° N latitude. Its monthly temperatures vary from 16° C (61° F) to 35° C (95° F), with a range of 19 C° (34 F°). And, as with the tropical monsoon climate, the maximum temperatures occur in late spring prior to the onset of the rainy season.

* The letter w identifies a winter dry season.

The distribution of rainfall at Allahabad is tied to the monsoonal wind circulation. Summer is a period of air convergence into the Asiatic low pressure system. This lifting, unstable air produces extensive cloudiness and heavy rainfall. Of the 1026 mm (40 in) of precipitation that falls annually, 88 percent arrives in the four-month period between June and September. During the winter the pattern is reversed. Air descending out of the continental high pressure systems are low in relative humidity, have little tendency for lifting, and provide Allahabad with clear skies. The occasional light rains that occur in this season are associated with the passage of weakly developed cyclonic storms.

D CLIMATES
(THE HUMID, SEVERE-WINTER REALM)

The humid, severe-winter climate types are often referred to as the snow climates. They are warm enough to have a summer season, but cold enough in the winter to sustain a snow cover. The length of time that the snow remains on the ground varies from a few weeks at their subtropical boundaries to several months in their poleward reaches (Fig. 3.19). These climates are found only on the large landmasses of North America and Eurasia. Their conspicuous absence in the Southern Hemisphere is largely related to the relative absence of land in latitudes optimal for their development. The effect of continentality is strong. Continentality coupled with dramatic seasonal differences in insolation provide for sharply contrasting seasonal temperatures and a large range in monthly temperatures. Three subdivisions of this climatic realm are recognized: the *humid continental (hot-summer) region*, the *humid continental (warm-summer) region*, and the *continental subarctic region* (Fig. 3.20).

Humid Continental (Hot Summer) Climates (Dfa, Dwa)

The hot summer varient of the humid continental climates is found extending from the east coasts into the continental interiors, between latitudes 35 to 45° N. In North America this climate occurs wholly within the United States. Lying south of the Great Lakes, it borders the semiarid climate

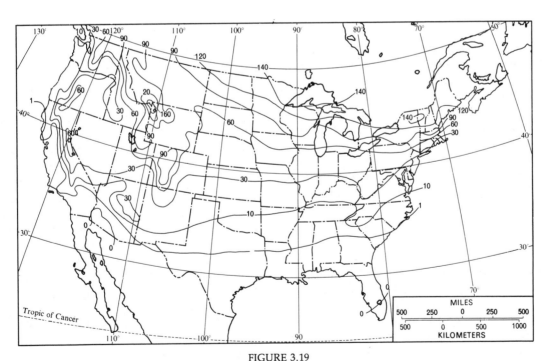

FIGURE 3.19
NUMBER OF DAYS WITH SNOW COVER IN THE UNITED STATES.

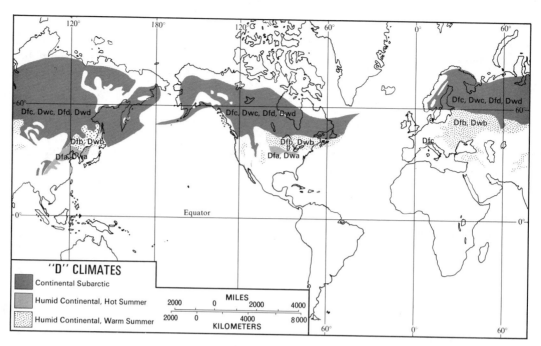

FIGURE 3.20
THE SEVERE WINTER CLIMATES OF THE MIDDLE LATITUDES.

to its west and continues essentially uninterrupted to the eastern seaboard; in Europe it is found in parts of Hungary, Romania, Yugoslavia, and Bulgaria; and in Asia in Manchuria, northeast China, Korea, and in Japan.

The supply of solar radiation to the humid continental hot summer climate regions varies considerably by season, the summer solstice (June 21st) receiving three times the energy available at the winter solstice (Dec. 22nd). This aspect of seasonal energy availability combined with the rapid heating and cooling qualities of landmasses produces widely divergent winter and summer temperatures. Winters are cold; summers are hot. January temperatures typically average below 0° C (32° F), and range from −18° C (0° F) to −1° C (30° F). July, on the other hand, usually varies between 18° C (65° F) and 24° C (75° F). An annual range of at least 28 C° (50 F°) or more, is not uncommon.

Summer in these climates is much like summer in the humid subtropics. A twenty-four hour period of weather in Peoria, Illinois, for example, may be hard to distinguish from the same period in Atlanta, Georgia. Humidity and temperatures are high, and the diurnal temperature range is small. This season is dominated by moist and un-

stable tropical air masses that provide for the region's rainfall maximums (Fig. 3.21). Most warm season rainfall occurs in cumulus clouds that are generated by convection. These rain clouds are frequently accompanied by lightning and thunder, and less frequently by hail and tornado funnels. Winter precipitation results from cyclonic storms and frontal activity; it may arrive in the form of rain, drizzle, sleet, or snow. Snowfall amounts vary. On the margins of the Great Plains less than 1 meter (3 feet) per year is the rule. In the snow belt of the Great Lakes Region, however, 2 meters (6 feet) per season is not uncommon and drifts may be three times that figure. Overall, annual precipitation averages between 500 to 1000 mm (20 to 40 in.). Even though this is a modest amount in comparison with the wet tropical regions, the lower moisture demands of the middle latitudes ensures that most years will have surplus water supplies in at least one season. On the average, two-thirds or more of the annual precipitation falls during the growing season.

Long and hot summer days combined with available moisture provide ideal conditions for rapid plant growth in the humid continental (hot summer) climates. The "corn belt" of the United States, often considered to be the world's greatest storehouse of farming wealth, lies within this region. Where soils are suitable, grains and legumes such as corn, soybeans, wheat, oats, and barley, are grown. These plants also support thriving dairy and livestock industries.

Humid Continental (Warm Summer) Climates (Dfb, Dwb)

The "warm summer" continental climates are controlled by the same processes that prevail in the "hot summer" type. The chief differences between the two result from the "warm summer" varient lying further poleward, roughly between latitudes 45 to 60° N. Once again, this type is only found in the Northern Hemisphere. In North America it lies astride the United States–Canadian border and fairly well encompasses the Great Lakes. In Eurasia, its eastern boundary extends from southern Norway to Poland and Czechoslovakia and eastward into central Siberia. It resumes again in Manchuria and crosses into northern Japan. In these locations winters are long and cold. For many stations the January average temperature is less than −15° C (5° F), with daily minimums that may fall to −45° C (−49° F). Snow remains on the ground between one to five months. Summer, on the other hand, is pleasantly cool. Although daily maxima may exceed 35° C (95° F), frequent incursions of polar air, accompanied by cloudiness, keep the mean monthly temperatures during the summer at modest levels; 18 to 20° C (64 to 68° F) is not uncommon for July. Not only is summertime relatively cool, it is also short. For most of the region, the frost-free period is less than 150 days per year. Compared to the "hot summer" type found to the south, the annual temperature range is significantly greater, about 39 C° (70 F°).

Precipitation totals are somewhat less than this climate's southern neighbor, but its seasonal distribution is about the same. A comparison of Figures 3.21 and 3.22 reveals the similarities in the two humid continental climates and also reflects the greater effect of continentality in the "warm summer" type.

This climate's relatively cool and short summers are not especially conducive to farming. Economic

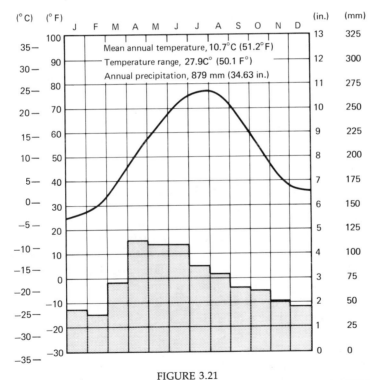

FIGURE 3.21
CLIMOGRAPH FOR PEORIA, ILLINOIS (HUMID CONTINENTAL, HOT SUMMER).

D CLIMATES (THE HUMID, SEVERE-WINTER REALM)

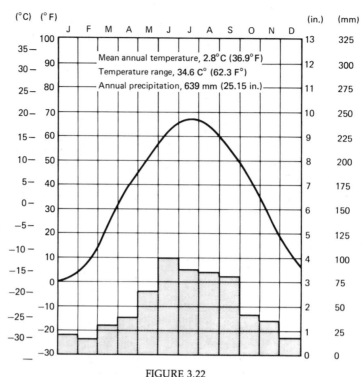

FIGURE 3.22
CLIMOGRAPH FOR INTERNATIONAL FALLS, MINNESOTA (HUMID CONTINENTAL, WARM SUMMER).

activity has always centered about these regions' vast coniferous forests. Trapping animals for pelts and utilizing the timber resources for lumber and pulp represent environmentally related occupations for the majority of the sparse population inhabiting these lands.

Continental Subarctic Climates (Dfc, d and Dwc, d)

Continental subarctic climates are the most extensive of all climatic groups. They occur between 50 to 70° latitudes and are found in North America, where they extend uninterrupted from Alaska to Labrador, and in Eurasia where they stretch from Norway to Asia's east coast.

Subarctic climates hold the world record for the greatest temperature ranges. As shown in Figure 3.23, Verkhoyansk, in east central Siberia (68° N) has a January temperature of −47° C (−53° F), a July average of 16° C (61° F), and an annual range of 63 C° (114 F°). Like the humid continental types, these very strong seasonal contrasts in temperature are linked to the broad expanse of the conti-

nent and to latitudinal position. On a typical day in July, the latitude of Verkhoyansk receives 87 times more solar energy than a day in January.

In the summertime days are long and average from 17 to 22 hours, maximum daily temperatures frequently exceed 25° C (77° F), and diurnal ranges amount to 10 or 15 C° (18 to 27 F°). Summer, however, is brief with three months, or less, having mean temperatures above 10° C (50° F). Although the growing season for crops and vegetables is short (50 to 90 days), the longer sunlight periods permit the cultivation of quick-maturing plants. Even so, frost may occur on any day during the warm season and wreck havoc with agricultural activities. Autumn is inconspicuous in these latitudes and winter arrives as early as October. The winter season dominates the region. Long nights (18 hours, or more) and large heat losses by ground radiation result in low surface temperatures and cold, dense air masses. For six to eight months temperatures average below 0° C (32° F) for most of the area, with many locations experiencing less than −15° C (5° F) for at least three to four months.

Precipitation in the subarctic climates is usually less than 500 mm (20 in.) annually. It is concentrated as rainfall in the summertime when polar front activity is at its strongest for these upper middle-latitude locations. During the winter, when the region is dominated by cold, dry and stable air masses, precipitation is meager. Yet, due to the very low temperatures, this moisture reaches the surface as snow and may accumulate to depths of one-half meter (1½ feet) or more. However, since the moisture content of snow relative to water has a ratio that may be as high as 15 to 1, total wintertime water supplies seldom amount to more than several centimeters.

One need only examine a map of world population density to see that subarctic climates are not highly favored for human settlement. Population is sparse. The very short growing season limits agricultural productivity; the region's great expanse of virgin forest is not impressive in either the size of its trees or the density of its stands; the area's soils are acidic and impoverished; and because of frozen subsoils the land is poorly drained.

Resources supporting small populations include trapping, fishing, mining, and some logging for pulpwood.

The permafrost problem of the subarctic climates is one that extends poleward into the summerless climatic realm. In these areas summer warmth is only capable of thawing the upper 1 to 4 meters (3 to 12 feet) of frozen soil, below which the ground is permanently frozen. The depth of the continuously frozen layer has been known to extend to 450 meters (1500 feet) below the surface. As might be expected, survival in permafrost regions requires unique adaptations to the environment. Homes must be built above the ground and heavily insulated to prevent heat losses that might melt the frozen subsoils. Otherwise, houses would slowly sink into the saturated soils. With vertical drainage of soil water impeded by the frozen subsurface layers, bog and swamp conditions prevail during the summer. These wetlands, in turn, are the homes of insects—especially mosquitoes and black flies—that make life extremely uncomfortable for both people and animals.

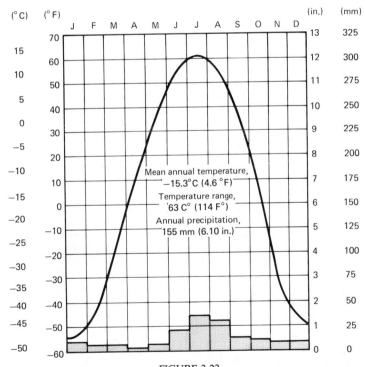

FIGURE 3.23
CLIMOGRAPH FOR VERKHOYANSK, USSR (CONTINENTAL SUBARCTIC).

D CLIMATES (THE HUMID, SEVERE-WINTER REALM)

E CLIMATES
(THE SUMMERLESS REALM)

At the opposite extreme from the winterless climates (A) of the tropics is the *polar* (E), or *summerless climatic realm.* These regions of monotonous cold lie poleward of the subarctic region. In both North America and Eurasia they are found to the north of the Arctic Circle, except on the eastern sides of the continents, where they dip farther toward the equator (Fig. 3.24). Also included within this climatic group are Northern Iceland, Greenland, islands of the Arctic and Antarctic Oceans, and the entire continent of Anarctica.

The primary requirement for E climates is that mean monthly temperature cannot equal or exceed 10° C (50° F). The 10° C warm–month isotherm closely coincides with the poleward boundary of forest and effectively defines the equatorward limits of the polar climates. Distinctive to all of the summerless climates is their lengthy alternating periods of illumination and darkness. At the poles six months pass during which the sun is always above the horizon followed by six months when it is completely out of sight. At the Arctic and

Antarctic Circles, sunlight duration varies from one 24-hour period on the summer solstice date to complete absence on the winter solstice date. Latitudes between 66½ to 90° N and S experience lengths of time in which the sun stays above the horizon, or is hidden below it, that range between one day and six months.

As a consequence of high latitude and the receipt of solar radiation at low sun angles, temperatures are perpetually low in the polar regions. The lowest *mean annual* and *mean summer monthly* temperatures on the earth have been recorded in these latitudes. Even though summer has a long illumination period, the intensity of sun energy is insufficient to be effective in appreciably raising air temperatures. Two primary subdivision of the polar realm are recognized: the *tundra* and *ice-cap climates.*

Tundra Climates (ET)

The tundra climate is a transitional type that lies between the polar ice caps of perennial ice and snow and the middle-latitude subarctic regions where contrasting seasons are the rule. In both the North American and Eurasian continents, tundra

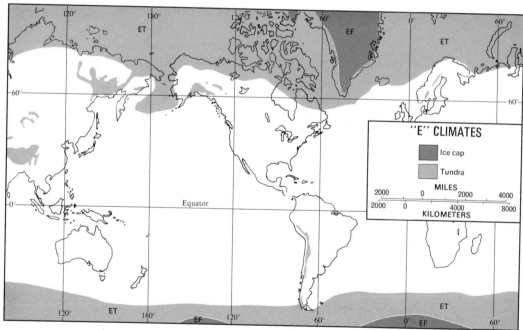

FIGURE 3.24
THE POLAR CLIMATES OF THE EARTH.

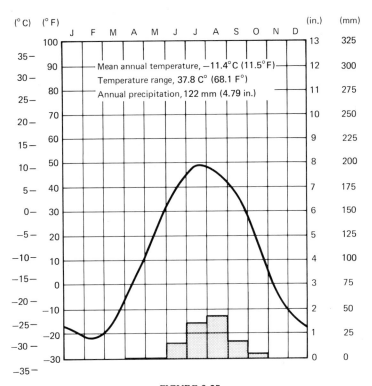

FIGURE 3.25
CLIMOGRAPH FOR COPPERMINE, NORTHWEST TERRITORY, CANADA
(TUNDRA).

extends from the subarctic boundary to the shores of the Arctic Sea. It is also found on islands of the Arctic and Antarctic oceans, in Northern Iceland, coastal Greenland, and in scattered small areas of the Antarctic Continent.

The fundamental differences between the tundra climate and the frigid ice-cap regime is that the former has two to six months of a warm season when temperatures are above freezing (Fig. 3.25). Summertime daily maxima are normally between 15 to 18° C (59 to 65° F), fairly typical of mild winter weather in the middle latitudes. Frost, however, may occur on any day of the year. Winter in the tundra is similar to that of the subarctic regions, with six to eight months averaging temperatures below 0° C (32° F) and extreme minima falling as low as −45° C (−49° F). The long nights result in large radiational heat losses accompanied by cold, dense, and clear air. Although the sun may be below the horizon during this period, moonlight, stars, twilight, and the electronic displays of the northern lights all shine brightly through the clear air to effectively reduce the im-

pact of the darkness. The temperature ranges are moderately large and average about 30 C° (54 F°) for inland stations, but not so great as are found in the subarctic climates.

Annual precipitation is usually less than 350 mm (14 in.). Most arrives as snow during the warm season when frontal activity is at its maximum. The winter season, with its cold, dense air masses of low-moisture content, is less conducive to precipitation forming conditions. The powdery snow piles into deep drifts, and, at times, is blown about by strong winds that recirculate the snow near the surface and reduce visibility to a few meters.

It is worth noting that coastlines and islands in polar latitudes, normally classified as tundra, have much milder winters than inland locations because of the "marine effect" (Fig. 3.26). For this reason, many climatologists prefer to consider these tundra climates, with a modest temperature range and minima monthly temperatures above −7° C (20° F), in a separate category. These are called *polar marine climates* (EM).

E CLIMATES (THE SUMMERLESS REALM)

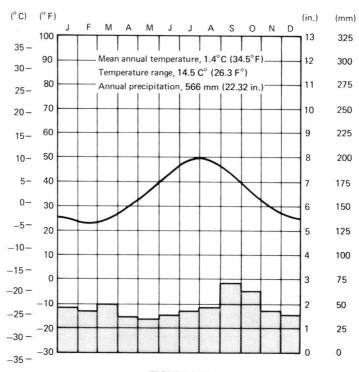

FIGURE 3.26
CLIMOGRAPH FOR MAKKAUR FYR, NORWAY (POLAR MARINE).

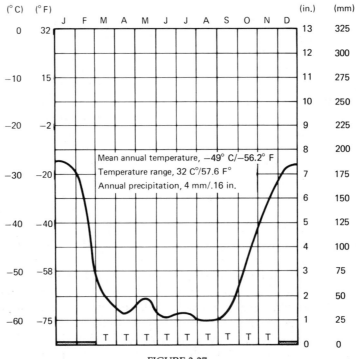

FIGURE 3.27
CLIMOGRAPH FOR AMUNDSEN-SCOTT, SOUTH POLE (ICE CAP).
T = trace

GLOBAL CLIMATES

84

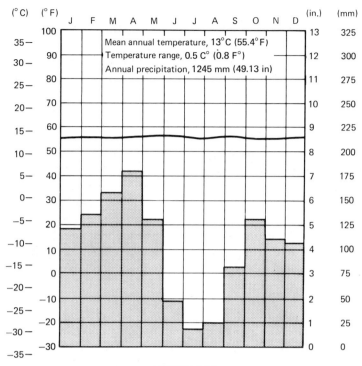

FIGURE 3.28
CLIMOGRAPH FOR QUITO, ECUADOR (ELEVATION: 2811 M OR 9280 FT).

Ice Cap Climates (EF)

This is the earth's coldest climatic type (Fig. 3.27). Every month of the year averages below 0° C (32° F) and snow, ice, or barren rock cover the surface. The principle areas of occurrence are found in Antarctica, the interior of Greenland, and in minor, fragmented highland locations.

The earth's lowest mean annual temperatures have been recorded in the interiors of Antarctica and Greenland. Summer, if it can be called that, with monthly means well below 0° C (32° F), is cold. Not only is the region receiving meager solar energy supplies during this season, but reflection of radiation from snow surfaces means that even less radiation is available for heating. Summer monthly means of −11° C (12° F) in Eismette, Greenland, and −34° C (−29° F) in Vostok, Antarctica, are normal. Winter is colder, with monthly means ranging from −20° C (−4° F), to −65° C (−85° F).

Precipitation in the interior of the ice cap climates is light, generally less than 125 mm (5 in.) annually, and originates from invading cyclonic storms. Under normal conditions the cold atmosphere, low humidity, and stable air of the region all combine to inhibit moisture release. Yet, due to low evaporation and melting, ice masses thousands of meters in depth are nourished and maintained by these scanty supplies of precipitation.

HIGHLAND CLIMATES*

As explained in Chapter 2, increasing elevation results in a reduction in air temperature. Mountains are favored vacationing sites during the summer for those who want to escape the heat. In winter, skiers flock to the mountains where temperatures are cooler and snow accumulates. From these two simple examples it may be concluded that, as regards to temperature, an increase in altitude is tantamount to moving poleward in lati-

* Although any highland climate can be classified according to the procedures already laid down, vertical changes in climatic types can often be abrupt. Since they often express complex patterns difficult to present on small-scale maps, these climatic types are often grouped together and shown on the map by a symbol "H".

tude. This analogy is almost correct, except that elevation does not control the seasonal weather factors, such as length of daylight period, that are associated with latitude. Climate does change as altitude increases, but it always retains some characteristics of the dominant lowland climate within which it lies. Quito, Ecuador, is located almost at the equator at an elevation of 2,811 meters (9300 feet). As illustrated in Figure 3.28, monthly temperatures are cool enough to classify Quito as a marine west coast climate (Cfb). Yet, its temperature range is not typical of the C climate because of the uniformity of solar radiation receipts that are characteristic of the tropical rain forest.

Because upland areas occur at all latitudes there is no single, predictable series of climate types that appear with increased altitude. In the humid

tropics, a sequence might start with tropical rain forest (Af), and be followed by humid mild winter types (Cfb and Cfc), tundra (ET), and lastly by ice cap (EF). In the humid subtropics (Cfa), the pattern could continue with Cfb, followed by humid continental (Dfb) or continental subarctic (Dfc), tundra (ET), and ice cap (EF). A diagram is provided in Figure 3.29 that illustrates the commonly expected sequence of mountain climate-types as they vary with latitude. In addition to altitude, however, slope orientation can add to the complexity of climatic patterns in mountains; thus only climatic realms are shown in the hypothetical diagram.

Throughout the discussion of climate, frequent reference has been made to plant associations of the earth. Parameters utilized in the Köppen clas-

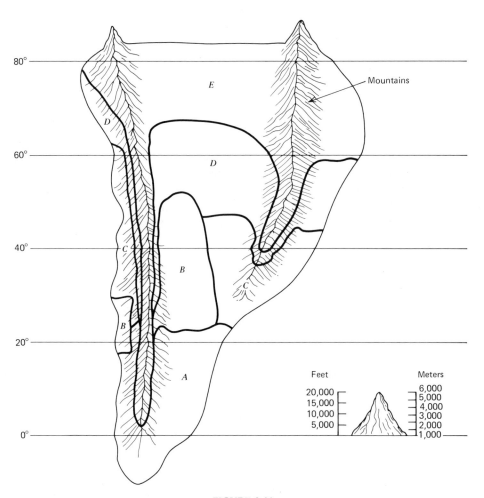

FIGURE 3.29
HYPOTHETICAL SEQUENCE OF CLIMATIC REALMS IN RELATION TO ALTITUDE.

sification were chosen because of their relevance to the global distribution of plants. There is little doubt that strong relationship exists between atmospheric characteristics and vegetation. Light, water, and temperature are elements crucial to the survival of plant life. Chapter 4 deals with the nature of earth's vegetation.

KEY TERMS AND CONCEPTS

climate
Wladimer Köppen
climatic region
humid tropical realm
humid
winterless (tropical) climates
tropical rainforest climate
diurnal range
tropical savanna climate
seasonal aridity
moisture deficient realm
arid climate
semiarid climate
humid, mild-winter climates
humid subtropical climate

growing season
Mediterranean climate
marine west coast climate
subtropical monsoon climate
humid, severe-winter climates
humid continental, hot-summer climates
humid continental, warm summer climates
continental subarctic climates
summerless (polar) climates
tundra climate
polar marine climate
ice cap climate
highland climates

DISCUSSION QUESTIONS

1. Define climate. How does climate differ from weather?
2. What is a climatic region? On what basis are climatic regions delimited in this textbook?
3. What is a climograph? How is it useful?
4. Describe the various types of tropical climates.

5. Which climatic region represents the northern limit of forest growth?
6. How do humid subtropical regions differ from humid continental regions?
7. Explain why the continental and latitudinal positions of the humid subtropical and Mediterranean climates cause them to be different.
8. What is the primary difference between subarctic and tundra climates?
9. Describe the various types of polar climates.
10. The marine west coast climate has relatively mild temperatures for its latitudinal position. Why?
11. Explain the seasonal distribution of precipitation in the tropical savanna climate in terms of wind and pressure systems.
12. In which climatic region is precipitation most unpredictable?
13. Which climatic type is transitional in nature, sometimes having characteristics typical of desert regimes and at other times humid?
14. Why is precipitation in Antarctica and Greenland very low, although these continents are considered to have humid climates?
15. What is meant by the term seasonal aridity?

REFERENCES FOR FURTHER STUDY

Boucher, Keith., *Global Climate*, Halsted Press, New York, 1975.

Chang, Jen-Hu., *Climate and Agriculture: An Ecological Survey*, Aldine Publishing Co., Chicago, 1968.

Köppen, Wladimir, and Geiger, R., *Klima der Erde* (map), Justus Perthes, Darmstadt, Germany; America distributor: A. J. Nystrom and Co., Chicago, 1954.

Mather, John R., *Climatology: Fundamentals and Applications*, McGraw-Hill Book Co., New York, 1974.

Sellers, William D., *Physical Climatology*, University of Chicago Press, Chicago, 1965.

Trewartha, Glenn T., *The Earth's Problem Climates*, The University of Wisconsin Press, Madison, Wisc., 1961.

4
THE NATURE OF VEGETATION

THE SOLID SURFACE OF THE EARTH IS COMPRISED OF diverse landscapes. Physical and cultural features provide areas with unique personalities. Aside from mankind's cultural imprints the most obvious element of a location's setting is its cover of vegetation. Blends of various plants distinguish one physical environment from another. Consider, for example, the contrasting scenes of the southern California deserts with the rainforests of the Olympic Peninsula (Fig. 4.1). In addition to providing character and aesthetic value to place, world vegetation also provides mankind with food, fiber, and materials for shelter, and serves a host of secondary needs as sources of fuel and medical supplies. The character of plant communities, their worldwide distribution, and their role in the environment are thus fitting topics of study for the geographer.

Vegetation may be considered as a mosaic of plants in which the individual plant is the basic structural unit. The typical green plant is a living organism that is fixed in place and possessed of cellulose cell walls. Through *photosynthesis,* the process by which green plants use solar energy to convert nutrients into living tissue, plants convert simple substances—chiefly carbon dioxide and water—into starch, sugars, and other complex materials. By assimilating solar radiation and storing it for future use, green plants sustain virtually all living systems.

Single plant species rarely occur as the only life form on the landscape. Rather, mixtures of plants are the rule. Moreover, faunal groups that ultimately depend upon plant populations for their survival coexist with these vegetation assemblages. A natural system of interacting flora and fauna is called an *ecosystem.* Although ecosystems contain all environmental organisms, our immediate concern will be communities wherein plant members live in a fashion that appears to be mutually beneficial to one another. For example, mosses and mushrooms may thrive because of the shade provided by a forest canopy. They, in turn, decompose surface and soil organic matter and aid in the recycling of mineral nutrients. The specific types of plants that occupy a community result from many complex interacting factors including: the plants' environmental tolerance limits; their ability to compete for space, water, and sunlight; and their evolutionary history. Two maxims hold true for sites where the vegetation has been relatively undisturbed by man.

1. The *biota* (life forms), available within an area that are best suited to both the regional environment and the specific location will be present.

   Each plant and animal has distinct habitat requirements. We do not expect to find palm trees, banana plants, monkeys, or elephants in Antarctica; nor do we look for ferns and

FIGURE 4.1A
DESERT VEGETATION IN SOUTHERN CALIFORNIA.

FIGURE 4.1B
LUSH VEGETATION IN THE RAIN FOREST OF OLYMPIC
NATIONAL FOREST, WASHINGTON.

mosses in clearings that receive large quantities of sunlight. Within vegetation regions of common evolutionary history, biota thrive in locations that are optimal for their development. Some plants demand shade, others intense sunlight. Some achieve maximum productivity in acid soils, while others prefer alkaline conditions. Some have high temperature requirements, other function well when it is cold. Some need large quantities of water, while others sustain themselves on a meager supply of water.

2. Plant communities tend to change in composition with time, until they achieve a stable association (called *climax vegetation*) within their regional environment.

The occupance of a site by plants normally procedes through a predictable evolutionary sequence referred to as *plant succession*. The sequence begins with the establishment of very simple plant communities and evolves through more complex types, ultimately leading to an association of plants that sus-

tain themselves on a quasi-permanent basis. Assume, for example, that vegetation is encroaching upon a newly formed site, such as a sand dune, a lava flow, or a recent stream deposit. The first plants to invade the new surface must be hardy and capable of surviving environmental extremes. With no ground cover sunlight is intense, surface temperatures fluctuate widely, and there is no protection either from the drying effect of the wind or from rainfall impact. The plants that survive this stage modify their environment as they inhabit the area. Their roots penetrate the soil, making it more permeable; their remains contribute organic matter to both the soil and the ground surface. Microorganisms and soil animals, such as worms and ants, find food resources in the decaying organic matter. They subsequently become permanent residents of the soil and serve to further modify its characteristics, both physically and chemically. As soil and surface climatic conditions become favorable for other plant species, those will invade the pioneer association. If the new vegetation forms are tall and provide shade, surface temperatures and sunlight intensity are reduced. Thus, the original pioneers are at a disadvantage. They cannot compete for space and sunlight and are eventually replaced by a successional community. This process may repeat itself several times, until succession has resulted in a relatively stable community of plants that are representative of a long-term balance between vegetation and its environment. These plant association are called *climax communities* and are likely to persist without appreciable change for an indefinite period.

GLOBAL VEGETATION

Plant communities exist at scales ranging from microscopic associations to regional complexes that may cover large portions of a continent. Concern throughout the remainder of this chapter will focus upon the latter forms. These *vegetation formations* are visible evidence of the dynamic interplay that occurs between plants and their environments. Knowledge of plant-environmental interrelationships, in turn, aid in understanding the manner in which man uses vegetative resources; the land-use systems he employs to maximize the economic potential of these resources; and the patterns of human settlement established in consequence.

Within any vegetation formation, groups of plants may be found that are not typical of the regional floral assemblage. These are species adapted to sites that differ in some respects from the macro-environment of the plant formation. Waterlogged depressions, excessively drained soils, and slopes that sharply reduce or enhance the receipt of solar radiation are examples of sites that could very well support localized variation in plant types. Nevertheless, on a macro-scale, vegetation formations have a remarkable degree of homogeneity, a response to certain environmental variables of like character being consistent over large expanses of the earth. In particular, a strong degree of interrelatedness exists between vegetation, climate, and soils. Comparisons of the maps illustrating the distributions of these variables should be made since the discussion of each plant formation takes them into account.

All of the earth's natural vegetation can be subdivided into four major structural subdivisions: *forest, grass, desert shrub,* and *tundra* (Fig. 4.2). Herbaceous plants (usually small, tender plants that lack woody stems) other than grasses dominate the tundra where temperatures are too cool for forest and, to some degree, for grass. Deserts are sparsely vegetated because of deficient moisture supplies. Forests are confined to regions relatively well supplied with water and where the growing season is sufficiently long to permit the maturation of trees. Grasses, on the other hand, cross a wide variety of environmental boundaries. Grasses do not normally tolerate the shade of a forest canopy, but can thrive in forest clearings. Grass dominates as a surface cover in semiarid to subhumid climates, and also represents an important part of the desert shrub biota. Although the distribution of plant formations is largely controlled by environment, vegetation classes are not mutually exclusive.

The Forest Formation

Forests consist of areas in which trees are dominant, even though vines, shrubs, grasses, and other ground coverings may be present. Since the total number of plants within any association is too numerous to enumerate, attention will focus upon the dominant members of the community. The primary requirements for trees include large quantities of available water during the growing season, and a suitable temperature and sunlight regime. Physical differences that are found among the earth's forest formations are a response to variations in these variables. When soil moisture is unlimited, and temperature and sunlight unrestricted, most trees do not experience a dormancy period. Growth is continuous, leaves are broad (wide in relation to length) in order to facilitate transpiration, and foliage lasts throughout the year. Although these *broadleaf* trees do drop their old nonfunctional leaves from time to time, the plant is never completely denuded, but *evergreen*. However, few areas on the earth have such ideal conditions for plant growth. A lack of seasonal rainfall or the prevalence of freezing conditions that immobilize soil water are the norm for large segments of the earth. These more common situations have necessitated the evolution of mecha-

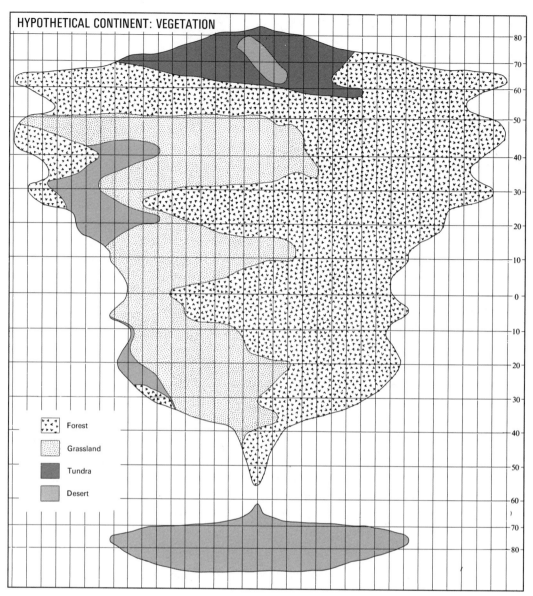

Forest

Grassland

Tundra

Desert

FIGURE 4.2

SCHEMATIC CONTINENT WITH VEGETATION FORMATIONS. This diagram represents all of the earth's landmasses squeezed together at their respective latitudes. The distribution patterns of the four major structural subdivisions of vegetation are shown in their appropriate latitudinal position and are proportionally representative of the areas they occupy.

nisms by which trees may survive stress periods. Some trees are *deciduous,* capable of shedding their leaves; some produce thick, leathery leaves that conserve moisture and reduce susceptibility to environmental extremes; some have developed appendages that are narrow in relation to length (*needleleaves*) to curtail water loss; and most maintain a growth cycle that includes a dormant, or rest, period. Based upon these characteristics, the

forests of the globe may conveniently be classified into six major formations:

Tropical Forests
1. Tropical Rainforest
2. Tropical Deciduous Forest*

---

* Areas identified as tropical deciduous forest actually encompass several plant associations, including semievergreen and semideciduous.

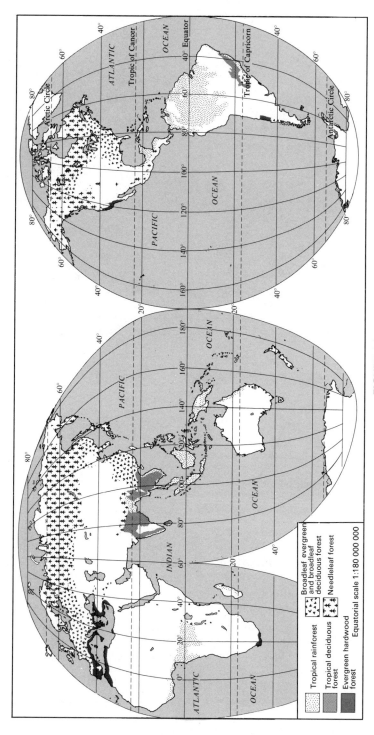

FIGURE 4.3
THE FORESTS OF THE EARTH.

Tropical rainforest

Tropical deciduous
forest

Evergreen hardwood
forest

Broadleaf evergreen
and broadleaf
deciduous forest

Needleleaf forest

Equatorial scale 1:180 000 000

GLOBAL VEGETATION
93

FIGURE 4.4
THE SELVA OF THE MIDDLE AMAZON, BRAZIL.

Middle Latitude Forests
3. Sclerophyll Woodland and Shrub
4. Broadleaf Evergreen Forest*
5. Broadleaf Deciduous Forest
6. Needleleaf Forest

Tropical Rainforest (Selva). The most luxuriant and complex forest on earth is found within those parts of the tropics where rainfall is plentiful and well distributed throughout the year (*Af climates*). These conditions are most extensively developed in the Amazon and Congo River basins; along the rainy coast of tropical West Africa; and on numerous islands, peninsulas and coasts ranging from Indochina through the Indonesian Archipelago (Fig. 4.3).

The tropical rainforest is often referred to as the

selva (Fig. 4.4). A most conspicuous feature of this plant formation is the large number of tree species that have evolved under millenia of uninterrupted warm and moist climatic conditions. Whereas mid-latitude forests may contain 10 to 12 species of trees per hectare, the number may exceed 100 in the selva. Each tree is structurally adapted to unique sunlight and moisture constrictions. The tallest members demand intense solar radiation; lower stratum species require shade. Thus, tree canopies are stratified by species tolerance to sunlight and by plant height (Fig. 4.5). In an ideal setting three levels of forest stratification can be recognized: the tallest trees are widely spaced and characterized by broad umbrella-shaped crowns reaching average heights of 35 to 40 meters (115 to 131 feet); the first understory is about 25 meters (75 feet) high; and the lowest tier of least sun-tolerant trees have canopies that are around 10 to 15 meters (30-35 feet) from the ground. Lower strata

---

* Other names for this formation include temperate rainforest, temperate evergreen forest, and laurel forest.

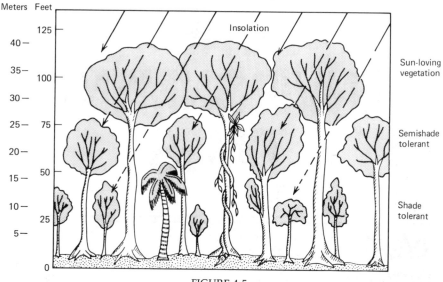

Stratification of Rainforest Vegetation

FIGURE 4.5
STRUCTURE OF A MATURE TROPICAL RAINFOREST.

species are more closely spaced than the giants that overshadow them and tend to develop conical-shaped crowns.

The trunks of tropical rainforest trees are uniquely different from their counterparts in the mid-latitudes. Such trees usually have thin bark. With no frost danger and little moisture stress, there is no need for a thick, insulating outer skin. Trunks tend to be slender, free of lower branches, and devoid of the annual growth rings typically found in forests where a dormant season is the rule. Roots are shallow. Because the soil is continuously wet, roots need not reach deeply to meet the tree's moisture and nutrient demands. Enormous buttressed bases are distinctive of the larger trees (Fig. 4.6). These ribbonlike flanges flare outward from the base to provide both support and stability for the tall trunks.

Lush and luxuriant are words that characterize the foliage of the tropical rainforest. This is a broadleaf evergreen plant formation. Leaves tend to be large and have a leathery texture. Their size provides for numerous pore spaces to transpire huge quantities of water efficiently. Their texture ensures protection from the intense midday sun and from the debilitating effect of long periods of high temperature. There is no distinct time when the forest as a whole is devoid of foliage. For most trees, shedding of leaves occurs as sporadic events

FIGURE 4.6
A GIANT BUTTRESSED TREE OF THE SELVA.

THE ENVIRONMENT TIMES

Volume 4, No. 9   December 1979

Rain forest destruction stirs warning of disaster

Reprinted by permission of Knight-Ridder Newspapers

WASHINGTON — The price of a double burger at your favorite fast food palace is among several things that may be contributing to one of the most fearsome environmental crises in history.

That is a central thesis of Norman Myers in his new book, "The Sinking Ark."

In much the same way that Rachel Carson's "The Silent Spring" alerted the world to the dangers of pesticides, Myers' book is a somber warning about the systematic destruction of vast tracts of the world's tropical rain forests.

"These forests cover only 7 percent of the earth's surface," Myers said the other day, "yet they contain at least 40 percent of all species, especially plants and non-vertebrates."

At the present rate of de-struction, he said, the forests could be effectively erased from the Earth in the next 50 years. And the destruction of the rain forests means a parallel destruction of plant and animal life.

It is a nightmare scenario, and some detractors are bound to argue that Myers has overstated his case.

But Myers—a global citizen who was born in Britain, educated in California and now resides in Kenya, in East Africa—is regarded by the environmental and scientific communities as a first-rate conservation scientist who knows as much about economics as most economists.

"The Sinking Ark" meticulously documents the environmental consequences of rain forest destruction, and then goes beyond the ordinary conservationist polemic by tracing the economic processes that have led to the predicament.

Which brings us back the price of a hamburger at the golden arches.

Many tropical forests, particularly in Latin America, are being cleared to provide grazing land for beef cattle.

The meat, instead of feeding the impoverished masses of the Latin countries, is exported to Japan, the United States and West Europe—where the bulk of it is consumed in fast-food outlets.

Purchased in bulk by brokers and fast-food chains, the beef costs a few cents a pound less than home-grown American beef. So it is "rational" and "economic" for the fast-food chains to buy it.

Yet, Myers said, if American and other consumers were willing to pay "a few pennies more" for their hamburgers, this economic chain could be broken—and perhaps part of the devastation of the rain forests could be abated.

Cattle-grazing is only part of the problem, according to Myers. He says the plunder of rain forests in Indonesia and Borneo by state-owned and international timber companies is destroying the habitat of thousands of species there, and is adding to flood threats and the progressive impoverishment of millions of Asian people.

Slash-and-burn farming in the rain forests of central Africa and Brazil's Amazonia is perhaps the leading cause of the deforestation.

But it is the most difficult to deal with because of the

when old, nonfunctional members drop off. The less dominant deciduous species may stand barren at any time, but are normally surrounded by their more numerous verdant neighbors.

The complex canopy structure of the rainforest is very effective in preventing direct solar radiation from reaching the forest floor. Of all the energy arriving at the uppermost crowns of a mature stand, less than one percent is capable of penetrating to the soil surface. As a consequence, the forest floor is in deep shade and, except for mosses and ferns, undergrowth is meager. Dense surface growth (typically shown in the movies as *jungle*) occurs only where greater amounts of sunlight reach the forest floor. This occurs in burned-over areas, abandoned agricultural clearings, along riverbanks and coasts, and on very uneven terrain.

Trees dominate rainforest landscapes. Nevertheless, there is a rich variety of other plant forms, including: *lianas*, or thick, woody vines; *epiphytes*, air plants such as the orchid that germinate high in the branches of trees and draw their nutrients and moisture from the atmosphere; and *parasites*, plants that use the tree as a host for their survival needs. All these elements combine to add to the complexity, diversity, and colorfulness of this majestic forest association.

The selva is the earth's most heterogeneous forest formation. Its native species include valuable cabinet woods such as mahogany, rosewood, and ebony; the resinous rubber tree; commercial nut trees such as the cacao; and tropical fruits, such as the banana and pineapple plants, and the mango and papaya trees. Unfortunately, under natural conditions plants of any one species are widely scattered. Thus, intensive utilization of forest resources is limited. Lumbering operations cannot operate on an efficient "clearcut" basis, because of the numerous tree forms that lack commercial value. Instead, they must selectively cut out valuable woods that are widely spaced and that can repay the cost of labor and transportation to distant markets. In some areas the original forest has been removed and replanted to single species. This has been done with the cacao and rubber trees and the banana and pineapple plants, all of which are now grown in large plantations. Some effort also has been made toward the plantation growing of the valuable cabinet woods, but on a very limited scale. Environmental conservationists have expressed serious concern about the future of the tropical rainforests. As explained in *"Rainforest destruction stirs warning of disaster,"* these areas are being modified at a rate unprecedented in earth history.

Tropical Deciduous Forest. Poleward of the selva are climatic regions that have a pronounced dry season. In this area the duration of the dry season increases and total rainfall diminishes with increasing distance from the rainforest. As might be expected, the region's vegetation reflects this variation in moisture availability (Fig. 4.7). Where the dry season is short, the general character of the forest is not too dissimilar to that of the selva. The primary exception is that the trees behave according to a definite seasonal cycle. Trees will sprout flowers simultaneously at the onset of the wet season and rhythmically shed their leaves during the water–deficient period. With increased duration of the dry season the forest changes dramatically. Most species are deciduous and there is more openness to the forest stand.

The tropical deciduous forest is best represented in India and Southeast Asia, and consequently is often referred to as *monsoon forest.* It is, however, the native vegetation of large parts of Africa and South America as well, even though these latter areas have been largely altered by man

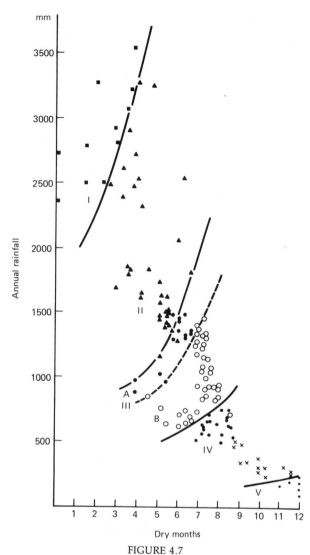

FIGURE 4.7
THE RELATIONSHIP BETWEEN FOREST VEGETATION AND ANNUAL RAINFALL (*ordinate*) AND LENGTH OF DRY SEASON IN MONTHS (*abscissa*) IN INDIA. I, evergreen; II, semi-evergreen-tropical rainforest; III, monsoon forest (A, moist; B, dry); IV, savanna (thornbush forest); V, desert.

Nevertheless, pure stands of any one tree type, such as the highly prized *teak,* are not uncommon.

Examination of Figure 4.3 reveals that the tropical deciduous forest occupies a large part of the tropics and lies in locations intermediate between the rainforest and savanna grasslands. The forest is not uniform throughout this large expanse. Along the margins of the rainforest, many lower stratum trees remain evergreen, while only the uppermost strata are deciduous. This unique association is sometimes separately classified as *semi-evergreen forest.* The drier portions of the region, on the other hand, often take on a very scraggly appearance. Referred to as *tropical scrub* or *thorn forest,* these plant communities are comprised of low-growing deciduous trees with trunks that are gnarled and thick barked. Grasses are more frequent as part of the surface cover, and the undergrowth may contain thorny plants and numerous bushes. Tropical scrub forest areas are verdant and lush during the rainy season, but are brown and appear lifeless during the dry part of the year.

Evergreen Hardwood Forest (Sclerophyll Forest). This evergreen plant association is found in subtropical regions where the rainy season occurs during the winter. In the summer, when potential plant growth is at its maximum, precipitation is scant and soils are dry. These conditions are found within the earth's Mediterranean (Cs) climates, as revealed by a comparison of Figures 4.3 and 3.13. To survive the moisture stress period of summer, a wide variety of plant adaptations enable vegetation to obtain and conserve water. These include tap roots 4 to 8.5 meters (13 to 28 feet) long that search for subsurface water; laterally developed roots that can absorb moisture from a large area; a thick insulative bark; and leaves that are thick, hard, and leathery. The leaves remain on the trees throughout the year and are referred to as *sclerophyllous,* a term that provides an alternative name for this plant community, namely, *sclerophyllous forest.*

The evergreen hardwood forest is a region that has been favored for settlement by man, and the plants seen on such a landscape today are only remnants of the original vegetation. From a reconstruction of the original forest based on relict plants near the Mediterranean Sea, it is believed that the original forests had trees 15 to 18 meters

and presently support tall savanna grasses. Maximum tree heights in the monsoon forest are less than in the selva, ranging from 12 to 35 meters (40 to 100 feet). Unlike the rainforest, trees here are more widely spaced, begin branching at lower levels, and have a thick, rough bark. The wide spacing permits large quantities of sunlight to penetrate to the surface. As a result, undergrowth is dense, true to the character of jungle, making the forest difficult to penetrate. Numerous species typify the composition of the monsoon forest.

THE NATURE OF VEGETATION

(45 to 55 feet) tall with a closed canopy, and were composed almost exclusively of live oak. Beneath the trees was a shrub layer 3 to 5 meters (10–15 feet) high, and a sparse surface stratum of herbs (about ½ meter in height). Today, however, the sclerophyll regions are mostly *woodland*: an open form of forest in which the tree canopy may only cover 25 to 60 percent of the surface, a result of man modifying the native plant cover. The once dominant trees have continually given way over a greater proportion of the surface to scrub forms of vegetation (locally known as *chaparral* in the American Southwest and *maquis* in Mediterranean Europe).

The trees of the evergreen hardwood forest are relatively low; they branch close to the ground; they have gnarled trunks; and they possess a thick, insulating bark. Being somewhat isolated from one another, each sclerophyllous region has its own dominant tree species. Around the Mediterranean Sea, live oak, cork oak, olive, and various pines are common; whereas in North America live oak and white oak dominate. Australia has numerous species of eucalyptus and acacia. South America's representative forms border on types that relate to the nearby deserts; and South Africa lies in a distinct floristic realm with tree species uniquely its own.

The seasonal rhythm of plant growth in the sclerophyllous forest association may seem strange to someone unfamiliar with the biota of the area. Vegetation becomes vibrant with life at the initiation of the autumnal rains. New leaves sprout and the plants take on a rich mosaic of green. Through the winter plants are verdant, reaching a peak in the early spring when flowering commences. During the summer their colors dull, and they appear in shades of gray or brown. Yet, though they may seem lifeless, there is no completely dormant season for these plants.

Broadleaf Evergreen Forest (Middle Latitude). The truest representatives of this forest community are found on the eastern side of continents, poleward of the tropical rainforest. In these areas the annual temperature range is small or moderate, and rainfall is both ample and well distributed throughout the year. Such conditions are ideally met in Southeast Asia, southeastern Brazil, eastern Australia, and northern and western New Zealand. This broadleaf evergreen association is stratified, as is its tropical counterpart. It differs from the latter, however, in having fewer species of trees, an increased frequency of pure stands containing one specie, smaller leaves, less dense canopies, and a thicker undergrowth.

Tree species in the broadleaf evergreen middle-latitude forest vary by region. In eastern Asia, there is little left of what must have once been a luxuriant plant cover. The demand for fuel in this densely populated part of the world has led to an almost complete removal of the native vegetation. Only a few relict species remain in very isolated locations. Thus, it is difficult to say with certainty what the climax community was like. In the southeastern United States (southern Florida and the Caribbean margins of the Gulf Coastal Plain), evergreen oaks, magnolia, and laurel are common. Australia's representative areas contain eucalyptus and acacia. In New Zealand, large tree ferns, podocarp trees, beeches and Kauri dominate the forest species (Fig. 4.8). The above plant com-

FIGURE 4.8
TREE FERNS NEAR GREYMOUTH, SOUTH ISLAND, NEW ZEALAND.

munities typically contain epiphytes, such as the Spanish moss of the southern United States, and varied forms of lower stratum vegetation that includes low-growing herbaceous plants, shrubs, bamboos, tree ferns, and palms.

Broadleaf Deciduous Forest (Middle Latitude). The middle-latitude deciduous forest is a plant community that is almost entirely restricted to the Northern Hemisphere. Aside from a few mountainous areas in South America and New Zealand, it does not occur south of the equator. Deciduous trees favor humid climates with relatively mild winters, and a growing season of four to six months. Due to large moisture demands during the growing season, high summer rainfall is especially conducive to their formation. These conditions are found in eastern North America and eastern Asia within the humid subtropical and humid continental climate regions, and in western and central Europe. Like the tropical deciduous forest, the trees of this plant formation seasonally shed their leaves. But, whereas defoliation in the tropics is related to moisture stress, leaf shedding in the middle latitudes is an adaptation to the cold season. The loss of the thin deciduous leaves protects this type of tree from water loss and freezing during the cold season.

The structure and composition of the deciduous forest is markedly different from the selva. The trees are shorter, 20 to 30 meters (65–100 feet) and thick barked; the canopy, although continuous, permits considerably more light to filter through to the surface, and variety is small, three to four tree species dominating an area. The distinct strata of trees, each comprised of unique species, so typical of the rainforest, are absent. Rather, a mature stand in this forest has a mixture of tree sizes of the same species—ranging from saplings to fully developed forms with trunks over a meter in diameter. Root systems are well developed and are often deep, thus negating the need for supporting, buttressed bases. The lowest stratum of the forest consists of a variety of perennial bushes and numerous herbaceous plants, including early-flowering annuals capable of completing their growth cycles before the tree canopy develops, and shade tolerant forms such as ferns. The trees of this plant community have specific habitat requirements. Each species is very competitive and constitutes a dominant life form in environments that lie within its narrow range of tolerance. Thus it is that this forest is frequently subdivided into associations that reflect the dominant species. In the United States, for example, three major associations are recognized. They are the oak-hickory, the birch-beech-maple and the oak-chestnut-yellow poplar (Fig. 4.9).

A remarkable characteristic of most species in a leaf shedding forest is that foliage changes color in early autumn when the green chlorophyll of the leaf tissue begins to degenerate. This signals the onset of cooler weather and a time when the trees are entering into a dormant (rest) period. During this time of metamorphosis, tree canopies exhibit spectacular color displays, ranging from yellows to scarlet.

Needleleaf Forest (Middle Latitudes). The needleleaf forests are largely composed of cone-bearing trees (conifers) that occur under widely differing conditions of moisture availability and temperature. These hardy plants have developed adaptations that enable them to occupy sites unsuitable for many other tree species. Their resistance to cold weather makes them dominant in subarctic climates; they constitute the chief vegetation in many fire-prone areas; and they are

FIGURE 4.9

MIXED DECIDUOUS FOREST OF MAPLE, BEECH, AND BASSWOOD.

FIGURE 4.10
DWARFED TREE FORMS ALONG THE THOMAS RIVER, YUKON, CANADA.

often found on sandy or gravelly surface materials where rapid soil drainage leads to moisture stress during the growing season. Common features of practically all conifers include a thick, insulative bark; broad, shallow root systems; slender and, in most cases, evergreen leaves; conical canopies; and seed cones.

The greater bulk of all the world's coniferous trees are found in two large expanses located between 45 to 75° N in North America and Eurasia. In both cases the forests extend completely across the continents. These are areas with a severe winter (lasting 6 to 8 months) and a short growing season of usually less than 120 days. This needleleaf forest, also called the *taiga* or *boreal forest*, does not contain large trees. In the southernmost part of the region, average heights may range from 10 to 15 meters (30–45 feet) and trunk diameters from 25 to 30 cm (10–12 in.). Farther poleward, tree heights and trunk diameters diminish to dwarf forms, often being less than 1 meter (3 feet)

tall and a few centimeters (1–1½ in.) wide (Fig. 4.10). What is most remarkable about these forest associations is the relatively few tree species present. In North America, white spruce and balsam fir occupy well-drained sites, and black spruce, jack pine, larch, and tamarack dominate in areas of poor drainage. The Eurasian association has a fairly simple composition in western Europe, where pine and spruce dominate the landscape. Farther east other species begin to take over, especially the Siberian fir, spruce, larch, and the dwarf Siberian pine. Of the foregoing, the larch is a special form of needleleaf vegetation. Like the bald cypress of the American Southeast, these needleleaf trees—although found in the coldest reaches of the subarctic climate—are deciduous and shed their leaves during the winter season.

The main characteristic of the boreal forest is uniformity. Over extensive areas, homogeneous stands of dense evergreen conifers overshadow the surface and allow little sunlight to penetrate for the support of undergrowth. Therefore, only small herbaceous plants and fungi that require little direct sunlight thrive. Under such shaded conditions, as well as relatively low temperatures over a significant portion of the year, microbial activity is low and surface organic debris decomposes slowly, resulting in a surface littered with needle leaves and semidecomposed branches.

The tallest stands of needleleaf trees are found in warm, humid regions, such as the marine west coast climates of North America and Europe, and in the humid subtropical coastal plains of the southeastern United States. The species found in these areas differ from those of the taiga and include several types that are highly prized for construction lumber and for pulpwood. In the American Northwest, species include the Douglas

fir with tall straight trunks of over 65 meters (200 feet), with diameters of 6 to 10 meters (18–30 feet); giant redwoods that attain heights exceeding 100 meters (325 feet); spruce, and a variety of pines (Fig. 4.11). The needleleaf forests of the southeastern United States are distinctly different from either the boreal forests or the conifers of the American Northwest. This area has been classified as a broadleaf deciduous forest (Fig. 4.3). What appears to be a misrepresentation on the map can be easily explained. Extensive areas of loblolly and slash pine dominate the uplands and bald cypress is found in the swamps and depressions of the region (Fig. 4.12). This current plant cover, however, is not believed to be representative of the region's climax community, which should be dominated by species of oak. Rather,

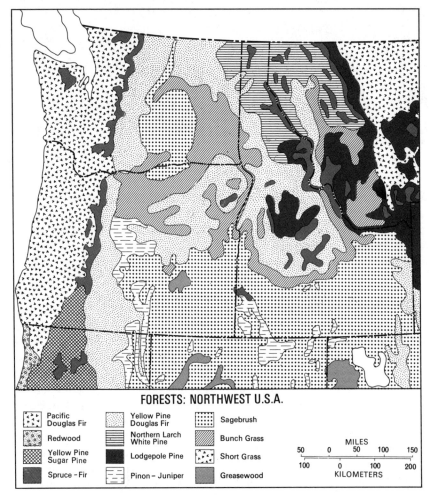

FIGURE 4.11
FOREST ASSOCIATIONS OF THE NORTHWESTERN UNITED STATES.

THE NATURE OF VEGETATION

FIGURE 4.12
A STAND OF LONGLEAF PINES IN THE CROATAN NATIONAL FOREST, NORTH CAROLINA.

naturally occurring fires coupled with deforestation by man have resulted in the regeneration of species associated with a subclimax community. These fast growing pine species are important raw materials for the plywood and pulpwood industries. As a consequence, their dominance upon the landscape in preference to climax species has been and most probably will continue to be promoted by foresters in the southeastern United States.

The Grassland Formation

Grasses and trees are antagonistic plant types. Where one is found, the other is usually excluded. But, while primary forest forms are sensitive to environmental vicissitudes, grasses are tolerant of widely varying climatic and soil conditions. Found from tropical to polar thermal regimes, and under moisture conditions varying from arid to humid, these hardy plants include species that range in

height from a few millimeters to giant bamboos that may exceed 20 meters (60 feet). Although grasses are occasionally found on sites in forested regions where conditions may be too wet, too dry, or too alkaline for trees, the largest continuous expanses of grassland occur in association with two distinct climatic types: the tropical savanna (Aw), and semiarid (BS) regions (Figs. 3.2, 3.9, and 4.13). In both of these climates, moisture deficiencies are common. Such environments provide favorable habitats for the grass family of plants that, on the whole, are less water demanding and more efficient in economizing moisture than are woody species. Like the forest associations that exhibit regional variation in both structure and composition of species, the grasslands may also be subdivided for classification purposes. Two major types are recognized: *Tropical Savanna Grassland* and *Middle-Latitude* Grassland.

Tropical Savanna Grassland. The tropical savannas are regions of homogeneous grassland upon which woody plants (dwarf trees and shrubs) are more or less evenly distributed. Tropical savannas are associated with the drier margins of the Aw climates and are largely confined to South America, Africa, Australia and a small portion of south Asia (Fig. 4.13). These plant communities are strikingly different from the grasslands of the middle latitudes. Nearly all of the savannas have a dominant cover of tall grasses that grow singly or in bunches, and most have short trees or bushes scattered throughout (Fig. 4.14).

The appearance of the savanna landscape changes dramatically with the seasons. With the advent of the rainy period, dwarf trees and bushes begin to flower and form leaves, young grass shoots burst with renewed life, and the overall plant cover seems lush and bountiful. Many unwary visitors who have viewed the savanna during this stage have departed with exaggerated ideas of the region's potential for agriculture and animal husbandry. As the rainy season progresses, the grasses increase in both stature and coarseness, attaining heights between 1 to 4 meters (3–12 feet). To maintain such heights individual grass blades must obviously become stiff and tough. Moreover, these native grasses tend to be low in nutritional value. Bleeding gums and malnutrition are common maladies of animals that graze on this forage. Consequently, savannas

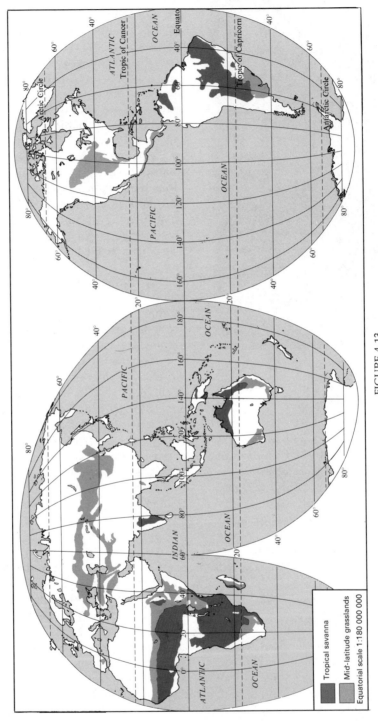

FIGURE 4.13
THE GRASSLANDS OF THE EARTH.

Tropical savanna

Mid-latitude grasslands

Equatorial scale 1:180 000 000

THE NATURE OF VEGETATION

104

FIGURE 4.14
SAVANNA GRASSLANDS OF KENYA, EAST AFRICA.

have not fostered the development of an important commercial livestock industry, even though grazing is the principal economy of the region. During the long dry season (4 to 6 months) the grasses turn yellow, brown, and brittle. The straw mat produced from the drooping aerial parts of the grass plants is highly flammable. And the savannas have had a history of being periodically burned over, by either natural or human agency. Fire, however, has little effect upon grasses. Their sensitive parts are protected below ground and after a surface burn they can easily regenerate. Tree growth, on the other hand, is limited by fire: only resistant species can maintain themselves in this highly competitive environment. As pointed-out below by Dr. Joseph E. Van Riper, the origins of the savanna are complex:

Students of tropical vegetation have uncovered as many problems in attempting to interpret plant succession in the savannas as have been raised with respect to the prairies and steppes. Some researchers claim that nearly all the savannas were once similar to tropical open woodland or semideciduous rain forest and that they have been transformed into savannas by periodic burning. Others admit that some of them were produced by burning but insist that, over wide areas, the seasonal droughts could never support more than a sparse stand of trees and that a dominance of the coarse grasses is to be expected. All are agreed, however, that the area of savannas is growing at the expense of the tropical forests and woodlands and that savannas have existed for many thousands of years.*

It is also noteworthy that the savanna regions contain unique plant associations found along the margins of streams. These *galeria†* forests and thickets of tall canes or reeds retain their greenness and foliage longer than the savanna plants and contain plants that are much more densely packed together. The reason for their occurrence is that water is available nearer to the surface in proximity to water courses for longer periods of time than is the case in the savanna proper. In an airplane flight over the region during the dry season, these vegetative forms often stand out as ribbons of greenery amidst the brown and yellow colors of the dry savanna grasses.

Middle Latitude Grassland. The middle latitude grasslands consist of prairie and steppe vegetation. *Prairie* is a continuous cover of tall grasses, averaging one meter (3 feet) high, whereas *steppe* is composed of shorter, bunch grasses. In each, flowering herbs intermingle with the grasses, providing the region with varied colors during the growing season. Both prairie and steppe are usually located between the forest cover of humid climates on one side, and the desert vegetation of arid climates on the other. They are found in Eurasia, extending from the Black Sea to almost as far as the Yellow Sea; in the North American interior plains and the Basin and Range province; and in the pampas of eastern Argentina.

There are very few remnants of undisturbed prairie vegetation. These tall grasslands have almost totally given way to the plow and now represent some of the world's most productive agricultural regions. The naturally fertile soils found here are deep, rich with humus and plant nutrients, and have structural and moisture retaining properties ideal for grain cultivation. In addition, the prairies occupy land surfaces that border on the forests and are in climates ranging from the subhumid to humid. Thus, moisture supplies for crops are more plentiful than in the steppes that

* Joseph E. Van Riper. *Man's Physical World*, McGraw-Hill Publishing Co., New York, 1962, p. 342.

† Where ground water is available for trees, along the banks of streams, the canopies stretch outward over the stream creating a tunnellike opening, that is likened to a corridor or gallery; hence the name galeria forest.

occupy the drier portions of the grasslands. Even so, drought is not uncommon; this explains, in part, the persistance of a grass cover in these regions. Yet, considerable debate has focused upon the question of whether the prairie is a climax plant community. Although trees and shrubs were almost totally absent when the area was first settled by Europeans, tree species planted in prairie soils have grown to maturity and can reproduce. Some argue that the ease with which trees establish themselves in the prairie signifies that the region may have once been covered by forest and that aboriginal populations modified the vegetation through the use of fire. Others suggest that recurrent droughts and a tendency for naturally occurring fires have deterred tree growth. And still others claim that grazing animals and climatic change are responsible for the lush grasses observed by our American pioneers. It is difficult to say with certainty that the prairie is truly a climax vegetation form or to verify its origins. Perhaps the answer lies in some mix of the number of probable causes suggested.

Steppe, also referred to as short-grass prairie, consists of short bunch grasses 25 to 50 cm (10–20 in.) high separated by bare surface. Ground cover is far from complete and low trees or scattered shrubs may dot the landscape. As mentioned above, this plant association occupies the drier portions of the grasslands. Although plant geographers are in general agreement that the steppe regions are climax grass-communities, all do not concede that the short bunch-grasses now observed are totally representative of the area's undisturbed vegetation. It is possible that overgrazing by either domesticated or wild animals has resulted in the selective removal of taller grass species that may have jointly occupied habitats with the present bunch grasses. Since grazing animals have always been present in the steppelands, the question regarding the original composition of the climax community will probably remain unanswered. Today, the steppe is used for commercial grazing and the extensive cultivation of grains. These are economic functions that are well suited to the region when proper land management practices are employed (Fig. 4.15).

The Desert Shrub Formation

Very little of the earth's land surface is absolutely barren of plant life. The few exceptions include ice-covered landscapes as in Antarctica and Greenland, regions of shifting sand dunes, and some rocky surfaces devoid of soil. In the remainder of the earth's deserts, the vegetation consists of widely spaced plants with considerable open space between them. As shown in Figure 4.16, the land area dominated by the desert plant formation is considerable, amounting to approximately 20 percent of the earth's total land area. Large expanses of this plant community are found in the Sahara, Namib, and Kalahari deserts of Africa; the plateaus of Arabia and Iran; and the Thar, Gobi, Sonoran, Australian, Atacama, and Patagonian desert regions.

Desert plants differ considerably from one part of the world to another. The most widespread aridland vegetation form, however, is desert shrub. *Sagebrush* and *creosote bush* are common species of this plant type found in the arid southwest of the United States. These shrubs have well developed vertical and lateral root systems and are capable of both maximizing the use of limited soil water during infrequent rainfall periods and of curtailing moisture loss through defoliation during periods of extreme aridity.

In addition to the desert shrub, other members of the aridland plant community include species of evergreen shrubs and low trees; leafless evergreens, such as the thorny cacti; a variety of salt tolerant plants; *ephemerals,* or short-lived plants; and grasses. The evergreen trees and shrubs are found mainly on the margins of the Mediterranean climates from whence plant invasions of sclerophyllous vegetation have occurred. The leafless evergreens are succulent, non-woody plants that utilize their stems as breathing surfaces. Their lack of leaves reduces photosynthesis and water loss, and their spines protect them from foraging animals. The large water storage capacity of these plants enable them to survive lengthy inter-precipitation periods. Salt tolerant plants (*halophytes*) occupy those poorly drained sites in desert regimes that have saline or alkaline soils. Ephemerals are usually small in size and include some grasses, tubers, and flowering annuals. They are able to remain dormant during indefinitely long dry periods and to rapidly germinate and mature during brief moist spells.

Within the desert shrub formation there is considerable regional variation as regards the disparity between water supply (precipitation) and atmospheric water demand. In some areas water

FIGURE 4.15

BLACK ANGUS CATTLE GRAZING IN COMMANCHE NATIONAL GRASSLANDS, COLORADO. Development of water facilities, fencing of pastures, reseeding with high-quality grasses, and good grazing practices are some of the management practices that help provide high quality grazing along with protection of the range.

supply may consist of only a light shower once every several years. In other locations, however, precipitation data reveal a moister regime wherein vegetation formations more typical of semiarid climates should prevail. To explain this particular anomaly researchers have coined the term *desertification*, meaning a process in which desert formation is caused by human misuse of the land.

One of the most recent examples of broad-scale desertification has taken place in the Sahel region of Africa. *Sahel* is an arabic word, interpreted as "border," that identifies the southern fringes of the Sahara Desert (Fig. 4.17). This area lies within a semiarid climate and receives rainfall in only four months of the year. Throughout history the region has experienced intense, periodic droughts. For centuries the widely dispersed native populations of the Sahel grazed scattered herds of cattle

on its vegetation cover and raised crops of millet and sorghum to meet basic food needs.

During the early 1960s rainfall was greater than expected for the Sahel. Accompanying these moister conditions was an intensification of agricultural activity and a substantial increase in the numbers of the region's people and animals. Then, towards the 1970s, the rains subsided and the region suffered from extreme drought. Large herds of animals competed for the scant growth of vegetation. Overgrazing and destruction of vegetation resources resulted. With the removal of the plant cover the soil of the area dried, cracked, became subject to wind erosion, and attained a character much like that of the deserts to the north.

Widespread suffering of the human and animal population accompanied the physical transfor-

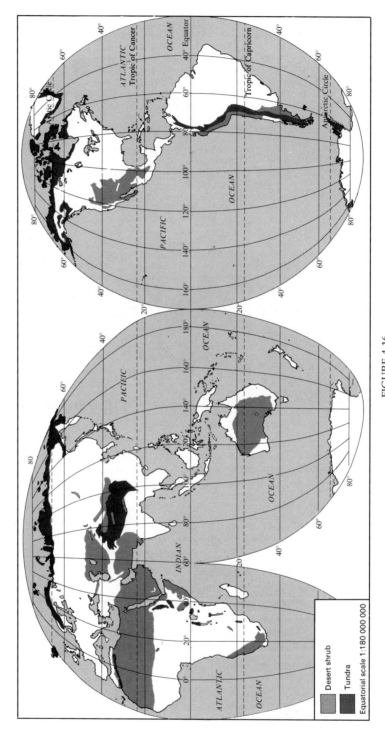

FIGURE 4.16
THE DESERT AND TUNDRA REGIONS OF THE EARTH.

Desert shrub

Tundra

Equatorial scale 1:180 000 000

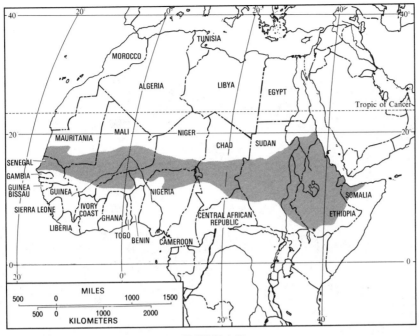

FIGURE 4.17
THE SAHEL.

mation of the landscape. It has been estimated that in 1974 more than 50,000 people died either through starvation because of crop failure, or from years of cumulative malnutrition that made them susceptible to diseases. Although an accurate count is not available, estimates place animal losses in the hundreds of thousands.

During the drought, the Sahara Desert expanded southward, largely because of man-induced disruption of the natural vegetation cover. Is the drought over? Will the Sahel vegetation recover? These are questions to which there are no absolute answers at the present time. What has been learned from observing man's use of such marginal lands, however, is that without careful use of their resources, environmental deterioration and desertification is bound to occur.

The Tundra Formation

The tundra, a vegetation community characterized by a low, but relatively complete ground cover, borders the polar reaches of the North American and Eurasian continents (Figs. 4.16 and 4.18). This region has a growing season of less than two months, experiences summer frosts, and has poorly drained, acidic soils underlain by per-

manently frozen ground. The most common plants are grasses, sedges, mosses, lichens, and occasional low-growing shrubs. In addition, there are also species of *forbs* (broadleaf herbs) that flower during the warm season, and a few dwarf species of trees. Of the latter, the arctic birches and willows are representative types which, at maturity, attain maximum heights of 15 to 60 cm (6–24 in.). Engulfed in snow during the long winter, the short trees are protected from strong winds that would dry out and abrade with driving snow larger plant forms.

Since much of the tundra was recently covered by glacial ice, some plant researchers have questioned whether the present vegetation represents a true climax community for the region. This is an issue that is apt to remain unresolved for some time to come. Economic activity related to tundra vegetation is minor and is restricted to grazing of the domesticated reindeer and to commercial hunting.

Highland Vegetation

The vegetation of mountainous areas has a vertical zonation that is associated with altitudinal climatic differences. Although in some respects it

FIGURE 4.18
TUNDRA VEGETATION NEAR THE SHORE OF HUDSON BAY, CAPE CHURCHILL, MANITOBA, CANADA.

is possible to draw parallels between latitudinal and altitudinal sequences of plant associations, highland vegetation patterns are more complex. Local position and exposure are critical to solar energy receipts, air temperature and wind patterns, and to precipitation types and amounts. Minor variation in either position or exposure can lead, in turn, to strong contrasts in these atmospheric variables and foster the development of strikingly different plant communities.

As explained in the previous chapter, it is difficult to generalize about highland climates. The same statement holds true for highland vegetation. Thus, only a single simplified, example of a vertical plant sequence is included (Fig. 4.19). In the central Sierra Nevada mountains of California, precipitation normally increases with elevation

along the western flanks to about 2500 meters (8200 ft), where total amounts approximate 1270 mm (50 in.). The accompanying vegetation forms range from the low, sclerophyllous (Mediterranean) types of the dry lowlands to the giant sequoia of the humid uplands. Above the 2500 meter (8000 ft) level, precipitation totals decrease and temperatures are substantially below those of the adjacent lowlands. Representative trees diminish in height and become species that are tolerant of cold temperatures. At about 3500 meters (11,500 feet) the tree line is reached, an altitude above which climatic conditions are too extreme for tree growth and where the vegetation community is dominated by grasses. The eastern flanks of the Sierra Nevada are generally drier and warmer than their western counterparts. Forest life forms ex-

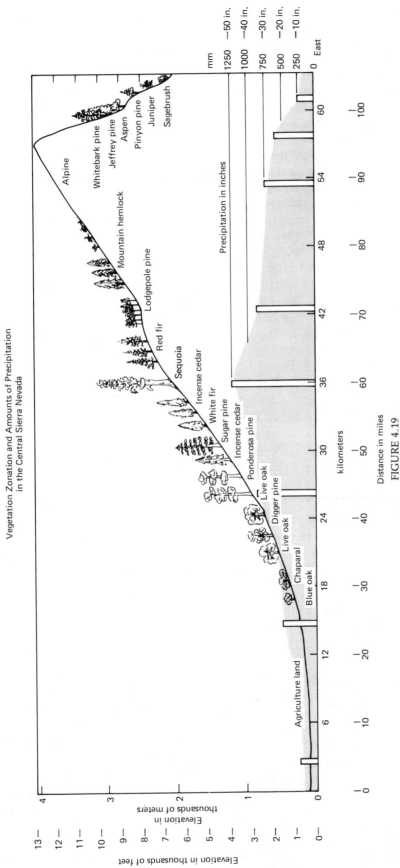

Vegetation Zonation and Amounts of Precipitation
in the Central Sierra Nevada

FIGURE 4.19

RELATIONSHIP BETWEEN ALTITUDE, PRECIPITATION, AND VEGETATION.

tend to higher elevations, and desert plants, especially sagebrush can be found extending above the 2000 meter (6600 feet) contour.

Throughout the last two chapters it has been suggested that the distribution of vegetation types and the physiologic characteristics of plants are strongly related to climate. Although this is true, it must be added that vegetation also affects climate. The canopies of trees shade forest floors and reflect short-wave radiation that might otherwise be absorbed to heat the surface. Plants transpire moisture that is withdrawn from beneath the surface into the air, thereby affecting atmospheric humidity. And, large closely-spaced plants reduce the speed of low-level winds. Thus, it is clear that vegetation and climate are not only strongly related, but are also strongly interrelated. The processes operating in one affect the characteristics of the other. To this dynamic interplay, yet a third environmental factor is of significance: soil. The ultimate character of soil is largely determined by climatic and biotic factors. Soil, in turn, can influence the type of vegetation found within an area and also serves as a storehouse of subsurface water available for release to the atmosphere. Soils of the earth are the topic of discussion in Chapter 5.

KEY TERMS AND CONCEPTS

photosynthesis
plant community (association)
biota
climax vegetation
plant succession
pioneer vegetation
vegetation formation
tropical rainforest (selva)
jungle
liana
epiphyte
parasite
tropical deciduous forest
monsoon forest

semi-evergreen forest
tropical scrub forest
thorn forest
evergreen hardwood (sclerophyll) forest
chaparral (maquis)
broadleaf evergreen forest (middle latitude)
broadleaf deciduous forest (middle latitude)
needleleaf forest (middle latitude)
taiga (boreal forest)
grassland associations

tropical savanna
galeria forest
middle latitude grassland
prairie

steppe
desert shrub formation
tundra formation
highland vegetation

DISCUSSION QUESTIONS

1. What is plant succession? How is plant succession effective in altering the character of the earth's surface?
2. Define vegetation formation. Which forest formation is dominant along the equator?
3. Describe the differences and similarities between broadleaf evergreen trees and broadleaf deciduous trees. In what areas of the earth would you find examples of each type? What climatic factors influence where each type occurs?
4. Why are buttressed-based trunks a common feature of tall selva trees?
5. Explain the environmental conditions that give rise to jungle growth.
6. How does the tropical deciduous forest differ from the selva?
7. Why has the sclerophyllous forest been so thoroughly altered by man?
8. What environmental factors make needleleaf evergreen trees more competitive than broadleaf deciduous trees in the taiga region?
9. Discuss the climatic characteristics typical of the savanna grasslands and the middle-latitude steppes. What do these regions have in common?
10. Explain the role of fire in maintaining the savanna region in a grass cover.
11. How does the areal extent of a galeria forest differ from that of a forest association that is closely related to climatic factors (for example the tropical rainforest)?
12. Explain the difference between prairie and steppe.
13. What is an ephemeral?
14. Describe the characteristics of tundra vegetation.

REFERENCES FOR FURTHER STUDY

Dansereau, Pierre, *Biogeography: An Ecological Perspective*, Ronald Press, New York, 1957.

Eyre, S. R., *Vegetation and Soils*, Revised Second Edition, Aldine Publishing Co., 1968.

Kuchler, A. W., *Potential Natural Vegetation of the Conterminous United States* (map and manual), American Geographical Society, Special Publication No. 36, 1964.

Odum, Eugene P., *Fundamentals of Ecology*, W. B. Saunders Co., Philadelphia, 1973.

Polunin, Nicholas, *Introduction to Plant Geography*, McGraw-Hill Book Co., New York, 1960.

Walter, Heinrich, *Vegetation of the Earth*, The English Universities Press Ltd., London, 1973.

THE NATURE OF SOIL

S OIL IS ONE OF MANKIND'S FUNDAMENTAL RESOURCES. The food we eat, the lumber we use, and the plants that respire oxygen to the atmosphere all receive their nutrients from the soil. The soil has also served as a crucial element in the cultural evolution of mankind. As humankind learned to till the soil and cultivate plants, sedentary patterns of living began to be established. The transition from a hunting and gathering economy to an economy based on agriculture marked the beginning of civilization. Cooperation in irrigating and cultivating crops required new social institutions, while surplus agricultural production enabled populations to increase in number. In areas of intensive agriculture cities evolved, occupational specializations developed, and social stratification came into existence.

When proper management practices are employed in soil utilization, productivity can be sustained endlessly. There are soils along the eastern seaboard of the United States that have been successfully cultivated for more than 200 years. When soil conservation is not practised disaster often results. Each year a growing world population demands more products from limited soil resources. Yet, losses of soil through erosion remain a serious threat to productive capability. Equally serious is the loss of soil fertility; that is, the destruction of the capacity to grow crops. To meet the food and product needs of the future, scientific knowledge and management skills must be ap-

plied to soil management and conservation. To face this challenge, it is necessary to understand soil characteristics and how they vary over the globe.

SOIL COMPONENTS

Soil may be defined as those surface materials of the earth capable of supporting the growth of land plants. Soil is a mixture of mineral and organic materials that serves as a home for dense populations of animals and plants, most of which are too small to be seen with the naked eye. Although normally thought of as an unchanging part of the landscape, soil constantly evolves, acquiring new characteristics.

Soils differ in appearance from one region to another. The prairie soils of Iowa are deep and dark colored; those of the North Carolina Piedmont are bright hues of red. But, regardless of their widely different appearances, every soil is made up of four essential parts: (1) *mineral matter*, (2) *organic material* (plant and animal), (3) *soil water*, and (4) *soil atmosphere*.

The Mineral Component

Minerals make up the bulk of most soils. Not only do minerals provide a base upon which plants can anchor and support themselves, but they also act as a reservoir of nutrients, supplying elements necessary for plant growth. Minerals are made available for soil formation by weathering processes. *Weathering* refers to all processes acting at or near the earth's surface to cause physical disruption and chemical decomposition of rock. Such commonly used terms as "sandy soil," or "clayey soil" signify the kinds of soil particles that are present after weathering. All mineral soil particles are classified by size. There are three groups. The largest soil particles are sand, intermediate particles are silt, and the smallest particles are clay. Particles larger than sand are not considered true soil.

Most soils contain mineral particles of more than one size. The proportion of particles in each size-class defines *soil texture*. Figure 5.1 illustrates the concept of soil texture. Note that a soil classified as sand must contain at least 85 percent sand. This means that the soil consists of relatively large particles. Since sandy materials tend not to stick together, these soils are very loose, easily moved by the farmer's plow, and can readily absorb large quantities of rainfall which quickly

migrates downward to be temporarily stored beneath the surface. A soil dominated by clay, on the other hand, sticks together, adheres to the farmer's plow (making tillage more arduous), and absorbs rainfall very slowly. A soil mixture of two or more particle sizes has characteristics that are intermediate between these two extremes.

The size and percentage of soil particles, and the manner in which they are arranged, determine the ease with which plant roots can penetrate the soil; the soil's moisture storing capacity; the degree of soil aeration, or exposure to air; and the rate of chemical reactions within the soil. The arrangement of soil particles is known as *soil structure*. Some soils, such as those consisting entirely of sand, may show no arrangement of their particles because they lack binding elements. In other soils, the particles may be grouped together to take on a platelike appearance, or be prismatic, columnar, blocky, or even granular in form (Fig. 5.2). These forms are the result of interacting factors, both environmental and human, that include the chemical nature of the materials, the amounts of clay and organic matter present, wetting and

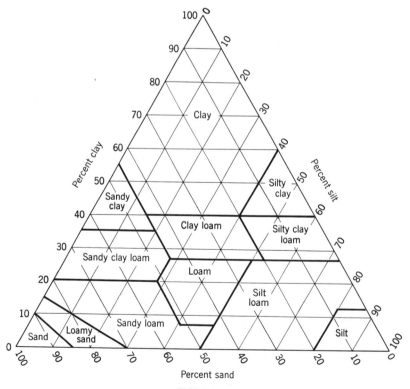

FIGURE 5.1
SOIL TEXTURE CLASSES.

THE NATURE OF SOIL

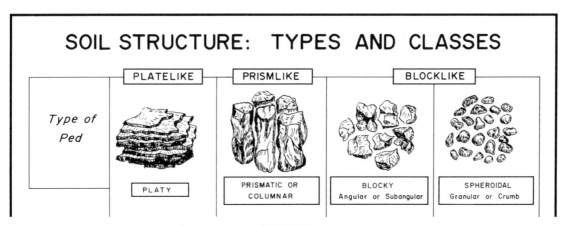

FIGURE 5.2
EXAMPLES OF SOIL STRUCTURE

drying patterns, freezing and thawing cycles, and cultivation practices.

The Organic Component

Soil organic matter includes living plants and animals, leaves, twigs, or stalks lying upon the soil, and remnants of decomposing plants and animals. The soil contains dense populations of life forms. Vegetation, especially such higher forms as grasses and trees, produces organic matter from inorganic materials and performs the very important function of returning soluble minerals from the root zone to the surface of the soil. This role is particularly significant with regard to calcium, magnesium, potassium, and sodium that are gradually removed, or leached, from the soil by *gravitational water*—a form of surplus precipitation that percolates through the soil to lower levels under the influence of gravity. These minerals are commonly known as *bases* because they tend to give a more basic and less acid chemical reaction to the soil, as the reading on "Soil Reaction" demonstrates. The more a plant takes up bases, the more it offsets the removal process. In general, grassland soils are less acid than forest soils because the grasses are heavy feeders on bases. Animals also constitute part of the soil's organic content. Worms, insects, gophers, moles, and woodchucks dig and burrow through the soil. As

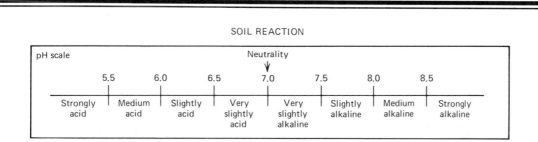

To provide a measure of soil acidity or alkalinity, scientists have devised the pH scale that reflects the concentration of hydrogen ions in a soil solution. Since hydrogen most often replaces bases that have been removed by gravitational water, the degree to which it is present is indicative of acid conditions. As shown in the above diagram, soils can range in reaction from strongly acid through strongly alkaline. Acid conditions are usually found in humid climates where surplus precipitation washes the soluble bases from the soil. In arid climates, where little water is available to remove bases, hydrogen content is low, pH is high, and the soil is often saturated with bases. Because most of man's agricultural crops demand large supplies of the basic minerals, pH values are a general indication of soil fertility.

they pulverize, mix, and transport mineral and organic matter, they serve to modify the soil's structure, aeration, and drainage properties.

Microscopic plants and animals, however, are more numerous in soil than either large plants or animals. Sometimes numbering several million within a single teaspoon of fertile topsoil, microorganisms obtain their energy by decomposing plant residues. In feeding upon plant remains, they minimize the accumulation of organic debris within the soil, and in the process aid in recycling plant nutrients. Decomposition, moreover, results in the production of *humus*, the relatively stable, dark-colored remains of organic matter. Soils rich in humus have colors that range from brown to dark brown, whereas soils deficient in humus have colors related to their mineral make up.

The Water and Atmospheric Components

The soil's pore space is occupied by either air or water. The quantities of air and water present in soil bear an inverse relationship to one another. When soil is saturated with water after a heavy rain, its air content may be near or at zero. On the other hand, if a drought persists for several months air increases as water percentage diminishes. The amount and size of soil pores determine, to a large extent, the water-storing capability of a soil and its oxygen availability.

Soil Water. When a soil particle is moistened for the first time, a minute film of *hygroscopic* water is absorbed on its surface. This form of water, amounting to as much as 5 percent of soil weight, is found in all of the earth's soils, including those of the desert. Hygroscopic water is so firmly attached to the particle surface, however, that it cannot be used by plants and can be removed only by oven drying the soil under intense temperatures.

When rainfall enters the soil it surrounds each hygroscopic film with a layer of *capillary* water. The amount of water available for plants is determined by the quantity of capillary water present in the soil, even though not all such water may be utilized. When the capillary layer is thick, plants withdraw moisture from it with ease. Shrinkage of the capillary layer by water removal, on the other hand, causes the remaining moisture to be held by the soil with increased tension. The energy required by plants to further withdraw moisture eventually becomes too great and they wilt. This level of moisture depletion is referred to as the *wilting point* (Fig. 5.3).

Following a period of prolonged rain or irrigation, the soil may be saturated. Much of this water is found in large pore spaces that will rapidly drain once the water supply ceases. Because this form of water stays in the soil for only a short time, it is little used by plants and is referred to as *gravitational* or *free* water. As we shall see, gravitational water is instrumental in transporting soluble minerals and producing texture differences within the soil.

After gravitational water drains, the soil is considered to be at *field capacity*; that is, it is holding its maximum amount of capillary water. As noted earlier, the amount of water held and available to plants will vary according to soil texture. A soil containing large quantities of water may supply

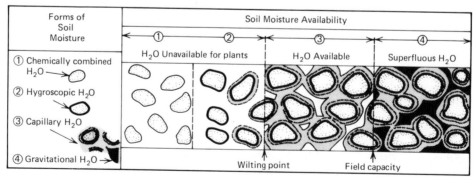

FIGURE 5.3
FORMS OF SOIL MOISTURE. The amount of water available to plants is equal to the "field capacity" minus "wilting point."

only meager amounts to plants. The *available water* is determined by noting the difference between field capacity and wilting point, as illustrated in Figure 5.4. When fine soil particles dominate a soil, capillary and hygroscopic water both increase. The proportion of water available to plants to the total water held in the soil, will, however, diminish.

Soil Atmosphere. Soil air exists in those pore spaces not occupied by water. Unlike atmospheric air, however, soil air is not only rich in carbon dioxide (with amounts of 10 to 100 percent greater than above the ground surface), but also has lower quantities of oxygen. This is because plant roots and the organisms living in the soil remove oxygen from, and respire carbon dioxide into, pore spaces. Most crops cannot grow if the carbon dioxide level in the root zone is too high or if oxygen availability is too low. A constant exchange must take place, between the soil and free atmosphere; the free atmosphere supplying oxygen and the soil diffusing carbon dioxide into the free atmosphere. For soils to be well aerated, there must be a sufficient number of open pore spaces to permit essential gases to move easily in and out of them. Saturated soils obviously have little space available for air.

SOIL FORMATION

When soil scientists analyze a soil in its natural setting, they concern themselves with the variation in the soil's composition and character with depth; that is, they examine a *soil profile*. A soil profile extends from the surface of the earth through the subsurface layers and into the nonsoil or parent material below. Knowledge of soil properties as they vary with depth provides information on plant root environments, nutrient and moisture availability, susceptibility to erosion, internal chemical reactions, and other data. The character of each soil profile is the evolutionary product of its environment.

Soil is a naturally occurring body that experiences continual change. Its properties are produced by the combined effects of climate and biotic activity upon parent material, as conditioned by topography over periods of time.

Earth materials are subject to weathering processes that change them physically and chemi-

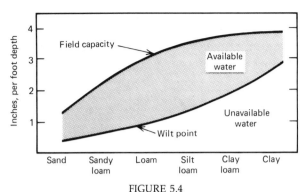

FIGURE 5.4
SOIL WATER HOLDING CAPACITY RELATIVE TO SOIL TEXTURE.

cally. Such materials, given sufficient time and favorable atmospheric conditions, can have established upon and within them a community of plants and animals. The subsequent accumulation of organic residues establishes a new complex between the earth's crust and the atmosphere—a mixture of material consisting of organic and mineral matter. These conditions stimulate the development of an expanded biological community: microorganisms including bacteria, protozoa, and a host of other minute plants and animals become increasingly abundant as they feed on organic remains and release nutrients for further plant growth. The soil during this first stage of development consists of a surface layer, darkened in color from decaying organic matter overlying the parent material.

If soil-forming processes continue and the rate of soil loss by erosion is less than the rate of rock weathering, the soil will deepen. As deepening takes place, soil properties begin to vary with depth. The surface layer (topsoil), under the influence of gravitational water, will have *eluviated* (removed) from it soluble minerals and very fine sized mineral particles. The latter move only a few feet at most before they become lodged, creating an *illuvial* (accumulation) layer. The presence of both an eluvial and illuvial layer, known as the soil's *A* and *B horizons* respectively, indicates an advanced stage of soil evolution. (If a soil contains only an A horizon resting upon partially altered parent material—the soil's C horizon—it is said to be young or *immature*; time has been insufficient for soil forming processes to produce horizon differences.) A soil attains a *mature* stage of development with the formation of the illuvial B horizon (Fig. 5.5). The concept of the mature soil with horizon differences is important to the understanding of the global variation of regionally distributed soils.

SOIL CLASSIFICATION

An analysis of elements from several profiles of a farmer's field would reveal that each profile is unique—no two soils are exactly alike. Just as there are some differences in identical twins, soils also vary over even short distances. Soils that are affected by similar environmental controls, on the other hand, share common properties. Thus, it is possible to classify soils into groups.

Various types of soil classification have probably been employed since the beginning of agriculture. There is evidence that the Chinese recognized and named different kinds of soils, more than 4000 years ago. The classification used was based largely on soil color and structure. Within the last 150 years or so, as data from soil analysis became more abundant, new and more sophisticated classification methodologies have been proposed.

In the mid-nineteenth century, soil classification on the basis of parent material was attempted. The assumption was that geology determined soil characteristics. This approach applied well to young soils, but was inadequate for many mature sites where climate and vegetation had substantially modified the parent material.

This so-called geologic approach was supplanted by revolutionary concepts that emerged from Russia during the late 1800s. Soil was considered a unique independent body that exhibited distinct traits as a result of the interactions of climate, parent material, flora and fauna, geomorphic (landform) factors, and time. Subsequently disseminated to the United States, these ideas on soil formation were incorporated into a

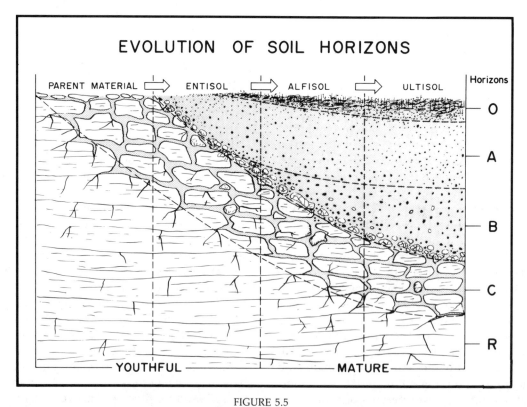

FIGURE 5.5
THE EVOLUTION OF SOIL HORIZONS. The identified layers are: O, surface organic matter; A, eluvial horizon; B, illuvial horizon; C, decomposing parent rock; R, original rock.

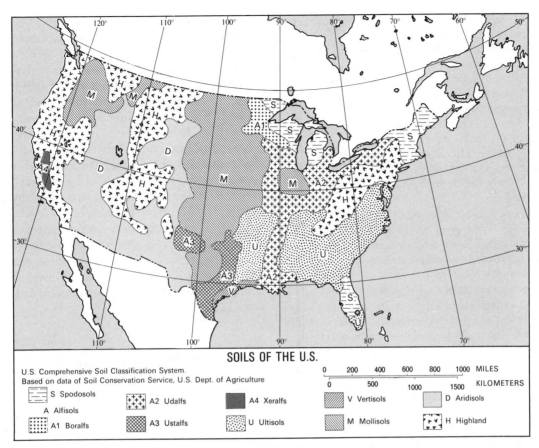

FIGURE 5.6
GREAT SOIL GROUPS OF THE UNITED STATES.

new classification scheme during the 1920s by the chief of the U. S. Soil Survey (C.F. Marbut). At the broadest level of generalization two classes were defined: (1) *Pedalfers*—soils of the humid climates, and (2) *Pedocals*—soils of moisture deficient regions. These two classes of soils were subdivided into lower categories, each designed to provide more specific information about soil types. Geographic emphasis was placed primarily upon Category IV (Great Soil Groups), the mature soils that expressed zonal distributions.

The new approach was a significant improvement over previously used methodologies; however, its limitations soon became apparent. New soils were identified that would not fit into the system; and the classification did not account for those soils that had been severely eroded or drastically reworked by man. Thus, it was necessary to continue the search for a better means of classifying and understanding soils.

THE SOIL TAXONOMY—CLASSIFICATION CATEGORIES

Six levels of generalization have been established for classifying soils according to the system used in the United States. At the broadest level of generalization (the order level) only a few similarities are present. In the lowest category (soil series) there is relatively complete homogeneity in both soil features and their genesis. The criteria for separation of the various classes are briefly summarized below:

I. Order Level. A given soil must possess common properties indicating similarity in kind and strength of soil forming processes, such as the presence or absence of major diagnostic horizons.*

* Each soil exhibits distinct features that are characteristic of the environment in which it has formed. These features are said to be diagnostic, in that they distinguish the soils in one category from those in another.

II. Suborder Level. This class has genetic homogeneity. Subdivision of orders are according to the presence or absence of properties associated with wetness, soil moisture regimes, major parent material, and vegetational effects as indicated by key properties.

III. Great Group Level. Differentiation is based upon similar kind, arrangement, and degree of expression of horizons, base status; soil temperature and moisture regimes; and the presence or absence of diagnostic layers.

IV. Subgroup Level. Provides for the central concept of the great group and for properties indicating intergradations to other great groups, suborders, and orders, and to extragradation to "not soil."

V. Family Level. This class identifies features of importance to plant root growth, such as broad textural characteristics, mineralogical composition, and soil temperature and reaction.

VI. Series Level. Soils are separated on the kind and arrangement of horizons; color, texture, structure, consistency, and reaction of horizons; and the chemical and mineralogical properties of the horizons.

The current period of soil classification in the United States has a quantitative orientation and is designed to be sufficiently broad to include soils not yet identified. The new *soil taxonomy* (classification according to natural relationships) dates from the 1950s. It was refined and tested for about 20 years prior to its publication. Six levels of generalization were established. Soils are placed within a category only if there is evidence of common horizons or features. The remainder of this chapter is concerned with the broadest level of generalization, and is a useful beginning for developing an understanding of the interrelationships that exist between global soils and large-scale realms of vegetation and climate (Fig. 5.6). Known as the *order level*, this level of generalization contains 10 soil categories, each with an identifying name ending in sol—from *solum*, Latin for soil (Table 5.1).

GLOBAL SOILS

Soil can be thought of as a record of the interaction of environmental factors over time. The distributional pattern that soils display on a global scale, as provided on a hypothetical continent, il-

TABLE 5.1 SOIL ORDERS: NAME DERIVATION, AREAL SIGNIFICANCE, RELATIONSHIP TO WEATHERING ACTIVITIES, AND PRINCIPAL COMPOSITION

Soil Order	Derivation of Root Word	% of Total World Soils[a]	Rank (Total Area)	Degree of Weathering[b]	Primary Composition
Entisols	Recent.	12.5	4	Low	
Vertisols	L: *verto* = to turn.	2.1	9		
Inceptisols	L: *inceptum* = inception, beginning.	15.8	2		
Aridisols	L: *aridus* = dry.	19.2	1		
Mollisols	L: *mollis* = soft.	9.0	6		Mineral soils
Spodosols	Gr: *spodos* = wood ash.	5.4	8		
Alfisols	Alf: combined from aluminum and iron.	14.7	3		
Ultisols	L: *ullimos* = ultimate.	8.5	7		
Oxisols	Oxi: from oxide	9.2	5		
Histosols	Gr: *histos* = tissue.	.8	10	High	Organic soils

[a] An additional 2.8 percent of the world total includes ice fields, unclassified lands, and others.

[b] Each mineral soil occupies a unique position within a hierarchy that is based upon degree of weathering; that is, degree of both horizon development and chemical alteration of the original parent material.

lustrate this point (Fig. 5.7). Soil features that recur throughout broad regions of like climate and vegetation suggest that interrelationships do exist between these elements of the physical environment. The remainder of this chapter expands upon this generalization by examining the 10 soil orders of the earth within their environmental context (Fig. 5.8).

Entisols (Recent Soils) and Inceptisols (Incipient Soils)

Entisols are immature soils that show little evidence of soil formation. Typically, they have an A horizon resting directly upon partially weathered parent material (a C horizon). Entisols are true soils and should not be confused with re-

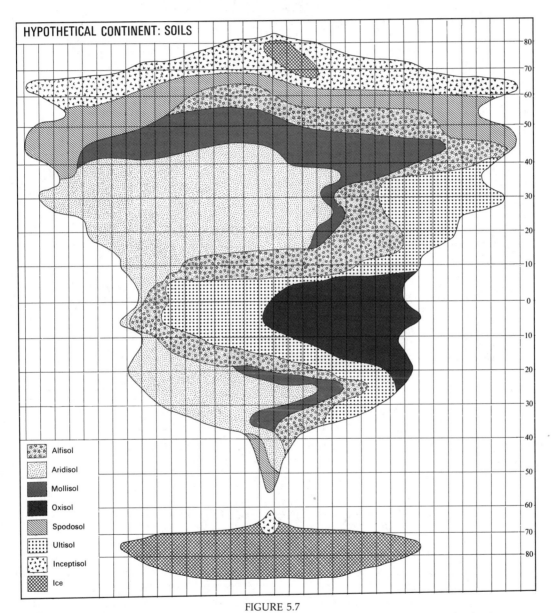

FIGURE 5.7

SCHEMATIC CONTINENT WITH SOIL ORDERS. This diagram represents all of the earth's landmasses squeezed together at their respective latitudes. The distribution patterns of regional soils are shown in their appropriate latitudinal position and are proportionally representative of the area each occupies.

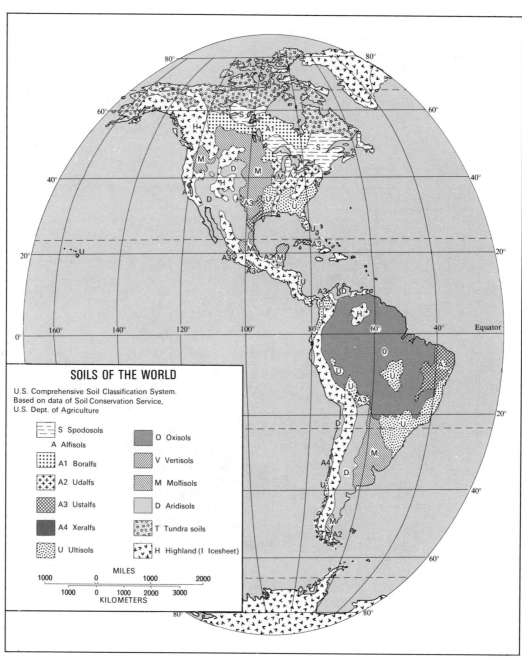

FIGURE 5.8
GLOBAL DISTRIBUTION OF SOIL TYPES.   (Courtesy of the Soil Conservation Service.)

cently weathered parent material incapable of supporting plant life. Various factors are responsible for the lack of horizon development or mineral alteration. In some, time has been a dominating element. These entisols are found on newly exposed surface deposits that have not been in place long enough for soil forming processes to operate to their fullest. They may be found on steep, actively eroding slopes, on floodplains, and on recent glacial outwash plains. There are types of entisols, however, that are very old. Consisting primarily of quartz or other minerals that do not alter readily, they simply do not form horizons. Entisols may exist in almost any moisture or tem-

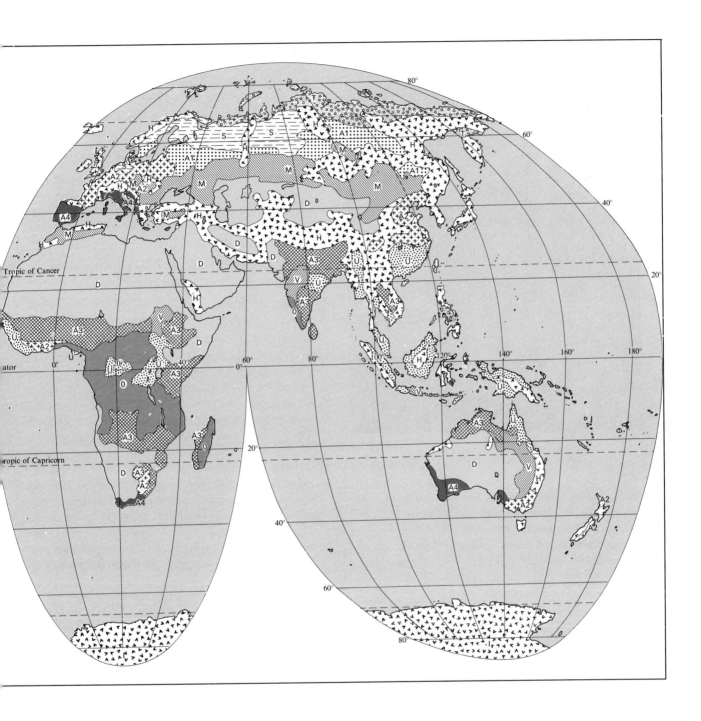

perature region, on any type of parent material, and under any form of vegetation (Fig. 5.9).

The *inceptisols*, like the entisols, include soils of widely differing environments in which a variety of soil forming processes operate. Some are weathered sufficiently to produce slightly altered horizons, yet they normally lack a well developed B horizon. In short, inceptisols are beginning to exhibit characteristics associated with their soil forming environment, but these features are too weakly developed for inceptisols to be considered mature soils.

Entisols and inceptisols have many uses and pose diverse problems to land managers. Most of

poor drainage are the chief causes of crop failure. Other entisols and inceptisols of importance are those developed in sandy deposits. In most cases these types of soil are of limited agricultural use. Sandy soils tend to have a low moisture-holding capacity and low organic content. Made up chiefly of quartz, they are also deficient in nutrients. Thus, potential crop yields are usually low. Their primary use has been for livestock grazing, although in favorable climates some have been modified by man and now support truck farms and citrus groves.

Vertisols (Expanding-Contracting Soils)

Vertisols are clayey soils that swell upon wetting and shrink upon drying. When dry, they produce open cracks that may be more than 25 mm (1 inch) wide and 50 cm (20 inches) deep. The open cracks permit surface materials that are dislodged by rainfall, animals, or the like, to fall to lower depths in the soil. When rain occurs, water runs into the cracks, moistening the soil both from above and below. With moistening, the clays expand and the cracks close, trapping the displaced particles at lower levels (Fig. 5.10). Because of the trapped particles, the lower portion of the soil now has an increased volume of material, and pressure is exerted in all directions to move excess soil material away. The soil, however, can only respond by moving upward. This repetitive process, in which soil moves to lower levels and is

Figure 5.9
AN ENTISOL PROFILE, CENTRAL COLORADO.

these soils have formed in wet regimes, and especially in areas adjacent to streams. Under natural conditions most support a forest cover of water-tolerant trees. When these lands are cleared for agricultural purposes, flooding and

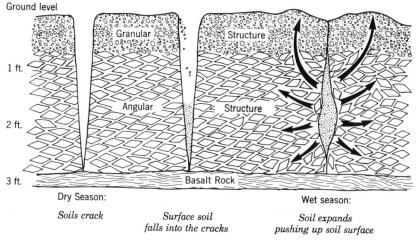

FIGURE 5.10
AN ILLUSTRATION OF THE WETTING-DRYING CYCLES OF VERTISOLS.

subsequently forced back toward the surface, means that the soil will eventually churn over, or invert itself.

Vertisols are primarily found between latitudes 45° North and South, and are most extensively developed in Australia, India, and the Sudan. On a worldwide basis they are used for cotton, wheat, corn, sorghum, rice, sugarcane, and pasture. The major problem in their use relates to their high shrink-swell capacity. The stress created by the alternate expansion and contraction of their clays has been known to crack building foundations and road surfaces.

Aridisols (Desert Soils)

Aridisols are pale-colored mineral soils that are low in organic content. They are shallow and exhibit features associated with an arid climate of meager rainfall and high temperatures. The largest expanses of aridisols are found in the Sahara, Namib, Atacama, Sonoran, Australian, Thar, and Gobi deserts.

The dominant process in the creation of an aridisol involves the accumulation of calcium and magnesium carbonates that produces a *calcic horizon* (subsurface layer of carbonate enrichment). The sequence of events producing this feature is referred to as *calcification* and begins with rainfall uniting with free carbon dioxide as it passes through the atmosphere. This mixture, a mild carbonic acid, enters the soil and dissolves calcium and magnesium from the surface materials. Drainage of gravitational water then transports the solution to the subsoil. Subsequent dehydration of the soil by evapotranspiration causes an increasing concentration of soluble minerals in the soil water, eventually reaching the saturation level. Any further moisture withdrawal results in the precipitation of the minerals held in solution. Each rain period repeats the process and thickens the carbonate layer in the direction of the land surface. In some desert areas these processes are sufficiently active to cement the carbonates into a solid rocklike mass (Fig. 5.11).

Land use in arid regions is generally confined to either grazing or the intensive production of crops under irrigation. Grazing predominates over most of the world's aridisols, and ranges from simple nomadic herding in less developed regions to large-scale commercial ranching in technologi-

FIGURE 5.11
AN ARIDISOL PROFILE, CENTRAL ARIZONA.

cally advanced areas. Grazing operations, however, frequently require some degree of surface irrigation for the production of supplemental feed and for winter grazing. Thus, these two land-use functions are not necessarily exclusive.

The land manager in an arid environment must be prepared to face problems that include not only scarcity of water, but also saline accumulations in the soil that may be toxic to plants; nutrient deficiencies; cementlike calcic horizons that restrict water movement; turbulent winds; precipitation that often arrives in heavy downpours; and the threat of drought.

Mollisols (Grassland Soils)

Mollisols are relatively fertile, dark colored, humus-rich soils. Their name is derived from the Latin word *mollis,* meaning soft, a term which is characteristic of their surface horizon. They are thought to be formed primarily by the underground decomposition of organic matter. Found in regions lying between the arid and humid climates of the middle latitudes, they may be thought of as existing in a climatic transition zone. Natural vegetation ranges from short, widely spaced bunchgrass along the arid margins to a rich, luxuriant tallgrass prairie along the humid boundaries. The most extensive stretches of mollisols are found in the North American Great Plains, from the Black Sea eastward beyond Lake Balkash, in Manchuria, and in the Argentine pampas.

A mollisol profile exhibits the pronounced effect of vegetation. The softness of the surface horizon, the soil's color, and the available nutrients are all related to the rooting, feeding habits, and life cycle of grasses.

Grasslands differ from forests in both the total amount of organic matter incorporated within the soil and in its distribution throughout the profile. In forested areas, the major sources of soil humus are derived from leaffall that accumulates on the surface. Grasses also produce an organic mat as their decaying aerial parts accumulate. In contrast to the rooting structure of trees, however, the dense, fibrous, masses of grass roots permeate and distribute themselves uniformly throughout the profile. As the plants die and decay in an almost continuous cycle, a large quantity of humus is produced (Fig. 5.12).

Grasses, in general, require greater quantities of basic mineral nutrients, particularly calcium, than do trees. As a result, their organic remains are richer in base nutrients, which upon decay of the plant are subsequently returned to the soil. The released bases are available for use by plants in a continuing cycle that under natural conditions maintains a relative high degree of soil fertility. It is only during infrequent lengthy wet periods that the soluble sodium and potassium are washed from the profile. The less soluble calcium salts have a tendency to accumulate in lower horizons because normal rainfall reaches only a few feet into the subsoil, where it gradually evaporates into the soil atmosphere or is utilized by plants.

FIGURE 5.12
A MOLLISOL PROFILE, SOUTHEASTERN SOUTH DAKOTA.

Under such dehydrating conditions calcium carbonate is carried into the subsoil in solution and may later become solidified into hard nodules.

SOIL FERTILITY AND PRODUCTIVITY

The supply of chemical elements necessary for plant life is generally regarded as a measure of *soil fertility*. However, the actual use of the term soil fertility by agronomists and soil scientists is more specific. They relate the term to the availability of chemical elements that support plants upon which mankind depends for food such as grains like wheat, corn, rice, and barley. By this definition, soil fertility does not depend on the ability of a soil to sustain trees and shrubs. *Soil productivity* refers to the capacity of soils to produce organic matter, regardless of vegetative form. The soils of the tropical rainforest and humid subtropical climates support prolific plant life, yet agriculturally their soils are naturally low in fertility. Desert regions have low productivity ratings because vegetation is sparse. Nonetheless, many of these soils are very fertile and capable of producing high-yielding plants, if supplied with water.

The fertility of mollisols and the climatic conditions that encouraged the growth of a natural cover of grass are conducive to the development of large-scale commercial grain farming and livestock grazing, especially along the desert margins.* The grains grown include wheat, rye, barley, corn, and sorghum. Management problems for the semiarid agriculturalist include variability of moisture resources, as well as the problem of having to maintain the favorable fertility and structure characteristics typical of these soils in their native state.

Spodosols (Soils with Mobile Iron and Aluminum)

Spodosols are mineral soils in which aluminum, iron, and organic matter have accumulated in the subsoil (*Spodic horizon*). These soils occur in humid areas ranging from the tropics to the tundra. They are best developed, however, in the subarctic realms of North America and Eurasia where winters are long and cold, where there is great seasonal variation in air temperature, and where precipitation is concentrated in the summer. Due to the absence of extensive land masses in the middle latitudes of the Southern Hemisphere, no parallel extensive region of spodosol development occurs south of the equator.

The processes that influence the development of Spodosols are known collectively as *podzolization*. The term is derived from the Russian word *podzol* which is translated "beneath ashes." This may seem a strange manner in which to describe a soil or soil-forming processes, until one recognizes that the original use of the term was to characterize the soil's color—ashen gray. In nineteenth-century Russia, the soils first classified as podzols had a surface mineral horizon from which most minerals, except quartz, had been removed by leaching. This frequently produced a sandy, bleached A horizon. The physical mixing of organic matter into this layer (by burrowing animals or percolating water) subsequently gave a grayish color to the horizon. Hence, the term *podzol*.

The vegetation of the subarctic consists of coniferous forest with mosses and lichens on the surface. The organic litter produced by these plants is low in calcium and nitrogen and responds slowly to decompositional processes. Consequently, a mat of organic debris develops upon the soil surface that deepens with time. Rainfall passing through this humus layer converts to humic acids that are capable of speeding-up chemical weathering activities. Under strongly acid conditions soluble minerals are rapidly washed from the profile; clays are removed from the A horizon and accumulate in the B horizon; and oxides of iron and aluminum migrate to lower levels where they form the spodic layer in the B horizon (Fig. 5.13).

The spodosols are the life-supporting medium for the world's largest continuous expanse of

FIGURE 5.13
A SPODOSOL PROFILE, NORTHERN NEW YORK.

---

* In many areas, there is evidence to indicate that early man, in controlling vegetation through the use of fire, modified the natural vegetation from forest cover to grass.

boreal forms of vegetation. These forested regions have, in recent years, experienced large-scale exploitation because their trees are a desirable raw material for newsprint. This demand for pulpwood, coupled with natural and man-induced fires, have resulted in the clearing of extensive stands of forest.

Some of the world's best known nurseries and market gardens, such as those in the western Netherlands, southeast England, Denmark, and northern Germany occur on spodosols. Many years of deep plowing, the use of fertilizers, and heavy manuring have created soils that are highly productive for human needs. These sandy soils lack the sticky character of clay soils, and their ease in cultivation has earned them the reputation of being "light." Since they are also highly permeable and contain relatively large pore spaces, plants can produce abundant fibrous roots with ease. However, this porous character can also be a disadvantage. Sandy soils have a low capacity to store water; hence, they warm up quickly in the spring and give plants an early start. On the other hand, drought periods have a rapid impact on vegetation sustained upon sands. This factor frequently requires the additional expense of supplemental irrigation for high-value crops. Intensively worked spodosols require considerable capital investment to insure maximum production, and they are profitable only where a market is available and the demand for produce is strong and dependable.

Alfisols (Fertile Forest Soils)

Alfisols have clay accumulations in the B horizon and a rich supply of soluble minerals, especially calcium and magnesium. No other soil order with mature development exists in so many diverse climatic and vegetation environments. Alfisols are found in warm and cool regions, in moisture realms ranging from humid to seasonally arid, and under flora that varies from broadleaf deciduous trees to the thorn trees and tall savanna grass of the tropics.

Alfisols are relatively fertile soils. They are formed on stable landscapes in which water has been instrumental in removing clays from the surface horizon and depositing them at a lower level. For this process to be effective, two pre-existing conditions must be fulfilled;

1. Either the parent material must contain very fine clays or subsequent weathering must be capable of producing them.
2. These very fine clays must be subject to dispersion. The presence of certain mineral salts, such as carbonates and free oxides, will tend to cluster clay particles, making them more difficult to displace. In such cases, the chemical weathering and removal of these salts must precede clay migration.

Even though alfisols may be acid in reaction, they possess a moderate to high reserve of bases.

FIGURE 5.14
AN ALFISOL PROFILE, CENTRAL MISSOURI.

This high percentage is linked to three primary causes: (1) the presence of base-rich parent material; (2) the presence of weatherable minerals, rich in bases, in the soil; or (3) the fact that bases are being added to the soil by wind and/or water (Fig. 5.14).

These fertile soils support some of the earth's most intensive forms of agriculture. In the United States, the well-known agricultural region called the Corn Belt occurs largely on alfisols (in association with mollisols). Over the years this area has yielded as much as two-thirds of the nation's corn, oats, and soybeans, and nearly one-half of its alfalfa. Land management systems include cash grain farming, dairying, and livestock feeding operations.

Ultisols (Forest Soils with Low Base Status)

Ultisols have clay accumulations in the B horizon and are particularly deficient in calcium and magnesium. They are more thoroughly weathered and have experienced greater mineral alteration than any other mid-latitude soil. Although their overall morphology and horizon sequence is similar to those of the alfisols, extensive leaching under acid conditions has resulted in the removal of most solubles (Fig. 5.15). Ultisols exhibit a very simple distribution pattern, being confined to humid subtropical climates and some youthful tropical landscapes.

The processes effective in developing the clay horizon that were discussed for alfisols are also operative in ultisols. In fact, except for differences in fertility, the overall characteristics of these two soils are similar. This has led some researchers to suggest that with time and further weathering alfisols may eventually degrade into ultisols.

Being the most weathered of all mid-latitude soils, ultisols are usually found on older surfaces. Extensive chemical weathering has led to the removal of solubles; yet contrary to expectations, nutrient availability does not increase with depth. The roots of trees extending several meters below the surface extract bases that are returned to the soil surface. Before these minerals can move downward, they are recycled by the roots. Hence, the plants effectively retard the loss of bases. Yet the maintenance of bases in the surface horizon occurs at the expense of the supply of bases in the

FIGURE 5.15
AN ULTISOL PROFILE, OUACHITA MOUNTAINS OF WESTERN ARKANSAS.

deeper horizons. Once the native plant cover is removed, as is necessary for planting crops, the meager store of nutrients is rapidly lost and crop yields decrease dramatically. Only through conscientious use of fertilizer can permanent agriculture be practiced on such soils.

The ultisols have deep and thoroughly weathered profiles. Yet they are found in a distinctly advantageous location in that they experience a lengthy growing season, with frost-free days ranging from 200 to 260 days in the United States. The growing season is normally accompanied by abundant rainfall.

Many forms of land use are found on these soils. The mild climate is favorable for the cultivation of crops, such as cotton and peanuts, that cannot be grown farther north; and with proper fertilization, these soils also sustain productive yields of corn, oats, tobacco, and forage crops. Mild winters permit the grazing of livestock throughout the year. In addition, the native forest cover represents a huge potential for the pulp and plywood industries, and extensive tracts of land have been set aside for tree plantations.

Oxisols (Oxidized Tropical Soils)

Oxisols are soils of the tropics that contain an *oxic horizon* and/or *plinthite* near the surface. The oxic horizon is a subsurface feature that consists of a mixture of oxides of iron and/or aluminum, quartz sands, and clay. Few solubles remain in the oxic horizon. Plinthite may occur within an oxic horizon or by itself. Plinthite is an iron-rich mixture of clay and quartz and commonly occurs as dark-red mottles that change irreversibly to hardened *ironstone* upon exposure to repeated wetting and drying. Plinthite is one form of the material that has previously been identified as *Laterite*.

These soils exist primarily in ancient landscapes in the humid tropics. Seldom found over broad contiguous areas, they are likely to occur in association with more youthful ultisols, entisols, and vertisols. There are, however, two main regions of oxisol concentration: in South America and in equatorial Africa. Smaller areas are found along the southern fringes of Asia.

The processes leading to the formation of oxic horizons are called *laterization*, in which most of the silica and soluble minerals are removed from the soil by chemical weathering and the action of gravitational water, leaving behind stable clays and oxides of aluminum and iron (Fig. 5.16). The latter tend to impart a reddish hue to the profile.

Oxisols require special management techniques to realize their yield potential. Under a rain forest cover, the fertility level of humid tropical soils is maintained in delicate equilibrium. Organic matter falling from the trees is rapidly decomposed, releasing nutrients and making them available for plant utilization in a continuous cycle that operates with the same small capital of nutrients. Once

FIGURE 5.16
AN OXISOL PROFILE, CENTRAL PUERTO RICO.

the rain forest is cleared, this balance is destroyed and soil fertility declines rapidly as oxidation and leaching are accelerated under high temperatures and rainfall in the absence of a protective mantle of vegetation.

Considering their limited fertility, most tropical soils support a surprising number of economic activities. Land use ranges from shifting cultivation on a subsistence level to permanent plantations, and includes grazing and lumbering operations.

Histosols (Organic Soils)

Unlike the other soil orders, the histosols are considered to be primarily organic, not mineral, in composition. They are commonly called bog, moor, peat, or muck, and contain well over half organic matter by volume. Unless drained, most histosols are saturated or nearly saturated with water most of the year. In certain cases they consist of only an organic mat floating on water. The presence of water is the common denominator in all histosols regardless of location.

These soils can form in virtually any climate, even in arid regions, provided water is available. They occur in the tundra and along the equator, and their vegetative cover consists of a wide variety of water-tolerant plants.

In addition to water, the controlling factors that regulate the accumulations of organic matter include the temperature regime, the character of the organic debris, the degree of microbe activity, and the length of time over which organic accumulation has taken place. The decomposition of organic remains is mostly a process of oxidation, in which oxygen is consumed. If this activity proceeds in a stagnant environment such as swamp water, the oxygen is soon depleted and the process thereby arrested. As a consequence, organic materials continue to accumulate.

Histosols can be utilized very profitably for intensive forms of crop production when they are drained in such a way as to permit the removal of excess water (Fig. 5.17). The wise management of water is critical to their use. In their original state, these soils are too wet either to support the operation of farm equipment or to produce crops. When they are drained the upper layers dry out and may oxidize and subside rapidly, literally vanishing into the air. The rate of oxidation and subsidence is very closely linked with the temperature regime in which a particular histosol occurs. In Florida, for example, subsidence takes place at an estimated rate of 5 to 7 cm (2 to 3 in.) each decade for histosols that have been drained. These rates are significantly reduced further poleward where temperatures are lower.

The ease with which plant roots can penetrate into the soft histosols permits intensive cultivation of a variety of crops. The major agricultural limitation in any given area is climatic. Cabbage, carrots, celery, cranberries, mint, onions, potatoes, and a variety of root crops have thrived on these soils.

Histosols are subject to fire damage when they

FIGURE 5.17
A CELERY CROP ON MICHIGAN HISTOSOLS.

become dry through excessive drainage or during prolonged drought periods. The dried organic material is a virtual tinderbox that can be ignited by careless human acts or by natural phenomena such as lightning. Once a fire is started these soils may smolder for months. The degree of soil destruction from uncontrolled fires can be extensive. One of the current theories regarding the formation of Drummond Lake in the Dismal Swamp of Virginia and North Carolina involves the notion of fire "burning out" a depression in relatively thick deposits of peat. The U.S. Department of Agriculture has recorded numerous accounts of such destruction in swamps and bogs during periods of intense drought.

Other problems relating to the management of histosols include erosion and bearing capacity. The very light character of the partially decomposed organic matter, which makes this soil ideal for plant root development, also contributes to its susceptibility to wind erosion. An intense storm with gusty winds can remove the surface organic horizon. Many farm managers try to counter this problem by maintaining trees as a windbreak on the margins of their fields. The problem of wind erosion increases significantly whenever the soil is left bare—after harvest and prior to spring planting.

Histosols have very low capacities to support weight. Soil subsidence and compaction under weight can cause stress that leads to the cracking and rupturing of roads and buildings. It is frequently necessary, therefore, to drive concrete pilings deep enough to reach the subsurface mineral strata. In the case of roads, the entire organic layer may have to be removed and replaced with a substitute such as sand and gravel. Similarly, the farm manager must exercise care in using equipment on these soils. The excessive weight of heavy machinery breaks down structure, compacts the soils, and reduces the ability of plant roots to penetrate.

Highland Soils

In the chapters on climate and vegetation it was noted that both of these environmental components were extremely variable in highland regions and that it is difficult to compare their regional and altitudinal relationships. Since the characteristics of soil are intimately related to both climate

and vegetation, considerable variation of soil types in mountainous terrain can be expected also. In general, lower montane sites have soils that are typical of their regional surroundings. As altitude increases, however, one must not only consider position and aspect in soil forming processes, but also include the steepness of slope. Steep slopes are not conducive to the development of mature soil profiles; erosion is rapid and soils tend to remain thin and stony. Thus, frequent interspersing of immature profiles occur, with entisols and inceptisols found intermixed with maturely developed soils. This is especially true as ridge lines are approached. Nevertheless, the dominant soil types associated with altitudinal zonation are capable of identification (Fig. 5.18).

ANOTHER PERSPECTIVE ON SOIL, CLIMATE, AND VEGETATION

Throughout this chapter it has frequently been said that soil, climate, and vegetation are interrelated. Atmospheric energy and water foster vegetative growth, and soil provides plants with their nutrients. Soil is a product of both climate and vegetation interacting on geologic materials through time. Climate is determined, in part, by characteristics of the surface cover. Thus: barren surfaces have more extreme temperatures than those mantled with vegetation; plants transpire water to the atmosphere and alter its humidity status; and a forest cover can appreciably reduce surface wind speed. With this knowledge, a novice might conclude that the earth is made up of natural regions, each composed of a distinct plant association, a uniform climate, and containing a single soil order. A comparison of the maps representing these environmental variables, however, reveals that this view is only realistic for portions of the globe.

By using a hypothetical continent an effort is made to show those areas where expected combinations of soil, climate, and vegetation occur (Fig. 5.19). Rather than being broad regions abutting one another, as is the situation when each variable is examined separately, the combinations appear as separated nuclei. Note also that over 50

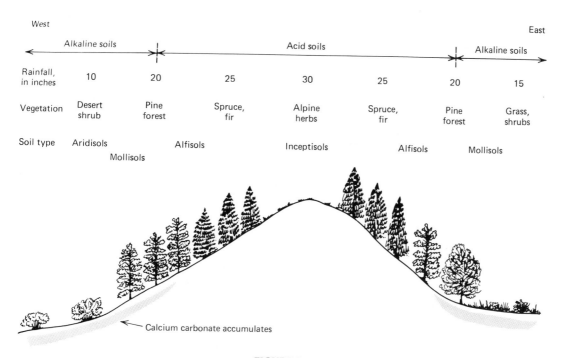

	West						East
	Alkaline soils		*Acid soils*				*Alkaline soils*
Rainfall, in inches	10	20	25	30	25	20	15
Vegetation	Desert shrub	Pine forest	Spruce, fir	Alpine herbs	Spruce, fir	Pine forest	Grass, shrubs
Soil type	Aridisols Mollisols	Alfisols		Inceptisols		Alfisols	Mollisols

← Calcium carbonate accumulates

FIGURE 5.18

GENERALIZED RELATIONSHIP BETWEEN ANNUAL PRECIPITATION, VEGETATION, AND SOILS IN MOUNTAINS OF THE WESTERN UNITED STATES.

percent of the continent is blank. Blank areas are those not perfectly conforming to our scheme of how nature is ordered. This apparent lack of continuity between nuclei may be explained as follows:

1. *Sharp boundaries are inconsistent with nature.* Soil, climate, and vegetation regions tend to gradually merge with those adjacent to them. What appears on the map as boundaries are in reality lines drawn through zones of transition. Each of the three variables also responds differently to environmental inputs. Thus, it is only logical that boundary lines should be irregular and overlap.

2. *Classification Systems have shortcomings.* (a) Although the classification of soils and vegetation is based upon physically observed traits, lack of a reliable data-base in some parts of the globe may lead to the erroneous delimitation of boundaries for each variable. (b) The climatic classification system utilized here is only a mental construct. Perhaps the criteria do not include a sufficient number of variables to differentiate ac-

curately climatic types that relate to the distribution of other environmental factors.

3. *Man disrupts nature.* It is difficult to asses the extent to which man has altered the orig inal vegetation of the earth. Some bounda ries representing the separation of natur plant formations are educated guesses, best.

It has been shown that soil, climate, and vegetation are interrelated. Yet, their expected distributions over the earth do not exactly coincide. Rather, distinct areas appear as nuclei, or cores, of regions within which all three predictable variables are most likely to occur. Between the nuclei, characteristics of one physical region gradually merge with those of another. These areas have only one to two expected variables present. They are, in fact, transition zones—areas in which one region's characteristics gradually merge with those of another. This is where boundaries are established between physical regions and where more complex intermediate regions may be identified.

Our analysis of the earth's physical environ-

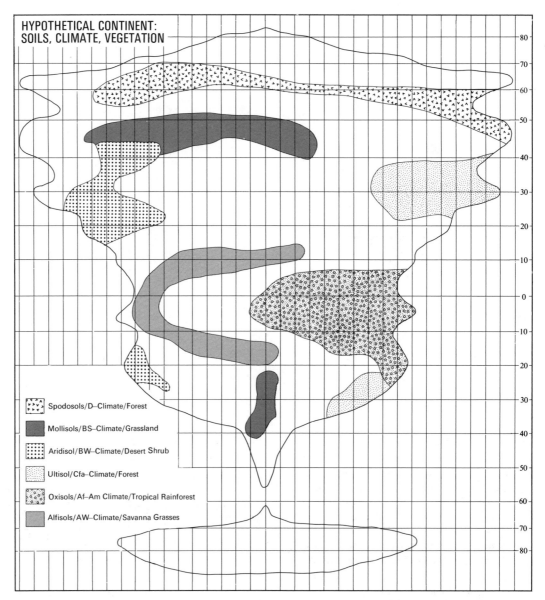

HYPOTHETICAL CONTINENT:
SOILS, CLIMATE, VEGETATION

Spodosols/D–Climate/Forest

Mollisols/BS–Climate/Grassland

Aridisol/BW–Climate/Desert Shrub

Ultisol/Cfa–Climate/Forest

Oxisols/Af–Am Climate/Tropical Rainforest

Alfisols/AW–Climate/Savanna Grasses

FIGURE 5.19
SCHEMATIC CONTINENT WITH APPROXIMATE AREAS IN WHICH EXPECTED OCCURRENCES OF SOIL,
CLIMATE AND VEGETATION ARE FOUND. Because of their complex relationships with other variables, the
Alfisols of the middle latitudes have been excluded.

ments has proceeded through three stages: first, the examination of the atmosphere; second, a study of global vegetation; and third, a view of planetary soils. In the following two chapters attention is focused upon the shape of the earth's crust; that is, on planetary landforms. These four environmental components provide areas with their physical characteristics that constitute the resource base available for human use.

KEY TERMS AND CONCEPTS

soil	soil structure
mineral component	organic component
weathering	humus
soil texture	bases

soil reaction

hygroscopic water

capillary water

wilting point

gravitational (free)
    water

field capacity

plant available water

atmospheric
    component

soil profile

Marbut's *Soil
    Classification System*

the soil taxonomy

entisols

inceptisols

vertisols

aridisols

calcic horizon

calcification

mollisols

soil fertility

soil productivity

spodosols

spodic horizon

podzolization

alfisols

ultisols

oxisols

plinthite

oxic horizon

ironstone

laterization

histosols

highland soils

areal combinations of
    soil, climate, and
    vegetation

11. What is the primary difference between an entisol and an inceptisol?

12. What unique feature(s) distinguishes the vertisol from all other soils?

13. Where are aridisols found on the globe? What unique problems are associated with their use?

14. Why are mollisols referred to as soft soils? Why do they have a high humus content?

15. What is a spodic horizon? Under what environmental conditions is it most likely to form?

16. Describe the similarities and differences between the alfisols and ultisols. How could you explain the fact that the two types may exist side-by-side in a humid subtropical climate?

17. Describe the formation of ironstone. How might ironstone restrict agricultural activity?

18. What makes a histosol different from the other nine soils orders?

19. Devise a general statement regarding the areal agreement that exists between soils, climate, and vegetation.

DISCUSSION QUESTIONS

1. Define soil.

2. Name and describe the four essential soil components.

3. Why is weathering of importance to soil formation?

4. Why is soil texture and structure of significance to soil use?

5. Define: gravitational water, hygrosopic water, capillary water, wilting point, field capacity, and plant available water.

6. Why are bases of importance to farmers?

7. What is a soil profile? Why is an understanding of soil horizons important to agricultural planning? How does one distinguish an immature from a mature profile?

8. What are the five factors that ultimately determine a soil's properties?

9. Why is it necessary to classify soils? How does the soil classification system utilized in this book differentiate soil types?

10. Describe an entisol that might be found in any climatic regime?

REFERENCES FOR FURTHER STUDY

Birkeland, P. W., *Pedology, Weathering,* and *Geomorphological Research,* Oxford University Press, New York, 1974.

Buol, Stanley W., Frances D. Hole, and R. J. McCracken, *Soil Genesis and Soil Classification,* Iowa State University Press, Ames, Iowa, 1973.

Foth, H. D., and L. M. Turk, *Fundamentals of Soil Science,* Fifth Edition, John Wiley and Sons, New York, 1972.

Soil Survey Staff, *Soil Taxonomy: A Basic System of Soil Classification for Making and Interpreting Soil Surveys,* Soil Conservation Service, U.S. Department of Agriculture, Agricultural Handbook No. 436, Government Printing Office, Washington, D.C., 1975.

Steila, Donald, *The Geography of Soils,* Prentice-Hall Inc., Englewood Cliffs, N.J., 1976.

Thompson, Louis M., and Frederick R. Troeh, *Soils, and Soil Fertility,* Third Edition, McGraw-Hill Book Co., New York, 1957.

6
SCULPTURING THE LAND
Universal Forces

A PRINCIPAL COMPONENT OF AN AREA'S "PERSONALITY" is the shape of the land; that is, its array of landforms. Earth's surface shapes are varied; ranging from high and jagged mountain peaks to low and uniform swamplands (Fig. 6.1). On many of these man makes his home, carries on economic activities, and establishes cultural institutions. Man is also an active agent in creating new surface forms. Landscapes have been transformed to permit agricultural production in areas once covered by seas; massive quantities of earth materials have been redistributed in the search for valuable minerals; and mountains and hills have been lowered to suit mankind's needs. The majority of mankind, however, cannot afford exotic and expensive landscape modifications; the choice of enterprises is limited by the very nature of the land surface. Suppose, for example, that a farmer wishes to grow wheat in the stony uplands of the Rocky Mountains where sheep grazing is common. The upland farmer cannot compete economically with the lower production costs of land managers on the Great Plains. He may elect to grow wheat. Yet considering the product's market disadvantages he would be well advised to choose an alternative land-use activity, such as sheep grazing. From this simple illustration, we learn that the economic activities in which men engage, and the consequent patterns of population distributions over the earth, are intimately associated with the nature of the physical environment.

THE BASIS OF LANDFORMS— ROCK

Whether turning over loose soil in the garden or sifting sand grains on the beach, the substances encountered were once components of a solid mass of rock. Although most of the earth's landmasses are covered with fragmental materials, solid rock is never far beneath the surface. Rocks and rock formations constitute the basic structure of most landforms.

Rocks are made up of groups of minerals and differ from one another according to composition and orientation within landscapes. Based upon the manner in which they are formed, rocks can be classified into one of three broad groups: *igneous, sedimentary,* and *metamorphic.*

The ultimate origin of all rock is *magma,* underground reservoirs of molten rock (Fig. 6.2) When magma cools and solidifies *igneous rock* forms by the interlocking of mineral crystals; common types of igneous rock include granite, rhyolite, and basalt. Endowed as they are with metallic minerals such as aluminum, chrominum, copper, gold, nickel, lead, and zinc, igneous rocks are a virtual storehouse of metallic wealth. Nonetheless, the availability and quality of their metallic resources varies considerably from one region to another, and their use is directly dependent upon man's technological development and whether they can be mined economically.

A

FIGURE 6.1

EXAMPLES OF EARTH'S VARIED LANDFORMS. (*A*) Louisiana cypress swamp. (*B*) Sand dunes of Death Valley, California. (*C*) Mountains of Switzerland. (*D*) Mesa and buttes of Monument Valley, Arizona.

B

C

D

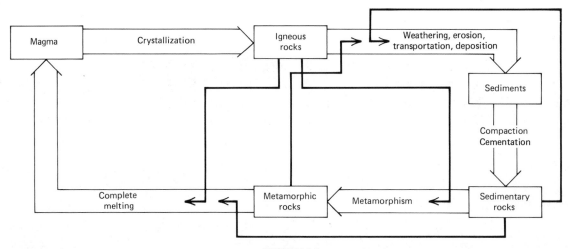

FIGURE 6.2

THE ROCK CYCLE. This diagram illustrates how fluid rock (magma) may become solid and then change in form. The total cycle involves going from magma to igneous rock to sediment, to sedimentary rock, to metamorphic rock, and then back to magma. Shorter cycles can also be initiated if a change in the processes affecting the rock material occurs.

When igneous rocks are exposed at the earth's surface forces of nature begin decomposing and disintegrating them. As they break up, rock fragments are swept away from their native sites by moving water, wind, ice, or gravity, and are deposited in river valleys, lakes, seas, and basins as *sediment*. Layers of sediment up to 20,000 meters (65,000 feet) thick are known to have accumulated. The weight of sediment accumulation causes compaction of rock particles. This, coupled with the cementation of rock particles by mineral substances trapped in spaces between them, transforms the sediments into *sedimentary rock*. Sedimentary rocks are the most ubiquitous rock type on the continental landmasses, forming about 75

percent of the continental surface. But they only form a thin veneer over the igneous rock that lies below. Examples of sedimentary rock are: sandstone, limestone, and shale. Like igneous rock, the resource potential of sedimentary rock is extensive. Sandstone is used for building material. Limestone is applied to farmer's fields to reduce soil acidity, is used to manufacture cement, and serves as a flux in the smelting of iron ore. Most fossil fuels, those fuels related to past biologic activity, are found within sedimentary rocks. Certain sandstones and shales contain petroleum which, with natural gas, is often found associated with oil reserves. These fuel resources have provided the major part of the energy required for

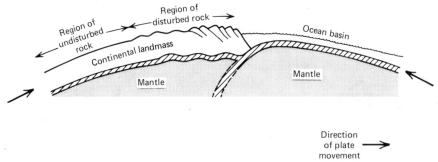

FIGURE 6.3

MOUNTAIN FORMING PROCESSES. Two lithospheric plates are being compressed because of convection within the earth's mantle. Because of convergence the surface is buckled, fractured, and uplifted. (For a review of the concepts of convection and convergence refer to Chapter 1.)

SCULPTURING THE LAND

heating homes and operating industries and vehicles. Also, many sedimentary rocks are very porous and contain huge reserves of groundwater that have collected through centuries. Today, many of these water bearing rocks are tapped to provide for domestic, agricultural, and industrial water needs.

The third group of rocks is the *metamorphics.* These form when igneous, sedimentary, or other metamorphic rocks experience high temperatures and high pressures. Under conditions where heat is intense, yet which is below the melting point of rock, new minerals begin to grow and replace some of the original rock's minerals, while some old minerals are realigned, thus changing the character of the original rock. Typical metamorphic activity changes limestone to marble, shale to slate, and granite to gneiss. These rock types are used as building construction materials, for statuaries, and for sidewalks.

LAND FORMING PROCESSES

Rock and rock fragments are shaped into landforms by processes that operate both within the earth and at its surface. Major landform features are formed by *internal processes* that provide the driving mechanism to move continents and bend, warp, fold, and uplift land surfaces (Fig. 6.3). These initial landforms are subsequently altered by *external processes,* both atmospheric and gravitational.

Internal Processes

The surface of the earth is wearing away. Minute sediments carried by each trickle of rainwater and dust particles that settle upon window sills are reminders that nature constantly redistributes earth materials. The combined effect of atmospheric and gravitational forces has been to wear down uplands and to fill in depressions; that is, to smooth out the globe's surface. These forces, operative for about 5 billion years, should by this time have eliminated all mountains, hills, and plateaus. Earth should have the appearance of a billiard ball were it not for internal processes that bring new rock to the surface.

The origin of all landforms produced by internal activities can be traced to one of two groups of processes: *vulcanism* and *diastrophism.* Vulcanism involves the redistribution of fluid rock, whereas diastrophism refers to the reorganization of solid rock materials. Although each of these two sets of processes create distinctly different geologic forms, they are frequently found operating together to create complex landform types.

Vulcanism. Vulcanism is the process by which magma moves from regions of high confining pressure and temperature toward areas of lower pressure and (usually) lower temperature. Movement may range from dramatically explosive events that create mountain-sized volcanoes to subterranean flows that produce new geologic forms beneath the surface (Fig. 6.4). Whether *extrusive* (reaching the surface) or *intrusive* (taking place beneath the surface), volcanic activity originates in a *magma chamber* that acts as a storage tank for molten rock. Although the processes involved are not well understood, it is likely that heat, pressure, and the chemical composition of the earth's interior are all important to the formation of magma chambers.

FIGURE 6.4
PARÍCUTIN VOLCANO, MEXICO.

Extrusive vulcanism occurs when a magma chamber has built-up sufficient pressure to force molten material to the surface through overlying rock. When pressure is released at a specific point, the end product will be a *volcanic cone*. If pressure is released along an elongated fracture or fault zone, magma may ooze out along a series of cracks or fissures hundreds of kilometers in length to create a *lava-flow plateau*.

Volcanoes are conical-shaped landforms consisting of *lava*—molten rock material that has reached the surface—and *volcanic* ash, a finely divided igneous rock blown under gas pressure from a volcano. Various types and sizes of volcanos occur, depending upon the viscosity of the lava and the degree of explosiveness associated with an eruption. The most common types of volcanos are illustrated in Figure 6.5 and are briefly described below:

Shield volcanos consist of highly fluid lava flows that have been extruded in a relatively nonviolent fashion. These flow from the magma chamber outlet in relatively thin sheets that produce a broad, gently sloping cone usually 4 to 10° from the horizontal. Although they are the most gently sloping of all volcanos, they are impressive in size. Mauna Loa, for example, is the largest of five shield volcanos that together comprise the island of Hawaii. Sitting upon the ocean floor 5000 meters (16,000 feet) below sea level, its summit rises to over 4000 meters (13,000 feet) above the water, totalling more than 9000 meters (29,000 feet) from base to top. In comparison to the height of Mount Everest (8850 meters; 29,040 feet), the earth's highest mountain, Mauna Loa certainly ranks as one of the tallest landform features on the globe.

Cinder cones are the smallest of volcanos, rarely attaining more than a few hundred meters in height. Often occurring in swarms that form volcano fields, they are the product of explosive eruptions of lava that are highly viscous. A large portion of the ejected lava solidifies while in the air and reaches the ground as solid fragments.

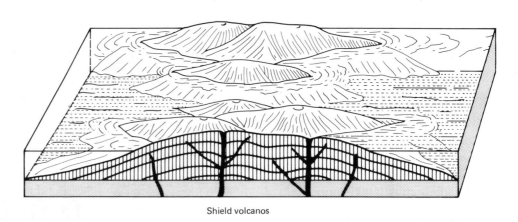

Shield volcanos

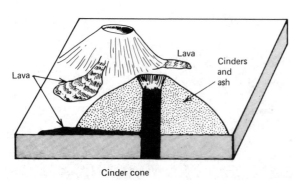

Cinder cone

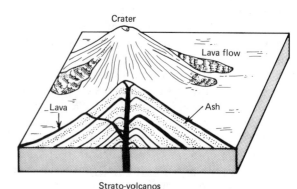

Strato-volcanos

FIGURE 6.5
BASIC VOLCANO FORMS.

SCULPTURING THE LAND

FIGURE 6.6
MT. FUJI, JAPAN.

These ejecta collect around the central vent in steep-sided conical piles, averaging about 34° from the horizontal. Parícutin is an excellent example of a cinder cone (Fig. 6.4).

Strato-volcanos are beautiful in size and shape. They may be thousands of meters high and consist of lava flows irregularly interspersed with layers of fragmented ejecta. Because the lava of their magma chamber seems to periodically or gradually vary in composition, eruptions likewise fluctuate. Some are quiet, while others are explosive. Thus, the volcano's form is intermediate between that of the shield volcano and cinder cone. The most famous strato-volcano is Mt. Fuji in Japan, with Mt. Vesuvius in Italy running a close second. Their splendor and symmetry has captured the attention of artists and photographers throughout the world (Fig. 6.6).

Lava flow plateaus are volcanic tablelands that form where fluid magma reaches the surface along extensive fractures in the crust. The extrusions are relatively nonviolent and intervals between flows of lava are difficult to predict. Individual flows range in thickness from about a meter (3 feet) to over 100 meters (300 feet), with numerous flows being common for each site. The fluid magma covers the surface in a succession of nearly horizontal sheets of lava that may accumulate to a thousand or more meters in thickness. In the Columbia Plateau of North America, for example, flows total more than 700 meters (2300 feet) thick over a wide area (Fig. 6.7).

FIGURE 6.7
AN EXAMPLE OF A LAVA FLOW PLATEAU: COLUMBIA PLATEAU, NORTHWESTERN UNITED STATES. The areas of gray represent only the remnants of the original flows, now reduced considerably in size by erosion.

Extrusive vulcanism is always spectacular, often violent, and sometimes catastrophic. Erupting volcanos are especially impressive sights. Each year, thousands of tourists visit volcanic parks throughout the world to view liquid rock emerging from the earth. Moreover, these windows to the earth's interior provide scientists with tangible evidence of what rocks and volcanic processes are like at great depths below the surface. Study-

ing a volcano at a safe distance, however, is much different than experiencing its terrifying impact at close range. Throughout history, devastating volcanic eruptions have occurred. In 79 A.D. *Mount Vesuvius* violently erupted, burying the city of Pompeii along with its inhabitants in white-hot ash. In 1902, Mount Pelée on the Island of Martinique came to life, sending dense swirling clouds of incandescent ash and intensely heated gases avalaching down the mountain's sides. The Peléan clouds raced across the city of St. Pierre, destroying the city and killing more than 30,000 persons. There is also evidence that the ancient Minoan civilization of Crete disappeared as a result of a catastrophic volcanic eruption. These examples of devastating magma eruptions are but a sample of the negative impacts that extrusive vulcanism has had on mankind's life. In a positive sense, volcanoes can be breathtakingly beautiful, and often serve as a recreational resource. Although their surfaces may remain barren and sterile long after their formation, most volcanic rocks eventually form fertile soils. In some countries of the world, such as Indonesia, these soils are intensely cultivated and support very dense populations.

It is estimated there are 450 active volcanos on

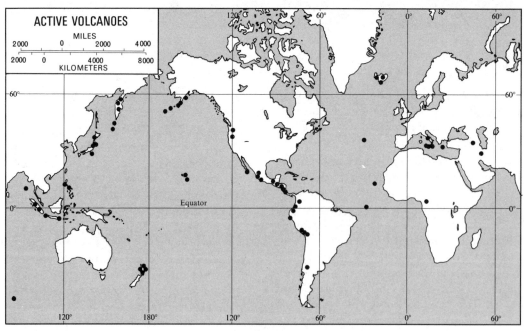

FIGURE 6.8
LOCATIONS OF SOME OF THE EARTH'S ACTIVE VOLCANOES.

the earth. These, as well as extinct volcanos, are found in active geologic zones where mountain building is taking place. Volcanos encircle the Pacific; they are found in a belt that passes through Indonesia, reappears in the Himalayas, and then extends through the Mediterranean region; and they are prominent along oceanic submarine ridges, where they appear as islands (Fig. 6.8). These are the areas where most earthquakes originate, where diastrophism is pronounced, and where intrusive vulcanism takes place.

Intrusive vulcanism occurs when magma is forced into rock formations that lie beneath the surface of the earth. As the magma cools and solidifies, various characteristic structures are formed (Fig. 6.9). The largest instrusive structure is a *batholith*, a massive body (perhaps part of a solidified magma chamber) that can extend over hundreds of square kilometers. Others include; *laccoliths,* lens-shaped features that lie in the same plane as the rock they intrude; *sills,* tabular bodies that lie in the same plane as the surrounding rock; and *dikes,* tabular bodies that cut across the strata they have intruded. Because these structures are covered by an overburden of rock during formation, they are not immediately established as landforms. Only after the overlying strata are stripped away by erosion do they appear as surface features.

Both extrusive and intrusive volcanic regions are areas where intense heat is generated to form "hot spots" (magma chambers) beneath the surface. In addition to producing igneous landform features these sources of heat energy may also manifest their presence at the surface in the form of *geysers,* fountains of groundwater and steam. Water, heated through contact with hot rock, may be forced through rock fissures and sprayed tens of meters into the air. The most notable areas in which geysers occur are: The United States, New Zealand, and Iceland. Old Faithful, in Yellowstone National Park, is North America's most famous geyser. Since its discovery in 1871, it has erupted about every 66 minutes and has a fountain well over 30 meters (90 feet) high.

In several countries engineers have found innovative ways to harness at least a portion of the heat energy released by magma chambers. In Iceland, warmed groundwater is used to heat homes and office buildings. Several nations have constructed geothermal electric generating stations that are powered by steam from the ground.

Diastrophism. Although the earth's crust may appear to be rigid and static, landscape evidence reveals it as everchanging. Repeatedly throughout earth history the crust has been raised, lowered, broken, bent, and wrinkled. This crustal defor-

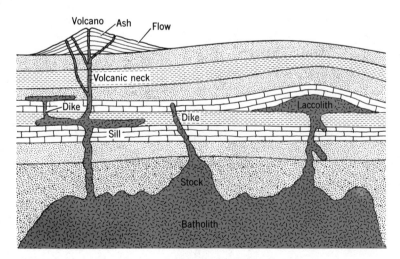

FIGURE 6.9
BLOCK DIAGRAM OF INTRUSIVE AND EXTRUSIVE IGNEOUS STRUCTURES. Although extrusive and intrusive igneous vulcanism have been discussed as separate igneous activities, they actually occur simultaneously. A variety of surface and subsurface igneous forms are shown in the diagram above.

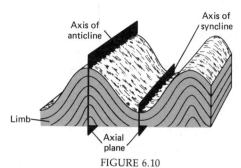

FIGURE 6.10

FOLDED TOPOGRAPHY. When compressional forces are applied to rock the result may be folding, especially if plastic deformation prevails. Consequently, there are formed a series of *synclines* (a troughlike fold in rock strata with the limbs of the fold oriented upward) and *anticlines* (uparched folds in the rock strata with downward dipping limbs).

mation is produced by internal processes collectively known as *diastrophism*.

Crustal deformation occurs when stresses applied to rock are greater than the rock's resistive strength. These forces stem from either compression (squeezing), tension (pulling apart), or torsion (twisting action). Depending upon the degree and duration of the applied stress, rocks behave in a predictable manner. Up to a certain point they are elastic; that is, they will deform under stress but regain their original shape and volume when the stress is alleviated. Under intense stress, however, they may become permanently deformed by either *folding* or *faulting* (fracturing).

Folding, or bending, occurs when rock is subjected to forces strong enough to induce plastic deformation. Just like putty, rock can be altered in form without internal cracking. If, for example, a sheet of paper lying upon a flat surface were pushed inward from opposite sides, it would adjust to compressional force by flexing upward in the middle. Rock strata will behave in a similar fashion when compressional forces are applied to them (Fig. 6.10).

The type of rock folding involved in landform development depends upon the degree, duration, and direction of compressional and/or vertical forces. Some folds may be symmetrical when opposing forces are relatively equal; others may be assymetrical when opposing forces are unequal; and still others recumbent or overturned when both direction and duration of forces are unequal, as illustrated in Figure 6.11. Folds vary not only in shape, but also in size. Some measure only a few *centimeters* (about an inch) in each direction, while others measure hundreds of kilometers (miles) in length and may involve vertical movement exceeding 6000 meters (20,000 feet). The Ridge and Valley Province of the Appalachian Mountains offers an excellent example of rock

Name	Description	
Symmetrical	Both limbs dip equally away from the axial plane.	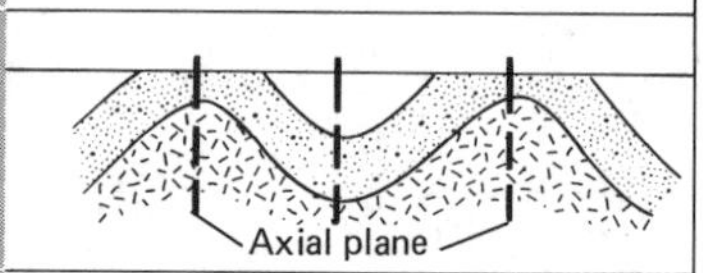
Asymmetrical	One limb of the fold dips more steeply than the other.	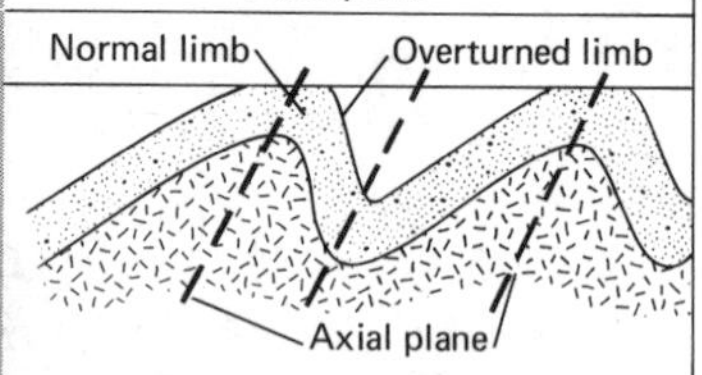
Overturned	Strata in one limb have been tilted beyond the vertical. Both limbs dip in the same direction, though not necessarily at the same angle.	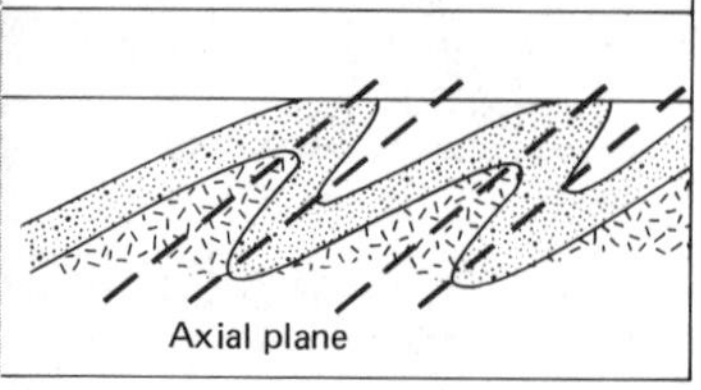

FIGURE 6.11
PRINCIPAL TYPES OF FOLDED STRATA.

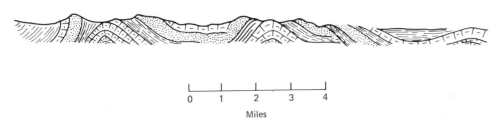

folding that resulted in mountainous terrain (Fig. 6.12). Linear folds as wide as 320 to 480 km (200 to 250 mi) from crest-to-crest have formed in the region.

Faulting takes place when brittle rocks, reacting to stress, exceed their elastic limits and rupture. If the resultant fracture is not accompanied by rock displacement, it is called a *joint*. If movement takes place along the zone of weakness to offset original rock units, the rupture is called a *fault* (Fig. 6.13). Movement along a fault is frequently abrupt, with large masses of earth material readjusting to stresses within the crust.

Faulting occurs primarily along inclined fractures. It is along these fractures that many veins

FIGURE 6.13
DAMAGE CAUSED BY THE EARTHQUAKE OF MARCH 27, 1964, IN ANCHORAGE, ALASKA.

of metallic ores have been deposited. Faults have been favored locations for miners to congregate and terminology regarding faults is influenced by the miners' pragmatic assessment of the fractures' characteristics. From the miner's perspective one wall of an inclined vein in a fault is beneath his feet, while the other is overhead (or overhangs him). Thus, in his terms, the *footwall* is the surface of the block of rock below an inclined fault; the *hanging wall* is the surface of the block of rock above an inclined fault. While these two terms do not apply to vertical faults, they do provide a means for describing the relative movement of most blocks of rock along faults, and therefore are useful in classifying fault-types. Common fault types are illustrated in Figure 6.14. A *normal fault*, generally steeply inclined, occurs where the hanging-wall block has moved relatively downward (Fig. 6.14*A*). Produced by the earth's outer crust being extended, or pulled apart, the motion along the fault is predominantly vertical and results in a *fault scarp* with a cliff-like surface, the height of which provides a measure of relative displacement of the affected rock masses. Along many faults, vertical displacements measure thousands of meters, and their lengths may run to hundreds of kilometers. Such colossal slippages are not the result of single events. Rather, they represent the cumulative effect of many small slippages over a great length of time. At most, individual movements along a fault are about 10 to 11 meters (35 feet). Usually they are much less, with about 6 meters (20 feet) fairly common. A *reverse fault* results from crustal shortening, or compression, wherein the hanging-wall moves upward (Fig. 6.14*B*). The motion is still predominantly vertical. Because an unstable overhanging scarp forms with reverse faults there is a constant danger of landslides accompanying fault movement. *Strike-slip faults* move in a horizontal direction, causing

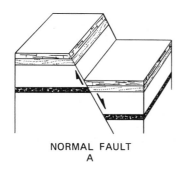

NORMAL FAULT
A

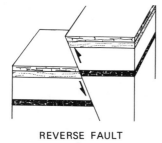

REVERSE FAULT
B

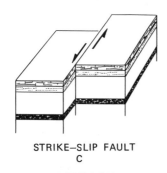

STRIKE—SLIP FAULT
C

FIGURE 6.14
PRINCIPAL TYPES OF FAULTS AND THE DIRECTION OF FORCES THAT CAUSE THEM.

nearby roads and streams to be offset by several meters (Fig. 6.14C).

Combinations of the common types of faults may occur; that is, mixes of horizontal and vertical movements. And, where one fault occurs there is bound to be more in the same region. A series of parallel faults produce a saw-tooth appearing terrain, and folded areas frequently are made up of segmented blocks that have been faulted.

The jarring motion of rock material moving along a fault normally generates a group of shock waves, an *earthquake*. Radiating outward from the disturbed zone, these shock waves cause the earth to shake and vibrate. Some waves pass through the interior of the earth. Others are restricted to the crust and travel around the earth. By studying

shockwave behavior on sensitive *seismographs*, considerable information has been learned about earthquake behavior and the interior of the earth. One of the most impressive facts to emerge from seismographic research is the sheer number of earthquakes that occur. Thousands occur every year, hundreds of which are strong enough to damage buildings. One or two a year may be potentially devastating.

An earthquake can be a traumatic experience for an observer. About 700 people were killed during the San Francisco earthquake of 1906, one of North America's worst natural catastrophes. Loss of property from the fires resulting from disrupted gas and water lines was estimated at $400,000,000. In 1923, a devastating earthquake in Japan left more than 140,000 dead or missing, 104,000 injured, and more than one-half million buildings destroyed. A vivid description of the aftermath of the Japanese earthquake is provided by Ordway:

Fires in . . . cities and towns quickly spread out of control because the city water systems had been disrupted by the earthquake. The intensely heated air over Tokyo rose in a violent updraft, formed clouds, and spawned powerful whirlwinds. Thus wind directions and velocities changed frequently, and the rapidly advancing flames encircled thousands of people, who were burned to death as they huddled on bridges, in parks, and in other open areas. Destruction by fire far exceeded that caused directly by the earthquake. Seismic sea waves (tsunamis) and landslides increased the magnitude of the catastrophe.[*]

Ranging from a few seconds to about a minute, earthquake shock waves are of surprisingly short duration compared to the damage they inflict. In the Japanese disaster, the major shock lasted a minute and a half. Major shocks are followed by a series of aftershocks of lower intensity and frequency that may continue for months.

Most modern buildings, if well built, can withstand relatively strong earthquakes. The impact of shock waves on structures, however, is directly related to the types of earth materials upon which they are built. Solid bedrock transmits shock waves and experiences less shaking than does soft, swampy land that quivers like a bowl of shaken jello (Fig. 6.15). Techniques for surviving an earthquake vary. In the country, the safest

[*] Richard J. Ordway. *Earth Science*, 2nd ed. New York: Van Nostrand Co., 1972. pp. 155–157.

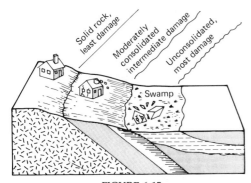

FIGURE 6.15
EARTHQUAKE DAMAGE SHOWN IN RELATION TO FOUNDATION MATERIAL.

place is open space away from buildings and trees. In the city, it's better to stay indoors, under doorways or stairs. Death rates increase when a frantic populace rushes into the streets to be crushed by the falling masonry of old, poorly built structures.

Although no region is immune, earthquakes tend to be concentrated along lines of crustal weakness. These zones correspond rather closely with areas of active vulcanism (Figs. 6.8 and 6.16). Faulting, folding, and vulcanism are all instru-

mental in developing most of the earth's mountains. These mountain building processes normally take place over hundreds of thousands of years. A vertical displacement of only one meter per hundred years would result in a mountain 1000 meters (over 3000 feet) high in only 100,000 years, a short time span in geologic terms.

The Continents

The surfaces of the continents are made up of vast interiors of gently rolling or flat land and higher mountainous and plateau regions, each associated with differing geologic provinces (Figs. 6.17 and 6.18).

Within each continent are *shield* regions of igneous and metamorphic rocks. These are broad rolling lowlands of the geologically oldest rocks, probably representing the remains of eroded roots of ancient mountain ranges. Adjacent to the shields are the *interior plains*, where a relatively thin cover of nearly flat sedimentary rock buries the downward dipping extensions of the shield. Together, the shield areas and the interior plains occupy the most stable part of a continent; that is, areas that

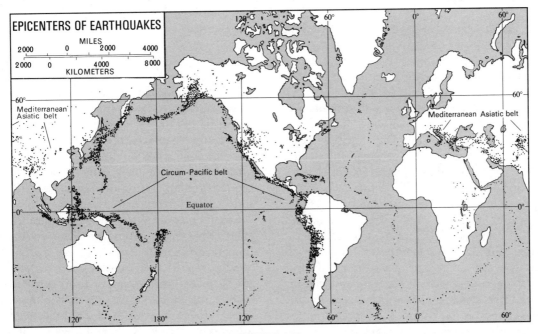

FIGURE 6.16
EARTHQUAKE ZONES AND LOCATIONS OF EARTHQUAKES RECORDED BETWEEN 1961 AND 1967. Each dot represents a single earthquake. The centers of the earthquakes fall into well defined belts that coincide with the margins of lithospheric plates.

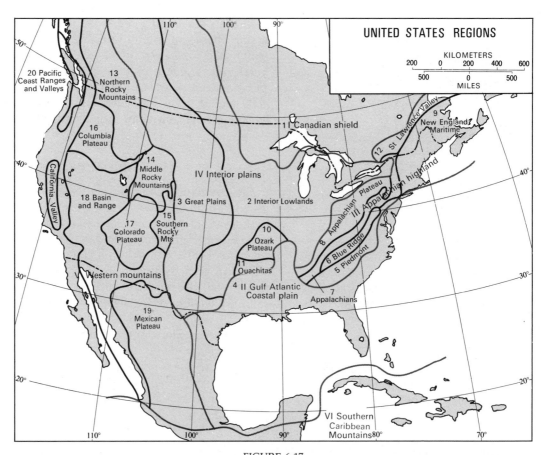

FIGURE 6.17

CONTINENTAL STRUCTURE OF THE UNITED STATES. The United States is part of a rather symmetrical continent, having a stable interior flanked by highlands.

have experienced little or no recent orogenic activity.

Continental uplands are formed of *plateaus, block mountains,* and *folded mountains. Plateaus* are highlands resulting from uniform uplift without folding or distortion of the original rock (Fig. 6.19). Originally, plateaus appear as tablelands; deep erosion, however, can dissect them to make their appearance distinctly mountainous even if their structure remains essentially horizontal. *Block mountains* are elevated rock masses produced by high-angle faulting. They may be found as single upthrust masses or in groups, the latter exhibiting a topography with a saw-tooth appearance (Fig. 6.20). Continental margins where vulcanism and earthquakes tend to be common are preferred locations for *folded mountains* to develop. Thus, we might expect folded mountains to be found in the circum-Pacific and Alpine-Himalayan belts of active volcanos and earthquake activity. This is exactly where the modern ranges occur (Fig. 6.18).

Older folded ranges, such as the Appalachians, or Europe's Caledonian and Hercynians, do not always conform to this pattern. The explanation for this apparent anomaly rests with geologic history; it is probable that when these worn ranges first formed they were on the margins of the stable interiors, but lost their exterior position by subsequent redistribution of crustal materials and by continental movements.

Coastal plains are common structural features of most continents. Extending from southern New England to Florida and along the Gulf Coast into Mexico, the Atlantic coastal plain of North America is an emerging part of the continental shelf. Unlike the stable interior plains, however, this region does not consist of a thin veneer of sedimentary materials. Rather, thousands of meters of sedimentary rock have accumulated in an apparent gigantic, elongated, crustal downwarp in the continental platform. Some earth scientists believe these accumulations represent the potential ma-

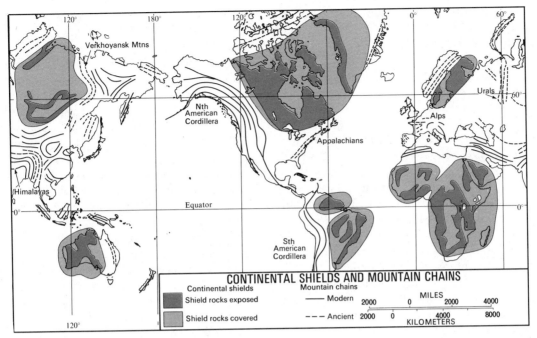

FIGURE 6.18
CONTINENTAL SHIELDS AND MOUNTAIN CHAINS OF THE EARTH.

terials for future mountain producing episodes.

Continents exhibit variable elevation from place to place. At one time it was thought that crustal rock was so strong and rigid that it could support all landforms, regardless of their height. Today's scientists believe that continents, and mountains contained within them are floating on the denser, more plastic mantle, like an iceberg floating in water. A floating iceberg projects much deeper below the water surface than it does above it. If the iceberg is very tall it also extends even deeper below the water surface. Similarly, continents are believed to have an irregular base, deepest under mountainous regions and shallowest where surface forms are low in elevation (Fig. 6.21). The concept of floating landmasses can help explain

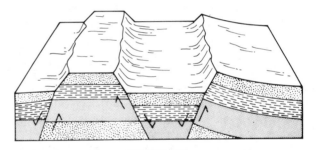

FIGURE 6.20
BLOCK MOUNTAINS PRODUCED BY HIGH-ANGLE FAULTING.

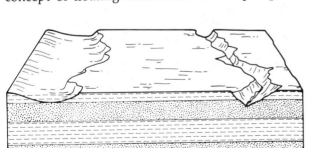

FIGURE 6.19
A BLOCK DIAGRAM OF A PLATEAU'S STRUCTURE.

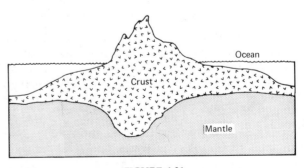

FIGURE 6.21
VERTICAL STRUCTURE OF THE CONTINENTS. The continental crust is deepest where it is also the highest.

LANDFORMING PROCESSES

certain vertical motions of the continents. Somewhat like a ship that settles low in the water when heavily laden and rides high when empty, continents (or parts thereof) can become overloaded and underloaded. During the Pleistocene Ice Age, for example, when large masses of ice accumulated on the continents, many areas were depressed, or sagged, from the weight of the ice. Since the ice has melted these same regions are rising. Similarly, where erosion has removed mountain tops, the lightened masses rise. Settling takes place where their sediments accumulate. Movements of this nature are not instrumental in mountain-building processes, but are a response of the crust toward achieving equilibrium, or balance. Movements that cause structural deformation of the continents are different and can be classified into two types. *Orogeny*, meaning mountain-making, involves the formation of faulted and folded mountains in relatively narrow, linear belts. *Epeirogeny*, meaning continent-making, is primarily vertical crustal movement that affects large portions of a continent. Epeirogeny alters the massive parts of the continents, warping shields and interior plains, forming block fault mountains, and lifting plateaus and coastal plains.

External Processes

Internal processes provide for the grand configuration of continents. They produce mountain ranges, elevate plateaus, and create subsurface forms that may ultimately reach the surface. Smaller-scale landforms, the infinite variety of individual hills and valleys, are formed by *external processes*.

The primary energy for external processes is supplied by the sun, because solar radiation is the driving mechanism for atmospheric motion. When converted into heat at the earth's surface, solar energy, in combination with air, moisture, and organic matter, unceasingly attacks the chemical and physical structure of the earth's crust, causing it to crumble and/or dissolve at varying rates in different locations. This decomposed earth is then directed by moving water (liquid or solid) and wind, under the influence of gravity, to lower elevations. The displaced debris ultimately accumulates in basins or is washed to sea by rivers.

Weathering and Mass Wasting. The earth crust is continually disintegrating and decomposing under the action of a set of processes known as *weathering*. *Physical weathering* causes the rock to crumble into smaller and smaller fragments, until only individual mineral crystals of the original rock remain. The most common agent of physical weathering is frost. Water that has penetrated cracks and crevices of rock expands 9 percent upon freezing and exerts considerable lateral pressure. This acts as a wedge to pry apart fragments from the original rock in cycles that may be daily or seasonal. In desert regions, salt crystals may form from solution. As water near the crust's surface evaporates, dissolved salts remain behind and crystallize. Continued evaporation causes the crystals to grow and to exert sufficient pressure to pry off particles from surrounding rock. Other forms of mechanical weathering include abrasion of rock by wind-borne sand; rupturing of rock by non-uniform heating, such as occurs during forest and brush fires; and weathering by plant roots that wedge their way into rock cracks and crevices. *Chemical weathering* occurs when atmospheric gases and water work together to alter rock minerals. While physical weathering reduces the size of the parent rock mass, chemical weathering changes its composition. Atmospheric water is normally slightly acid and is capable of converting many rock minerals to soluble or weakened forms. Biota can also contribute to physical and chemical weathering processes. Animals mix and break down surface debris through burrowing and pawing, and vegetative debris forms organic acids that can speedup the chemical decomposition of rock.

Physical and chemical weathering operate together to break down the exposed crust. Mechanical weathering tends to dominate in high altitudes and elevations where temperatures drop below freezing; chemical weathering is most active in warm, moist climates such as the humid tropics. Weathering has important implications for man, especially in the formation of soil upon which he depends for food and forest products. In addition, many valuable mineral deposits have been produced by weathering processes. These deposits may be concentrated in place by removal of waste products or by the accumulation of minerals released through weathering. Ore deposits of aluminum, copper, and iron have formed in this fashion. How rocks behave when weathered is also a critical factor in deciding which rock types should be used in the construction of buildings. Sandstone cemented by calcium carbonate or iron are unsatisfactory for use in humid climates, because the cementing agent will either dissolve or stain the rock. Badly fractured sandstone would not be suitable in cool climates, since water entering cracks would expand upon freezing and disintegrate the rock. Soluble limestone is also unfit for use in humid climates, but may be used in arid regions with success.

As crustal material weathers, gravity tends to move it downslope. The movement of rock debris toward lower elevations is called *mass wasting*. Mass wasting takes many forms. It may involve a single soil particle or hundreds of tons of earth material. The rate of movement may be imperceptively slow or amount to thousands of feet in a matter of seconds. Of the many processes that constitute mass wasting, some of the most important are *creep*, *flows*, and *slides*.

Creep is the very slow downhill movement of soil and rock material. It is a form of mass wasting too slow to be directly observed. To determine its impact upon the landscape indirect evidence must be employed, such as observations of the warping of utility lines and fences, or the tilting of telephone posts and tombstones (Fig. 6.22). Creep results from the force of gravity acting upon surface materials that have been dislocated. The disturbance may be caused by freezing of soil water, by the burrowing and trampling of animals, by the prying action of swaying trees, or by man's activities. Even though individual movements may involve only single grains moving a fraction of an inch, when the process is repeated over and over, huge volumes of material are capable of being redistributed.

FIGURE 6.22
COMMON EVIDENCES OF CREEP.

There are two basic types of flows, *earthflows* and *mudflows*. Earthflows occur most frequently in humid climates on relatively steep slopes. Involved is the movement over bedrock of unconsolidated rock debris and soil material, a process usually helped along when surface materials are saturated by heavy rains. Earthflows move slowly, yet perceptibly, and may transport from a few to several million cubic meters of earth material. Mudflows are more fluid than earthflows and are most common in arid and semiarid climates. The typical mudflow originates when a cloudburst sweeps unconsolidated or unstable rock materials into stream channels. The channels normally become blocked by the initial surge of this debris, until the growing pressure of water backing up in the channel is sufficient to break through. Then the mix of water and debris (containing as much as 90 percent solid matter) moves down valley, carrying along or pushing aside obstacles in its path. Many unwary campers have lost their lives in the southwestern United States by taking refuge in canyons prior to the onset of storms that have triggered mudflows.

Slides are the most rapid and destructive movements of rock and soil. Depending upon the amount of material displaced, the slide may be called a *debris slide* or a *rock slide (landslide)*. Debris slides are small areas of fast moving unconsolidated earth. They are normally found along steep or unstable slopes of stream valleys and shorelines. Minor disturbances or excessive moisture can trigger rapid movements that result in a surface of low hillocks and intervening depressions backed by a scarp face. Rock slides are larger in scale and the most catastropic of all mass movements. They represent the sudden, rapid slippage of varying amounts of bedrock and unconsolidated materials along planes of weakness within underlying rock.

The causes of landslides are numerous. Many are set in motion by earthquake activity, a classic example of which occurred in the canyon of the Madison River west of Yellowstone Park in August, 1959. Jarred loose by an earthquake, 30 million cubic meters (35 million cubic yards) of broken rock shot down the valley wall, covered a 1½ kilometer (1 mile) span of river and highway, and climbed 120 meters (400 feet) up the steep opposite valley wall (Fig. 6.23). A number of people lost their lives. The debris dumped by the slide formed a natural dam across the river, behind

which a lake formed that was several hundred
meters deep and several kilometers long. In this
example, the earth material that slid was lying
along fractures that slanted downward toward the
river and that were nearly parallel to one of the
canyon's steep walls. When the shaking accompanying the earthquake occurred, the mountain
face literally slid off.

Because gravity is pervasive over the globe, the
most universal of all landscape sculpturing agents
is the mass wasting of earth materials. But climate
and man aid the process by increasing or restrict-
ing the susceptibility of matter to movement.
Water is crucial to the amount of movement that
takes place. Since gravity can only move matter
when it overcomes the mass's resistence to motion, the presence of water as both a solvent and
a lubricant makes wetted earth materials increasingly prone to gravitational sliding. After heavy
and/or persistent rains the incidence of earthflows, mudflows, and slides increases dramatically. Human activities can also trigger mass
movements. Removing plant cover through defo-

FIGURE 6.24
HOUSE DAMAGE AS A RESULT OF A LANDSLIDE IN LA-
GUNA BEACH, CALIFORNIA.

restation or overgrazing makes surface materials
more prone to displacement. Large quantities of
earth materials removed from beneath the ground
in mining operations have led to the collapse of
tunnels and to subsequent surface subsidence.
Massive extractions of groundwater have also led
to compaction and subsidence of underground
materials. This form of settling and fracturing of
the ground can destroy homes and streets.

Mass movements of the land are not only dangerous; they are also costly. Thousands of people
have been buried in fast-moving debris. According to the California Division of Mines and Geology, as much as $10 billion in property damage
will occur throughout California between 1970
and 2000, due to landslides (Fig. 6.24). It is estimated that 90 percent of this projected loss could
be avoided by removing or stabilizing landslide
masses, and by regulating the kind and density
of development in landslide prone areas. Obviously, identification of areas prone to massive
earth movements and the wise planning of such
areas is a prerequisite to saving lives and preventing property damage.

FIGURE 6.23
THE MADISON RIVER LANDSLIDE, MADISON COUNTY,
MONTANA.

Streams and Stream Erosion. Precipitation falling on land surfaces is directed along several paths (Fig. 2.29). A large portion is returned to the atmosphere by evaporation or through the transpiration of plants. Another part immediately runs off the surface into streams and lakes; and a third sinks into openings in sediments and rocks to become groundwater, much of which later emerges at the surface at lower altitudes.

Runoff is a term encompassing the physical movement of surface waters, under the influence of gravity, toward lower elevations, and ultimately the sea.* Runoff, originating as precipitation, is initiated as broad sheets of water moving toward depressions and small rills. Called *overland flow*, these broad sheets of water are eventually carried to a stream. *Stream* denotes a channelized flow of any size, and includes small brooks and large rivers. With the exception of mass wasting, water running over the surface has had the most far-reaching effect upon earth landscapes. Running water also has the greatest impact upon people. It is the most important long-distance transporter of earth material and has carved its signature of erosion and deposition on surfaces in climatic realms ranging from arid to humid and tropical to polar. Ancient civilizations were founded along the fertile floodplains of rivers, and today rivers provide energy, irrigation water, recreational resources, and domestic water supplies, and also serve as transportation arteries.

Precipitation that falls on land surfaces is either absorbed by earth materials, involved in transpiration processes, or removed by runoff. As indicated in Figure 6.25, the direction of overland flow is determined by slope orientation. A combination of two opposing downward dipping slopes constitute a drainage unit, with the outer boundaries of an interconnected set of drainage units representing the limits of a *drainage basin*. Drainage basins, in turn, are organized into a larger complex—the *drainage system*. Drainage systems can be small, consisting of only one or two streams, or large, draining thousands of square kilometers and containing hundreds of tributary streams.

---

* In a few cases, where runoff is confined to landlocked basins, the movement of precipitation to the sea is abbreviated and the entrapped water is subjected to groundwater seepage or evapotranspiration. These cases, however, are exceptions.

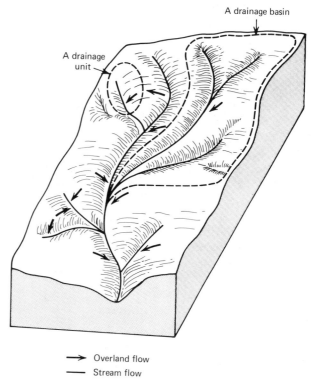

→ Overland flow
— Stream flow

FIGURE 6.25
A DRAINAGE SYSTEM.

Streams have a variety of shapes and sizes. Streams can be narrow or wide, deep or shallow. Some flow swiftly, others are sluggish. Their characteristics depend upon the energy of the stream and the type of materials through which they flow. A stream's volume, and the width and depth of its channel, normally increase in a downstream direction (Fig. 6.26). Viewed in cross section, its gradient (vertical drop per unit of horizontal distance) tends to be steep near the headwaters and to flatten out toward the mouth, presenting a profile that is concave upward.

Streams are one of man's basic resources. A large proportion of the world's population obtain its water directly from streams. Streams have long served as natural transportation arteries, as areas for recreational activity, as power sources, and as a medium for food production. Oceans are enriched by the nutrients and minerals carried to them by streams. Streams also serve to carry off mankind's wastes, a situation that tends to erode the quality of a resource upon which humans are so heavily dependent.

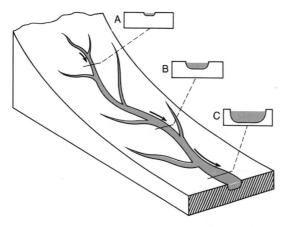

FIGURE 6.26
CHANGES IN THE DOWNSTREAM CHARACTERIS-
TICS OF A STREAM. Discharge is increased down-
stream by entrance of water from successive tributaries.
Width and depth of channel are shown by cross sec-
tions, A, B, and C.

Seen from the air, stream systems display distinct patterns that correlate with the degree of uniformity and resistance of the rock through which
they flow. In regions of relatively homogeneous
surface materials, a branching tree-like arrangement develops into a *dendritic pattern* (Fig. 6.27A).
Erosion on parallel bands of rock of varying resistance results in a *trellis pattern*. Here the primary
stream occupies a valley eroded into the soft rock;
tributaries flow down adjacent ridges, entering
the primary channel almost at right angles (Fig.
6.27B). *Radial patterns*, resembling the spokes of

a bicycle wheel, form where an elevated structure
such as a volcano or dome exits (Fig. 6.27C).

Streams alter the surface they traverse, carving
valleys that are the single most characteristic feature of the earth's landscapes. As shown in the
aerial photograph of Death Valley, the work of
streams is evident even in the deserts (Fig. 6.28).
As streams flow, they *transport* debris weathered
from valley slopes and channel bottoms, *erode* the
stream channel deeper into the land, and *deposit*
sediments either along their valleys or in lakes and
oceans.

Streams transport sediment as mineral matter
that has become soluble *(dissolved load)*, as fine-
sized material in suspension *(suspended load)*, and
as heavy matter that is rolled or pushed along the
channel bottom as *bed load*, a process called *saltation*. The Mississippi River well illustrates the
transport power of a stream. Each year it carries
750 million metric tons of material to the Gulf of
Mexico. Five hundred million are in suspension,
200 million in solution, and 50 million are moved
as bed load.

Erosion of stream valleys takes place along
zones of maximum stream turbulence (Fig. 6.29).
In these zones *hydraulic action*, *abrasion*, and *corrosion* combine to deepen and widen the channel
and its valley. Hydraulic action is the excavation
of unconsolidated material (gravel, sand, silt, and
clay) by the force of flowing water which exerts
an impact and drag effect upon the bed and banks
of the stream channel. This process enhances
mass-wasting where the stream undermines its

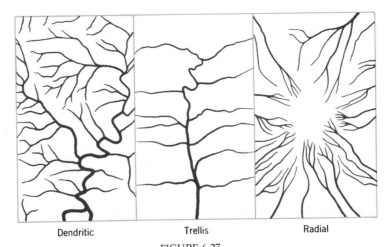

Dendritic   Trellis   Radial

FIGURE 6.27
STREAM DRAINAGE PATTERNS. (A) Dendritic. (B) Trellis. (C) Radial.

FIGURE 6.28
STREAMS CAN INTRICATELY ERODE AND SCULPTURE DESERT LANDFORMS, DEATH VALLEY, CALIFORNIA.

cycles? Imagine a gently dipping plain lifted above sealevel by tectonic forces. The initial drainage upon this landscape will be poor. Water slowly moves downslope toward depressions. The depressions, in turn, serve to collect overland flow and direct its movement toward lower elevations. Erosion is at a maximum along these paths of concentrated water flow. It is here that stream channels are established to conduct runoff to the oceans; and landscape sculpturing by moving water is set in process.

During the *youthful stage of erosion* streams erode vertically and downcut the surface. Waterfalls and rapids are common in areas of resistant rock; valley-walls slope directly to the channel, giving a V-shaped cross–valley profile; and branches, or tributaries, of the stream system extend themselves into the surrounding *interfluves*—those areas between the streams. The stream's gradient is usually steep, its course relatively straight, and level land within the stream valley is either minor in extent or absent. The V-shape of a youthful valley makes it ideal for dams for generation of hydroelectric power and storage of water for it is more economical to build a narrow, high dam than a broad, low one. Thus, more power is generated by the water of small mountain rivers than by the giant rivers of the plains. Youthful stream valleys have very little land avail-

banks. Without underlying support, large masses of valley slope materials may slump into the river and be carried away. Abrasion is mechanical wear. Rock fragments carried by the stream grind away at the surface of channel walls in an effect much like that of sandpaper rubbed over loose paint. Corrosion is the process by which soluble rock materials are dissolved into the stream channel. When the cutting action of a stream is combined with mass-wasting of its valley slopes, the landscape may be expected to be modified. Stream erosion studies have validated this conjecture and proved that there is also a gradual evolution in the morphology of both streams and of the topographic features they carve. The evolution takes place in a somewhat orderly fashion that may be arbitrarily divided into three stages: youth, maturity, and old age. Figure 6.30 illustrates these stages.

How does a stream progress through these

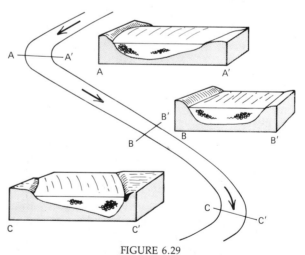

FIGURE 6.29
AREAS OF MAXIMUM STREAM TURBULENCE. Shown by curly symbols are areas of maximum stream turbulence. They occur where the change between the two opposing forces—the forward flow and the friction of the stream channel—is most marked.

LANDFORMING PROCESSES
159

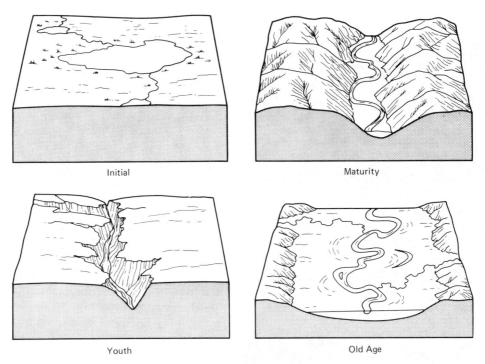

Initial

Maturity

Youth

Old Age

FIGURE 6.30
STAGES OF STREAM CHANNEL EVOLUTION.

able for agricultural or urban development. If such valleys are numerous in an area and lie close together, their interfluves possess an intricate sawtooth pattern. Such irregular surfaces make the building of roads costly and travel time-consuming. Nevertheless, these areas, with their deeply incised gorges and canyons, are some of the earth's most aesthetic landscapes and are highly valued as recreation resources.

The load of debris carried by a youthful stream seldom matches the stream's potential for removing earth materials. As a result, downcutting predominates. However, with the passage of time and the continued weathering of interfluves, the supply of rock to the channel increases. Eventually, the supply of debris will exactly equal the stream's transport capacity and vertical erosion will cease. In theory, as rock moves downstream it is replaced by freshly weathered rock from upper slopes. In other words, the profile of the stream eventually attains a balance, or equilibrium, in supply and removal of debris and becomes stable in its landscape position.

A stream that has achieved equilibrium is said to have a *graded profile*, a characteristic that ear-

marks the onset of its second evolutionary stage: *maturity*. During maturity vertical erosion, or downcutting, diminishes and lateral erosion becomes dominant. The effect of lateral erosion is valley broadening and floodplain development.

When a stream's erosive power is directed toward the sidewalls of its channel, a widening of the valley floor is inevitable. Erosion is most pronounced on the outside of river bends where water velocity is greatest. As the stream undercuts the valley wall on the outside of bends, it shifts its channel in the same direction. The sweeping bends of a stream channel are called *meanders*. Meanwhile, on the inside of the same bends where stream velocities are at a minimum, *alluvium* (stream deposited materials) collects to create a flat-floored valley bottom. This low, flat strip, or portions of it, is inundated in flood conditions. Hence it is called a *floodplain*. A floodplain is a narrow strip in its early evolution but widens considerably with the passage of time. Along with floodplain development comes an extension of tributary streams by headward erosion, a process leading to the intricate dissection of interfluves. Mature valleys have considerably more level land

SCULPTURING THE LAND

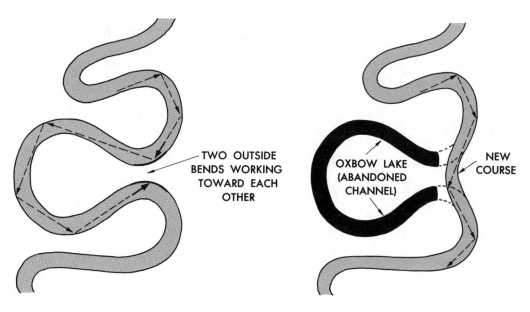

FIGURE 6.31

SHIFTING STREAM CHANNELS. The cutting outside edges of meanders extend the distance water must travel. As meanders expand, the neck of land within their bends is narrowed. Flood waters eventually cut across the neck, abandoning meanders, leaving channel scars, and forming *oxbow lakes.*

than youthful valleys. Historically, they have been more densely settled than the latter. As floodplains widen, increased amounts or arable land are available in proximity to water. This favors agricultural development. Mature streams, possessing a more gentle gradient than youthful rivers, are better suited to navigation and often serve as major transportation arteries. Highway and railway construction in maturely eroded stream valleys can be engineering marvels. Routes may have to traverse steep valley bluffs and frequently cross meandering channels. Interfluves of maturely eroded regions provide little level land, and agricultural activity must be undertaken with great care to prevent rapid soil erosion.

The *old-age stage of erosion* begins after a stream has cut a floodplain considerably larger than the width of its *meander belt.* At this stage, the river shifts its channel relatively easily through unconsolidated aluvium; essentially ceases to broaden its floodplain; and changes form by elongation through meander growth, by cutting off meanders, and by developing natural levees (Figs. 6.31 and 6.32). Slope wash, lateral erosion, and mass-wasting lower most of the interfluvial areas to remnants of inconsequential height. Theoretically, the floodplain and the interfluves would

eventually meld their forms into a *peneplain* ("near plain") if the erosion cycle were not interrupted (Fig. 6.33). Throughout earth history, however, uplifting usually re-initiates an earlier stage of erosional activity, called *rejuvenation,* long before a peneplain can form (Fig. 6.34).

The broad valleys of old-age rivers have large expanses of nearly level land and possess relatively fertile soils. Repeated deposition of fresh silt and organic matter over the floodplain replenish plant nutrients that normally wash out of the soil. Many of these areas are used for crop production. In addition, they support forest product industries as well as some grazing, water dependent industries, and urbanization. The major drawback to the uti-

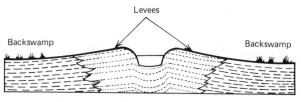

FIGURE 6.32

NATURAL LEVEES. These are banks of sand and silt that build up along the stream channels of old-age floodplains. They form during flood times, when water spilling over the river banks loses velocity and turbulence and deposits its load of sediment.

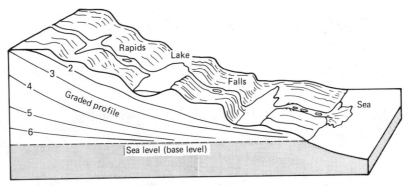

FIGURE 6.33

STREAM GRADATION. As streams erode the surface, they eliminate channel irregularities to form graded profiles. Thereafter, as the landscape is worn down, the
graded profile diminishes in steepness throughout its entire length. Sea level is the
ultimate surface to which the land can be reduced and is referred to as the *base-level*
of stream activity.

lization of these lands is the constant threat of
floods that may inundate thousands of square kilometers, destroy crops, damage property, and
result in the loss of human lives.

Floods are naturally recurring phenomena; yet
floodplains are deemed desirable locations for human settlement. Disasters are, therefore, inevitable. Nevertheless, through thoughtful floodplain
zoning and hydrologic planning, floodplain danger, especially in terms of human lives, has been
significantly reduced within the United States.
Floodplain zoning limits use of flood-prone areas
to functions, such as crop and forest growth and

the provision of recreational activities, that cannot
be permanently destroyed. Dense settlements are
not permitted in these zones. Hydrologic planning involves applying comprehensive hydrologic
principles to reduce flood threats. These measures
include reducing overland flow by revegetating
watersheds, building artificial levees to contain
the stream; and shortening the river course by cutting off meander loops (a process that speeds the
flow of water through the channel).

Stream deposited landforms. Some of the depositional activities of streams were identified in
the previous section. Floodplains, oxbow lakes,
and natural levees are all products of stream deposition within a river valley. Much of a stream's
load, however, is carried to the terminus of its
channel, where deltas and alluvial fans are formed.

When a river terminates in an area of relatively
still (oceanic or lake) water, the stream loses its
forward velocity and deposits its sediments. These
accumulate to form *deltas*, low-lying wedges of
land projecting into the sea (Fig. 6.35). Normally,
the finest silts and clays settle out at some distance
from the mouth of the stream channel, whereas
coarse sediments accumulate at the immediate
mouth of the channel. Deltas may extend over
many thousands of square kilometers and effectively increase continental margins. Some of the
world's most famous deltas are those of the Mississippi, Nile, Ganges, and Amazon rivers.

Alluvial fans are similar to deltas, except that
they form on land rather than in water. Fans are

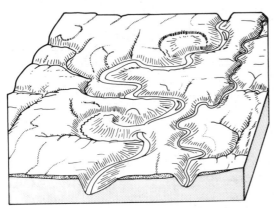

FIGURE 6.34

LANDSCAPE REJUVENATION. The diagram represents a rejuvenated landscape. The meandering course of a mature river
is evident on the surface. Yet—probably due to uplift—the
steam is now eroding vertically and producing steeply dipping
entrenched valleys, typical of the youthful erosional stage.

From Strahler *Modern Physical Geography* © 1978, John Wiley &
Sons, New York.

SCULPTURING THE LAND

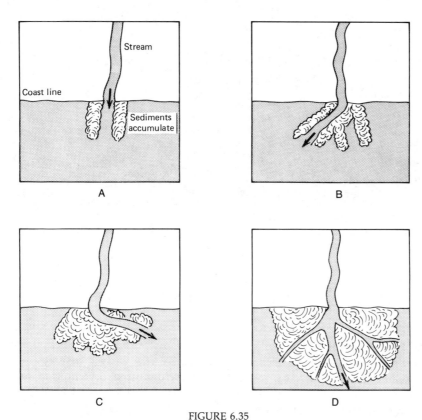

FIGURE 6.35

EVOLUTION OF A DELTA. A stream's reduced velocity on flowing into standing water causes it to deposit its load of sediment, thereby forming a delta. In the soft sediments it has deposited, the stream easily changes courses and subdivides into many branches, thus laterally extending the delta.

especially prominent features in arid regions where mountain streams reach a plain or broad valley. Carrying a considerable amount of sediment from sparsely vegetated slopes, the stream loses its forward direction and velocity upon reaching the valley floor and spreads out, depositing its load of debris in the process. These deposits accumulate in a cone-shape with the apex of the cone oriented toward the valley mouth (Fig. 6.36). In a plan view, these features resemble a Japanese fan. If the process continues for a sufficient length of time, alluvial fans from adjacent valleys will begin to coalesce, eventually filling in valley depressions as the uplands wear away.

Alluvial fans are formed of relatively porous materials. They lie at the lower reaches of a channel toward which streams flow, and consequently have served as collectors of vast quantities of water. Throughout the southwestern United States these extensive groundwater reserves have been used to supply irrigation and urban water needs.

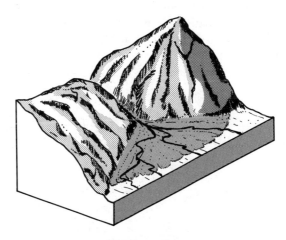

FIGURE 6.36
AN ALLUVIAL FAN.

In many instances groundwater withdrawal often outpaces groundwater recharge. This results in increased pumping expenses as deeper water reserves are tapped.

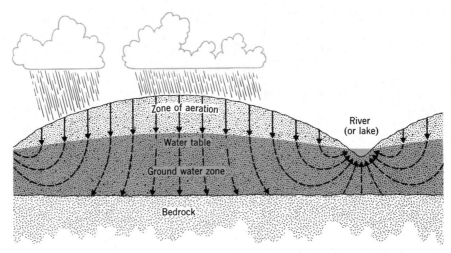

FIGURE 6.37
THE WATER TABLE. The water table is the three-dimensional surface of the zone of groundwater saturation. Its configuration approximates the shape of the ground surface, but is of more subdued relief.

Groundwater and Subsurface Erosion. Some of the precipitation that reaches land areas becomes *groundwater*. This is water that has passed through the soil and moved to lower levels, saturating spaces between sediments and rock. The upper limit of this saturated zone is called the *water table:* a three dimensional surface with a shape resembling the topography in which it is found, yet of less relief (Fig. 6.37). The depth from ground surface to water table varies from place to place. In deserts, it may be hundreds of meters. In swamps, the water table is at, or above, the surface.

Groundwater becomes a visible element on the surface in a variety of ways. It may serve as the primary source of water to feed a lake, sustain flow in a stream channel that has eroded its base into the water table, or be drawn to the surface by pumps in wells. A *well* is an opening drilled into the zone of groundwater saturation. In a well, a large opening is created where previously small pore-space filled with water existed. The bore of the well becomes filled with water by a horizontal flow from the saturated zones adjacent to it. Water removal from the well is replenished in like fashion.

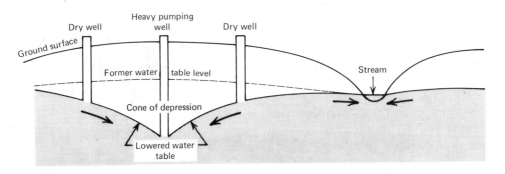

FIGURE 6.38
GROUNDWATER EXTRACTION EXCEEDING REPLENISHMENT. In areas of heavy pumping, the water table is drawn downward, creating a *cone of depression*. In this example, adjacent wells that are shallow no longer reach the water table and have become "dry." The stream eroded into the water table, however, is unaffected by the drawdown and is still being fed groundwater.

Groundwater reserves represent water that has been stored underground for months, years, or even centuries. In some of the desert areas of the western United States, water is being pumped that fell as rain at least 10,000 years ago. How long such a well can be pumped is largely determined by the rate of water withdrawal. If water is withdrawn at the rate it is being replaced, the well may last indefinitely. Excessive extraction, on the other hand, depresses the water table and may lead to the failure of a well, or it may sufficiently increase the cost of pumping to make the well's use un-

economical (Fig. 6.38). Excessive groundwater extraction may have a major impact upon the surface of the land. In many areas of the earth ground subsidence has accompanied a lowering of the water table. As water is removed from the tiny spaces between individual rock grains, earth materials become compacted, meaning that they occupy less space. When this happens the surface above the compacted layer subsides. Subsidence in Mexico City, for example, has amounted to as much as 7 meters (23 feet) in places, resulting in tilted buildings, reversal of the flow in sewage

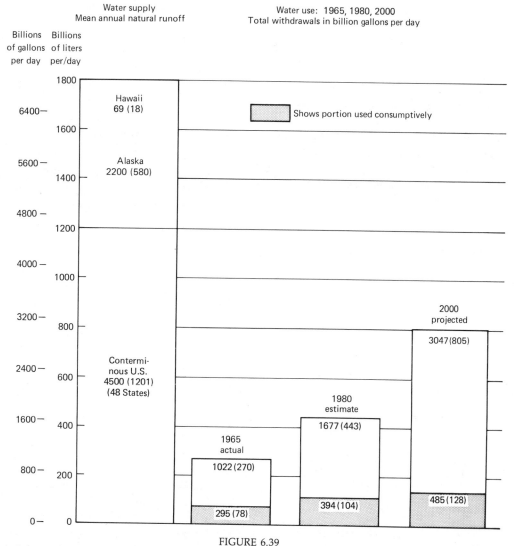

FIGURE 6.39
WATER SUPPLY AND PROJECTED WATER USE IN THE UNITED STATES. This graph showing water supply and use in billions of liters (gallons) per day suggests adequate water supplies throughout the year 2000. These average figures do not, however, account for regional imbalances in both water quantity and quality. (*Source*: Water Resources Council, 1968.)

lines, and ruptured well-casings. Throughout the Central Valley of California, a similar problem exists. Withdrawal of large amounts of groundwater for crop irrigation has been accompanied by a costly lowering of the water table, and land has subsided as much as 8 meters (26 feet). Such alterations of the land surface are irreversible. Only enlightened long-range planning can reduce the threats of property damage and the eventual loss of groundwater resources.

Water is one of mankind's most important natural resources. It is needed to sustain human life, plant life, industries, and cities. Fortunately, the United States is blessed with very large water supplies. Although local shortages and surpluses result from the irregular distribution of precipitation and its seasonal or annual variation, surface and groundwater reserves are ample to meet most current water needs (Fig. 6.39). The U.S. Geological Survey has estimated that present fresh water supplies in the continental United States amount to about 4500 billion liters/day (1200 billion gals/day). In 1980, well over 1600 billion liters/day (400 billion gals/day), or roughly 35 percent of total available fresh water, was being used to meet domestic, agricultural, and industrial water needs. Based on the latter figure, it might seem that the United States has little to fear in terms of a major water shortage. Unfortunately, this impression is not true. The demand for water increases with time, while the water supply is static. In 1965, water used was only 22 percent of water reserves.

Since then annual increased water use has averaged about 1 percent per year. Projected water needs for the year 2000 is over 3000 billion liters/day (800 billion gals/day), or about 67 percent of the supply, a predicted 1½ percent increase in use per year. Although water supplies for the next few decades appear adequate, long-range shortages seem likely, especially in densely populated regions and in areas of low rainfall. The identification of areas prone to water shortages, and planning for water augmentation programs, are basic to human welfare.

Springs and Artesian Systems. A *spring* results where the water table intersects the earth's surface over an impervious layer through which water cannot readily flow, or where fractures and solution channels intersect the surface (Fig. 6.40). *Artesian systems* are those in which groundwater rises in a well, or spring, to a level higher than that in which the groundwater is first encountered (Fig. 6.41). Three essentials are required for such systems to function:

1. A series of dipping sedimentary layers that include an *aquifer* (a porous stratus) lying between impervious layers.
2. Rainfall to infiltrate the aquifer where it is exposed at ground surface.
3. A fracture or a well that provides access through the impermeable roof rock, permitting the escape of water to the surface.

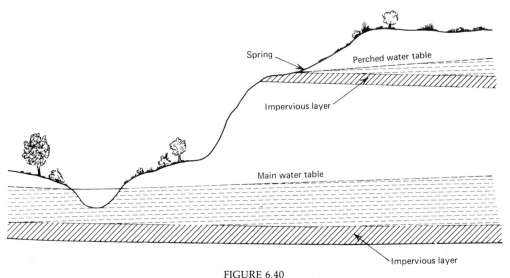

FIGURE 6.40
SPRINGS. Springs are frequently found in areas where the water table intersects the ground surface.

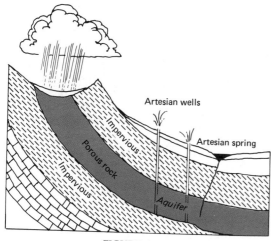

FIGURE 6.41
ARTESIAN WELLS AND SPRINGS. Three essential conditions for an artesian well or spring are an aquifer, an impermeable rock roof, and water pressure sufficient to make the water in any well or spring rise above the aquifer.

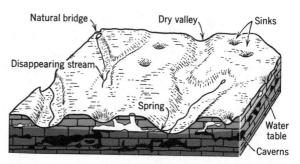

FIGURE 6.42
KARST LANDSCAPE AND ASSOCIATED SUBSURFACE FEATURES.

The name for artesian wells and springs comes from the French province of Artois, where the first such well in Europe was bored. This particular well began flowing in A.D. 1126 and is still active today. It stands as a testimony to the longevity of artesian systems when withdrawal of water does not exceed replenishment. When conditions are especially favorable, artesian systems have been known to produce spectacular fountains, lifting water as much as 60 meters (200 feet) above the surface.

Subterranean erosion. Groundwater, moving through earth materials, erodes by dissolving soluble rock. To explain this process, it is necessary to refer to an earlier discussion of the role of rainwater in altering landscapes. Recall that most precipitation is slightly acid. As raindrops fall through the atmosphere they unite with carbon dioxide to produce a mild carbonic acid. This solution then infiltrates the soil, where decaying plants add complex organic acids to it. Portions of the final product then gravitate toward lower depths to become groundwater. Acid groundwater, in turn, reacts with many rock-forming minerals, those most susceptable being the calcium and magnesium carbonates found in limestone and dolomite.

In areas of extensive limestone deposits the surface exhibits unique features associated with groundwater erosion (Fig. 6.42). These areas are said to have *karst topography*, a term originating from a well-eroded limestone region of Yugoslavia. Here, as in other limestone regions, water moves toward lower levels along joints and fractures within the bedrock. As it moves, it enlarges its pathways by dissolving the walls of its channels. (These subterranean decay zones form the fascinating caves and caverns that spelunkers and bats frequent.) The difference between limestone caves and caverns is one of size. The former are small, the latter large. Each, however, is often ornamented with calcium-rich evaporation deposits known as stalactites and stalagmites. *Stalactites* have the appearance of giant icicles, and hang from cavern ceilings; *stalagmites* stand as mounds building up from mineral drippings upon the cavity's floor.

Subterranean rock corrosion may eventually enlarge a cave or cavern to the point where the cavity begins to approach the surface. Then, and often suddenly, when the weight of overlying roof becomes too great to support, the surface materials fall inward to create a surface depression (Fig. 6.43). These may be dry, steep sided holes. If the area has a high water table, they may stand-out as ponds or lakes. It is estimated that in the limestone areas of Kentucky and southern Indiana there are 600,000 of these depressions varying in depth from one to more than 30 meters (3 to 100 feet). Precipitation in these areas is rapidly funneled below ground, often leaving the surface relatively dry and giving rise to stories of "disappearing rivers swallowed-up by the earth." Such waters are diverted undergound to flow either as subsurface streams in cavities, or as groundwater to other outlets.

FIGURE 6.43
A SOLUTION DEPRESSION. This feature, called a "sink hole," resulted from a cave-in of an underground cavity in Shelby County, Alabama.

Mankind, as well as other forms of life, has found functional uses for soluble landscapes. Birds, insects, and larger animals have found sanctuary, and even permanent shelter, in subterranean caves and caverns. In the caves of Altamira, Spain, and Lascaux, France, walls are painted with hunting scenes dating to between 25,000 to 10,000 B.C.. These paintings afford a glimpse of mankind's earliest art forms and provide evidence of his habitation of caves.

The actions of internal forces, gravity, and water in sculpturing the earth's surface can be considered univeral landscape-forming processes. They have therefore been grouped together in this chapter. In the next chapter, land sculpturing agents associated with more specialized conditions will be considered.

strato-volcanoes
lava-flow plateaus
intrusive vulcanism
batholith
laccolith
sill
dike
diastrophism
folding
faulting
footwall
hanging wall
normal fault
fault scarp
reverse fault
strike-slip fault
earthquake
continental structures
shield
interior plains
plateaus
block mountains
folded mountains
coastal plains
orogeny
epeirogeny
external processes
physical weathering
chemical weathering
creep
earthflows
mudflows
slides
debris slide

rock slide (landslide)
runoff
overland flow
stream
drainage basin
drainage system
dendritic stream
    pattern
trellis stream pattern
radial stream pattern
stream load
stream erosion
stream evolution
graded profile
meander
alluvium
floodplain
peneplain
oxbow lake
natural levee
base level
delta
alluvial fan
groundwater
water table
cone of depression
spring
artesian system
Karst topography
stalactites
stalagmites
solution channels
solution depressions

KEY TERMS AND CONCEPTS

igneous rock
sedimentary rock
metamorphic rock
vulcanism
extrusive vulcanism

magma chamber
lava
volcanic ash
shield volcanoes
cinder cones

DISCUSSION QUESTIONS

1. Name the three families of rock. How do they differ from one another? Why might it be said that the basis of landforms is rock?
2. Explain the difference between internal and external landforming processes.

3. Describe the various forms of extrusive and intrusive volcanism.

4. What evidence of folding exists on the earth's landscapes?

5. Define the various types of faulting. How has faulting of the crust affected mankind?

6. What is the relationship between faulting and earthquake activity?

7. Describe the primary structural components of continents.

8. What is the difference between physical and chemical weathering?

9. Discuss the significance of the various forms of mass-wasting upon human activity.

10. Define: runoff, overland flow, stream, drainage basin, drainage system, dissolved load, suspended load, bed load.

11. How does the structural characteristics of land surfaces determine stream patterns?

12. Explain how hydraulic action, abrasion, and corrosion effectively change a stream valley.

13. Describe the processes involved in the development of a floodplain.

14. Define: interfluve, graded profile, meander, alluvium, natural levee.

15. How are deltas and alluvial fans formed?

16. How is groundwater collected and stored? What is the water table and how is its depth significant in determining where to place a well?

17. What conditions are necessary for the formation of an artesian spring?

18. What is the predicted availability of water supplies for the year 2000, for the twenty first century?

19. Describe conditions favorable for the development of Karst topography.

REFERENCES FOR FURTHER STUDY

Carson, M. A., and M. J. Kirkby, *Hillslope Process and Form*, Cambridge University Press, Cambridge, 1972.

Derbyshire, E., ed., *Climatic Geomorphology*, Harper and Row, New York, 1973.

Hunt, Charles B., *Natural Regions of the United States and Canada*, W. H. Freeman and Co., San Francisco, 1974.

Leopold, Luna B., M. G. Wolman, and J. P. Miller, *Fluvial Processes in Geomorphology*, W. H. Freeman and Co., San Francisco, 1964.

Morisawa, M., *Streams: Their Dynamics and Morphology*, McGraw-Hill Book Co., New York, 1968.

Thornbury, W. D., *Principles of Geomorphology*, Second Edition, John Wiley and Sons, New York, 1969.

7
SCULPTURING THE LAND
Regional Forces

CHAPTER 6 DEALT WITH LANDFORMS CREATED BY INTERnal processes and with the modification of such landforms by the combined action of gravity and flowing water. These processes affect the majority of the earth's land area. This chapter deals with some agents of change found operating under special circumstances. These landscape altering processes include the work of glaciers, oceanic waves, and winds.

GLACIERS AND GLACIAL LANDFORMS

Glaciers are seas and rivers of moving ice whose existence require a cool thermal regime. Confined today to mountainous or polar regions, glaciers cover about 9 percent or 15 million square kilometers (5.8 million square miles) of the earth's surface. The geologic record indicates that in the past an area at least three times as large was buried under massive sheets of ice (Fig. 7.1). Why the earth experienced such large-scale periods of glaciation has been a topic of considerable speculation.

Scientists have proposed numerous theories to explain the origin of ice ages, or periods of widespread glaciation. Among these theories are:

- *A change in the eccentricity of the earth's orbit about the sun.* The earth's orbital eccentricity (oblateness) is not constant, but increases and then decreases on a 92,000 year cycle. This means that the distance between the earth and the sun at aphelion and perihelion varies. Perihelion (when the earth is nearest the sun) occurs in January. Thus, winters in the Northern Hemisphere are not as severe as they might otherwise be. If, however, an increase in eccentricity were accompanied by a change of the aphelion position (when the earth is farthest from the sun) so that aphelion occurs in January, winters in the Northern Hemisphere would be considerably more severe than they are now and would possibly support glacier development.

- *Shifting of continents.* As plate tectonic activity shifts about the continents, it may happen that landmasses are moved to more poleward locations where conditions favor ice sheets.

- *Changes in the content of atmospheric particulate matter.* During periods of active vulcanism, the atmosphere contains large quantities of volcanic dust. This results in more solar energy being absorbed in the upper atmosphere, while considerably large amounts of solar radiation are subjected to loss by scattering. Less energy reaching the surface of the earth means that the lower atmosphere is cooler. At the same time, the numerous dust particles serve as nuclei for the condensation (or sublimation) of water vapor; thus, processes producing snowfall are enhanced.

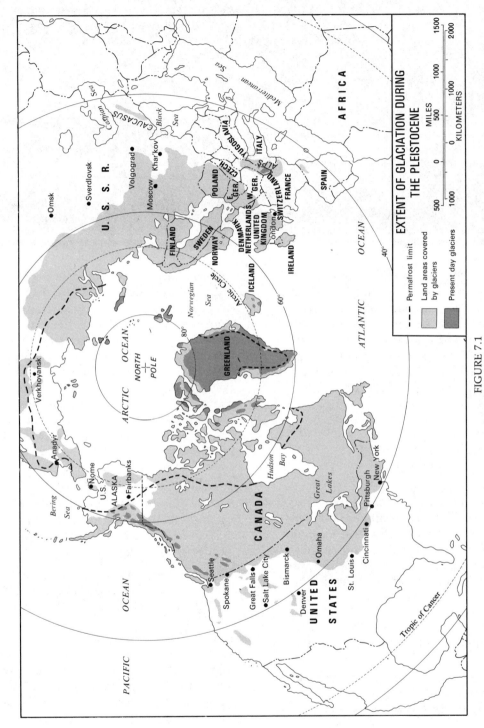

FIGURE 7.1

GLACIATION OF THE NORTHERN HEMISPHERE. During the last two million years, the northern hemisphere has experienced four periods of extensive glaciation. Light areas shown above indicate the maximum surface affected by ice. The gray show present-day glaciation.

■ *Increased elevation of continents.* A period of active mountain building and general land-mass uplift is conducive to glacier development. Their cooler environs and their role in promoting orographic precipitation make mountains ideal sites for snow accumulation.

There are other suggested causes for glaciation. Some include changes in oceanic circulation, others a change in the rate of energy emitted from the sun. So far, no single theory can account for all of the times ice has accumulated and spread over the continents, nor can any one theory explain all of the occasions upon which glaciers decreased in size or were absent from the land. Nevertheless, the search for the factors causing ice ages continues. Regardless of the theories individual earth scientists may favor, glacial ice and its role in modifying the landscape has had far-reaching effects. The threat of another ice age is realistic. About 10,000 years ago the areas in which Cleveland, Detroit, Chicago, and Milwaukee now stand were covered by ice. The social and economic ramifications of a return of glaciers to the mid-western United States would be profound. People would have to emigrate to new locations, homes and industries would be abandoned, and a wealth of regional investment would be lost.

Glaciers, by definition, are bodies of ice that are either flowing on the land or that exhibit evidence of past motion. They form in areas where winter accumulations of snow exceed the amount of snow lost in summer by evaporation and melting, and range in size from small ice masses contained within mountain valleys to extensive ice sheets thousands of square kilometers in extent. The continuation of yearly snowfall surpluses deepen the *snowfield* (wide covers, banks, and patches of snow). This snow, initially powdery, is converted into granular ice by melting and refreezing. When the thickness of granular ice becomes sufficiently deep, its weight compresses the lower layers into hard crystalline ice. Continued deepening eventually causes the lower layers of ice to behave plastically and the ice begins to move downslope or outward. Once ice movement has taken place, a full-fledged glacier is born.

Glaciers expand and shrink, deepen and thin, in response to the amount of nourishment by snow in their upper ends and to the wastage by melting and evaporation in their lower reaches. If snow accumulation exceeds wastage the front of the glacier advances; the front recedes when wastage predominates. Regardless of whether the glacier's front is advancing, retreating, or relatively stationary, the ice still moves forward. Velocity of glacial ice flow has a maximum measured rate of 30 meters (100 feet) per day. Though it is believed that glacial surges can produce even faster motion, most flows range from a few centimeters to several meters per day, a rate normally too slow for the casual observer to detect. Figure 7.2 illustrates the character of movement and the features of a small glacier contained within a mountain valley. Note that the mass of ice moves downslope like a solid river. Bearing in mind that the ice at the bottom and sides of the glacier is in direct contact with rock surfaces, it can be expected that the velocity at these boundaries is slower than at the center of the glacier, where there is less friction.

Glacial flow is initiated by the pull of gravity upon an ice mass. The lowest layer, below 60 meters (200 feet), behaves as though in a plastic state, constantly undergoing slow deformation and re-crystallization. The upper layer is a rigid zone. It does not flow, but is dragged along with the underlying ice. This ice is brittle and readily cracks to form deep fissues, or *crevasses*.

Some of the world's most spectacular landform features are found in mountainous regions that were once glaciated. In these upland areas, pre-glaciated landscapes normally contain stream valleys that are long, narrow, and V-shaped in cross-

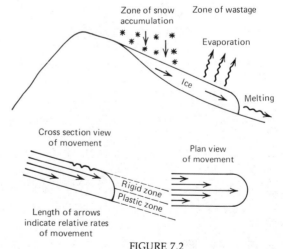

FIGURE 7.2
CHARACTERISTICS AND MOVEMENTS OF MOUNTAIN GLACIERS.

GLACIERS AND GLACIAL LANDFORMS

profile. These characteristics result from the downcutting action of streams that are well above their base level. During glaciation, the valleys are transformed into steep-sided, flat-floored, U-shaped *glacial troughs* by the action of *abrasion* and *plucking* (Fig. 7.3). When the debris carried by moving ice consists of coarse rock materials, scratches or deep grooves are carved into bedrock surfaces. When made up of fine-sized sediments, the debris serves a function parallel to that of a jeweler's rouge—it smooths and polishes exposed bedrock. In general, glaciers move like giant pieces of sandpaper, abrading all surfaces with which they come in contact. Plucking occurs when glaciers flow over fractured bedrock surfaces. The ice simply freezes to underlying blocks of jointed rock that are carried off as the ice moves forward. The combined action of abrasion and plucking effectively widens, deepens, and straightens out

mountain valleys. If a glacial valley leads to a shoreline and the grinding action of moving ice erodes the valley base below sea level, the trough will be occupied by ocean waters when the glacier melts and recedes. A trough thus "drowned" by the sea is called a *fiord* (fjord).

Just as a stream may have several tributaries feeding it, so a large valley glacier may have tributary valley glaciers. The degree of erosion in each valley depends upon the total mass of ice flowing within it. Obviously, tributary valleys are not eroded as deeply as the main valley. As a consequence, when the glaciers eventually recede, tributary valleys lie at elevations high above the main trough. These *hanging valleys*, if reoccupied by streams, produce cascading waterfalls (Fig. 7.4).

At the uppermost limits of valley glaciers are found *cirques*, amphitheater or bowl shaped depressions where snow accumulates and from

FIGURE 7.3
A GLACIAL TROUGH NEAR MOUNT SUMDUM, ALASKA.

SCULPTURING THE LAND

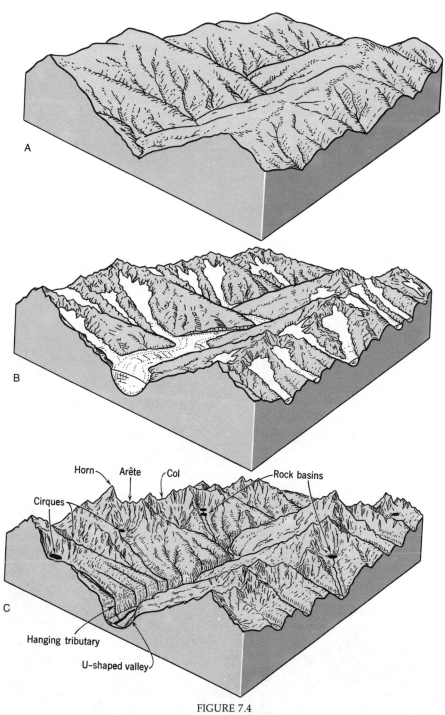

FIGURE 7.4
LANDFORMS CARVED BY MOUNTAIN GLACIERS. (*A*) Landscape prior to glaciation.
(*B*) Period of maximum glacial activity. (*C*) Landscape after glaciation.

GLACIERS AND GLACIAL LANDFORMS

which ice movement originates. Cirques are formed by the plucking action of the moving glacier ice against the headwalls of the valleys. After the ice has disappeared, the depression makes an ideal site for small lakes or ponds. The continued headward erosion of cirques lengthens the glacial valley. If cirques from opposite sides of a ridge continue their gnawing action, the intervening ridge takes on a knife-like, jagged, and serated appearance. These rugged ridges are known as *aretes* (Fig. 7.4). Similarly, when three or more cirques erode their way into a single mountain, its summit is reduced to a spire or jagged peak called a *horn*, the most classic example being the famous Matterhorn of Switzerland (Fig. 7.5). Both

aretes and horns have long challenged adventurous mountain climbers.

Glaciers are like gigantic conveyor belts that carry eroded rock debris from source regions to termini where it is dropped. The most impressive materials moved in this manner are *erratics*, individual blocks of rock, some weighing several tons, that are foreign to the bedrock on which they rest (Fig. 7.6). All remaining materials deposited by moving ice are classified either as *till*, an unsorted mixture of clays, sand, cobbles, and boulders; or *stratified drift*, materials sorted by melt waters of the glacier prior to their being deposited.

The most common till landforms are *moraines*

FIGURE 7.5
THE MATTERHORN OF SWITZERLAND.

FIGURE 7.6
A LARGE ERRATIC IN TULARE COUNTY, CALIFORNIA.

of which *end moraines* are most conspicuous. End moraines appear as prominent ridges that form at the margin of the ice mass—in areas where the glacial front remained in a relatively stable position for some time. The end moraine marking the furthest advance of the glacier is called a *terminal moraine;* moraines formed while the glacier is retreating are *recessional moraines.*

Mountain valley glaciers produce varied and often spectacular landform features. These features are restricted in their occurrence, however, to localized upland regions. On a much grander scale is the landscape modification associated with large ice sheets ranging in thickness from 1500 to over 3000 meters (5000 to over 10,000 feet) and covering millions of square kilometers. Antarctica and Greenland are the only ice sheets now present on the earth. Nevertheless, continental-sized glaciers occupied extensive parts of the Northern Hemisphere in recent geologic times. Landscapes remaining today bear witness to the tremendously powerful erosional and depositional activities of these massive ice sheets.

Large ice sheets differ from mountain glaciers. Instead of being confined to a valley, ice may accumulate over an extensive area in upper latitude regions and totally engulf all mountains and valleys within their realm. Upon reaching a sufficient depth, the weight of the ice causes the glacier to move outward, terminating along a line where a balance between snow accumulation and ice wasting is achieved.

The sheer weight of ice sheets, thousands of meters thick, leaves its mark on land areas under them. Like a ship settling deeper into the water as it is loaded with cargo, land masses tend to sink when loaded with large quantities of ice. Removing the cargo from a ship permits it to float higher in the water. Similarly, continents experience vertical lifting when relieved of ice pressure.

The magnitude of glacier erosive power is also related to the dimensions of an ice mass. While a mountain glacier erodes a valley's walls and bottom, ice sheets can level hills, subdue mountains, and gouge out enormous valleys. The Great Lakes of North America were initially formed when ice sheets gouged out large depressions in solid bedrock, while simultaneously leveling and laying bare of its soil mantle the interior of Canada and parts of New England.

The erosive power of mountain glaciers creates landscape features that are, generally, more impressive than those produced by their depositional activity. The reverse is normally true of ice sheets. An ice sheet's relict erosional surface tends to be subdued; its depositional features, in turn, are more conspicuous (Fig. 7.7). In particular, sand, silt, and gravel are deposited as extensive stratified deposits in advance of end moraines to form *outwash plains*. Within these plains numerous depressions, called *kettles*, are found. Kettles originate from ice-blocks buried in glacial drift. Subsequent melting of the ice leaves a depression, often filled with water. *Drumlins* and *eskers* are two depositional landforms that are unique to ice sheets. Both form under the ice. Drumlins normally occur in swarms and resemble huge inverted teaspoons that range in length from one to two kilometers (1-1½ miles) and that may be as much as 60 meters (200 feet) high. Drumlins form beneath the ice upon irregularities in resistant bedrock. Heavily choked with fine debris, the moving glacier builds up streamlined hills by plastering these irregularities with successive layers of clays and silt. Eskers are natural casts of former stream channels that existed within glacial tunnels. They are composed of washed sands and gravels and appear as long sinuous ridges, often tens of kilometers in length.

The impact of glaciation upon human activity

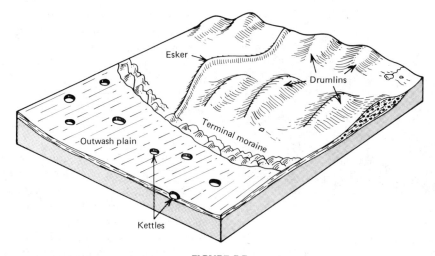

FIGURE 7.7

EXAMPLES OF DEPOSITIONAL LANDFORMS ASSOCIATED WITH CONTINENTAL GLACIERS.

is far-reaching. Rugged, ice-scoured mountainous regions have long served as barriers to the movement of goods and people. Throughout much of mankind's history these areas have been sparsely settled and remained as wilderness. The reasons for their being avoided in the past explain why they are regarded as assets today. Their pristine beauty and recreational potential attracts campers, hikers, mountain climbers, cliff hangers, skiers, and nature lovers. While glacially sculptured mountains discouraged settlement and deterred the movement of goods and people, ice-scoured troughs facilitated access to complex mountainous terrain and often afford the only level land available for agricultural development and dense population settlements. In addition, glacially derived deposits of sand and gravel have great economic potential for concrete and for paving. North America's greatest fresh water resource, the Great Lakes, is a direct result of glaciation. Depending upon the type of debris deposited by ice, some of the earth's most productive and most sterile soils have formed in glaciated regions. Moreover, ice itself influences human activities. *Icebergs* formed when the advancing front of a glacier pushes into the sea and breaks off, are a menace when they float into shipping lanes. If the ice present upon the earth today were to melt, sea level would rise and many coastal areas would be inundated. If the ice sheets grow, the threat is reversed. In either case mankind would find it necessary to make major adjustments.

OCEANIC MOTION AND COASTAL LANDFORMS

Artists and poets have been inspired by the restless motions of the seas. Love songs and poems often evoke images of oceanic movement. Yet, for the most part, observers of the ocean seldom perceive the complexities of movement taking place within a realm that ceaselessly alters the margins of landmasses.

Oceanic Motion

The primary movements of oceanic water can be conveniently divided into four categories: density currents, surface currents, tides, and waves.* *Density currents* normally refer to deep-ocean circulations that are governed by gravity and moved by density differences. Variations in seawater density are associated with two factors: temperature and salinity. When temperature decreases or salinity increases water becomes denser. These principles are depicted in Figure 7.8. Note that cold North Atlantic water tends to move equatorward. This dense water originated at the surface in the Gulf Stream. Gulf Stream water has a high salinity content due to high evaporation rates in low latitudes. As it flows poleward it is chilled

---

* These four categories exclude other forms of motion that are of localized importance, or effective during brief time intervals; turbidity currents and tidal waves (tsunamis) are examples.

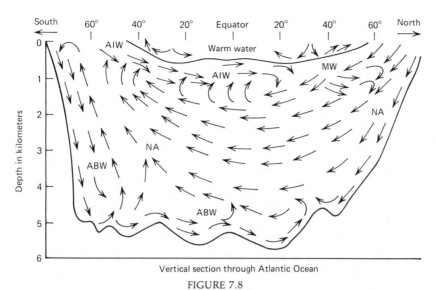

Vertical section through Atlantic Ocean

FIGURE 7.8

CIRCULATION OF DEEP WATER IN THE ATLANTIC OCEAN. The above diagram is oriented along a north-south transect. NA–North Atlantic; ABW–Antarctic bottom water; AIW–Antarctic intermediate water; MW–Mediterranian water.

and its density increases. Upon reaching the Arctic region near Greenland it sinks to the bottom of the North Atlantic basin. Thereafter, it flows southward, and can be traced as far as the Antarctic region.

The surface waters of the Antarctic region are the densest of all the oceans. In this area temperatures are sufficiently low to form sea ice. Sea salts are not included as a part of sea ice, thus rendering the remaining winter water saltier than normal. This extremely cold, saline Antarctic bottom water sinks to the bottom of the Atlantic basin to move northward, crossing the equator and extending well into the Northern Hemisphere, where it underrides cold (but less dense) North Atlantic water. It is estimated that once these cold dense waters sink and become part of the deep ocean circulations they will not reappear at the surface for 500 to 2000 years. The densest water in the world's oceans thus remains cold and deep for thousands of years.

Two additional discrete density flows are important to the general circulation of the Atlantic Ocean. Antarctic intermediate water forms when the salty Brazilian Current is chilled as it moves poleward toward Antarctica. The other, Mediterranean water, originates in the Mediterranean Sea. Because of the warm dry climate prevailing over the Mediterranean region, evaporation rates are high, causing high sea-water salinity. During the winter these saline waters cool and sink, and move westward along the sea bottom to pass into the Atlantic Ocean through the Strait of Gibralter. The water lost from the Mediterranean Sea is replaced by a less dense countercurrent that flows at the surface and enters through the Strait of Gibralter from the Atlantic Ocean.

Surface currents are of more direct importance to man than are deep ocean circulations. The earliest mariners quickly realized that their voyage time could be reduced by "sailing with the currents." More important, surface waters significantly influence global climates. Low latitudes are regions of energy surpluses, where more solar energy is received than is radiated back to space. The opposite is true of high latitudes. Yet the tropics are not becoming warmer, nor are polar regions becoming progressively colder. The reason is that heat is transferred from surplus areas to deficit areas by winds and ocean currents. Winds account for about 75 percent of the total heat transported poleward; oceanic water movements transport the remaining 25 percent.

Surface circulation patterns in the earth's oceans are largely caused by the frictional drag of air moving over water. A close relationship exists between atmospheric and oceanic movements. Consider the region of the Northeast Trade Winds that drag water toward the southwest. Ocean currents, like flowing air, are subject to deflection by the

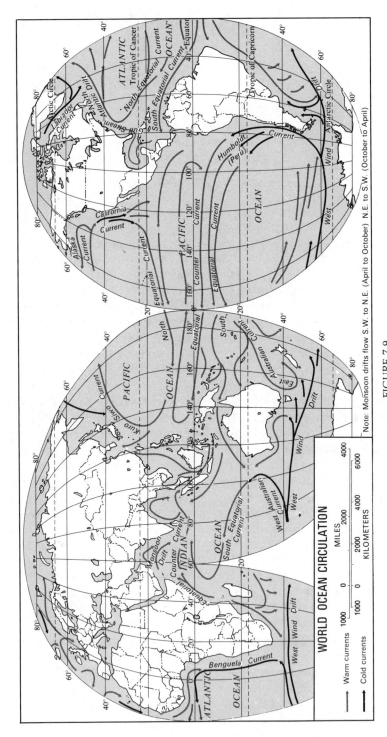

FIGURE 7.9
CURRENTS AND DRIFTS OF THE OCEANS.

Coriolis force. In this instance, the resultant flow of water is directed due west, to the right of its intended path. This flow, known as the *North Equatorial Current,* has a counterpart in the Southern Hemisphere, the *South Equatorial Current.* Between the two is a less significant reverse flow of water, the *Equatorial Countercurrent* (Fig. 7.9), formed as the North and South Equatorial Currents force water to "pile up" against South America's eastern coast. Because the trades are weak in the vicinity of the equator, a portion of this "piled up" water flows "downhill" toward the east, forming the countercurrent.

As the warm equatorial currents reach the eastern coasts of landmasses, most of their flow is forced to turn poleward. Coriolis deflection causes them to be diverted away from the continents and to travel on a more easterly course. By the time the middle latitudes are reached, the currents come under the influence of the prevailing westerlies wind system. Coriolis deflection and the driving power of the westerlies then combine to direct them due east. In the open southern hemispheric seas this produces a circumpolar current known as the *West Wind Drift.* In the Northern Hemisphere, land barriers once again cause the moving water to split. Some of this water will move equatorward as *cold currents* to complete a path back toward the tropics. As cold currents, normally found along west coasts of continents in subtropical latitudes, move toward the equator they tend to be deflected away from the coast by the Coriolis effect. This removal of surface water away from the continents causes colder water to rise to the surface from deeper levels of the ocean. This upwelling intensifies the climatic influences of cold currents along continental margins (Fig. 7.10). Deep ocean water also tends to be rich in nutrients, particularly in nitrates and phosphates. These support abundant microscopic plants which, in turn, are fed upon by fish populations. Some of the world's greatest fishing fields are found in areas of upwelling. Characteristically, upwelling is most common along the eastern sectors of the oceans, most notably along the shores of California, Peru, and West Africa. The remainder of the water reaching land barriers is directed poleward, carrying relatively warm water into higher, cooler latitudes (*warm currents*). Figure 7.9 illustrates the large-scale circulation patterns of the oceans. In the Northern Hemisphere the primary flow is rep-

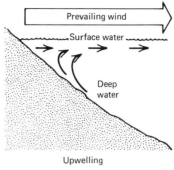

FIGURE 7.10

OCEANIC UPWELLING. Where prevailing winds blow surface water away from the coast, *upwelling,* the rising of water from deeper layers occurs. This sluggish upward flow form depths up to 300 m (1000 ft) brings cold water to the surface and creates a zone of low temperature and high incidence of fog.

resented by huge clockwise movements in each major ocean basin. In the Southern Hemisphere the motion is counterclockwise.

A third form of oceanic motion is represented by the periodic fluctuations in elevation of the sea's surface. This movement, called *tides,** arises principally from interactions between the moon and earth and, to a lesser extent, from interactions between the sun, moon, and earth. It was Sir Isaac Newton who first showed that tide producing mechanisms were a consequence of the law of gravitation. He demonstrated that there is a mutually attractive force between any two bodies. In the case of the moon and the earth, the ocean, which is free to move, is deformed by this force. In effect, the moon's gravitational attraction pulls toward itself the seawater on that side of the earth facing it, creating a tidal bulge in the earth's oceans. On the opposite side of the earth, where the earth-moon gravitational pull is least, the earth's centrifugal force provides for another "mounding up" of seawater. The positions of these two bulges (or high tides) are not stationary but rotate as the earth rotates. As a consequence, any spot on the sea experiences two high tides and two low tides each day. As shown in Fig.

---

* Tide is the rhythmic rise and fall of water bodies caused by gravitational effects of the moon and the sun. *Tidal waves (tsunamis)* are powerful ocean waves reaching great heights along coastlines. They differ from tides both in size and origin. Their great size is related to displacements in oceanic crust and they occur as nonperiodic events. A *tidal bore* is a rapidly moving front of tidal water moving through a narrow inlet or channel. These are rhythmic events associated with the tides. They are usually characterized as having significantly larger waves than are characteristic of most tides, yet the waves are smaller than those of a tsunami.

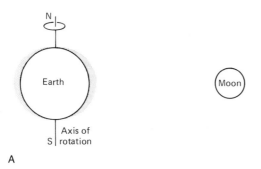

A

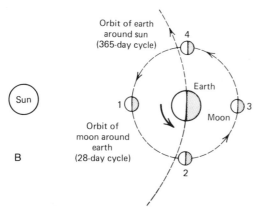

B

FIGURE 7.11

GRAVITATIONAL ATTRACTION OF THE MOON AND SUN RAISE TIDAL BULGES IN EARTH'S OCEANS. (*A*) Tidal bulges, relative to earth's axis of rotation with the moon. (*B*) When the moon and sun attract in the same direction (moon positions 1 and 3) the highest tides are experience. When them moon and sun have opposing positions (moon positions 2 and 4), the lowest tides occur.

7.11, the sun also influences the tides. At times the sun pulls in the same direction as the moon, increasing the height of the tidal bulge. At other times, it acts at right angles to the moon and results in diminished tidal elevations.

In deep ocean water tides are of little significance to either the movement of oceangoing vessels or to land erosion. Along coasts, however, they can seriously affect the movement of water craft and be notable agents of erosion and sediment transport (Fig. 7.12). Erosion is especially pronounced when tides move through narrow inlets, where their rapid currents scour out and remove sedimentary materials. These speedy, steep-fronted tide waves are called *tidal bores*. By the action of tidal bores, many harbors that would otherwise be blocked are cleared of accumulating debris each day.

Most of the work involved in altering coastlines is accomplished by wind generated waves. Figure 7.13 depicts the dimensional elements of a wave in deep water where it is unaffected by the ocean bottom. Contrary to popular opinion, there is little forward motion of water accompanying the passage of waves. Rather, the form of the wave travels through water without significantly displacing it.

A

B

FIGURE 7.12

THE BAY OF FUNDY. Located between Nova Scotia and New Brunswick, this bay experiences extreme tidal ranges and rapid tidal currents. (*A*) High tide. (*B*) The same place at low tide.

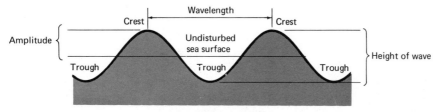

FIGURE 7.13
CHARACTERISTICS OF A DEEP SEA WAVE.

Each water molecule involved in transporting the wave revolves in a vertical loop, returning very nearly to its original position (Fig. 7.14). This looplike, or oscillating motion, of water molecules is effective to a depth that is approximately one half the length of a wave in the open ocean.

When deep ocean waves approaching the shore reach a water depth that is less than half their wavelength, wave form begins to change. In part this is because some wave energy is expended in moving bottom sediments back and forth, an action that slows down the wave. Meanwhile, the slightly faster seaward waves catch up to the dragging forward waves, piling into them, decreasing their wavelength, and causing the forward waves to grow in height. Eventually the steep advancing front of lifted water is incapable of being supported and the waves collapse, or break. These breaking waves instantaneously produce turbulent waters, commonly referred to as *surf* (Fig. 7.15).

The role of surf in changing coastlines is difficult to assess unless observed directly and under varied circumstances. In some instances breaking waves may form from dissipating swells, producing no more than minor beach surges that lazily lap against the shore. In others, violent storms generate rapidly moving waves hundreds of meters in length and tens of meters high. These smash into coasts with forces known to exceed 25 tons per square meter. Under such tremendous pressure, water is forced into every crack, crevice, and pore space of rock and sedimentary materials. This pressure is sufficient to displace blocks of rock and to increase the size of preexisting fractures. Added to this type of coastal erosion is the work of abrasian. Armed with rock fragments, smashing waves perform a sawing and grinding action on coastal rock formations, constantly sculpting new forms (Fig. 7.16).

Since deep ocean waves travel in a direction related to the dominant wind flow, they might be

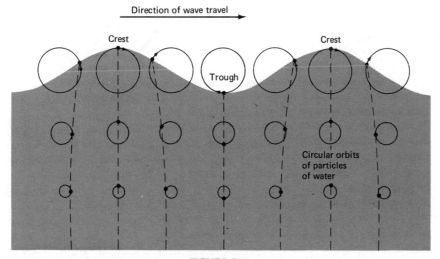

FIGURE 7.14
THE MOTION OF WATER MOLECULES IN DEEP SEA WAVES. In deep sea waves, water molecules travel in loops that return to their point of origin.

OCEANIC MOTION AND COASTAL LANDFORMS

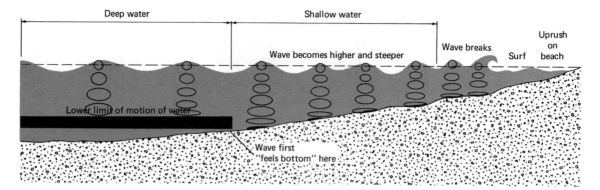

FIGURE 7.15

WAVE MODIFICATION. Waves change form as they travel into shallow water. Their wave length decreases, height increases, and eventually the wave collapses and forms surf.

expected to move at angles that only by chance coincide with the compass bearing of coastlines. This is correct for deep water waves. Yet, as the waves enter shallow water they are bent approximately parallel to the shore. This curvature, called *wave refraction*, results from the wave nearest the shore slowing down due to bottom drag while its end in deep water continues forward with regular speed to eventually swing around nearly parallel with the shore (Fig. 7.17).

Coastal Landforms

Wave erosion tends to straighten irregular coastlines. To understand this idea we will examine the characteristics of an idealized shore profile; that is, the elements of a surface along a line at right angles to shore. The usual elements of a coast are: wave cut cliffs, wave cut benches, wave built terraces, and beaches (Fig. 7.18). The first two result from erosion, the latter two from deposition.

Wave cut cliffs and *benches* (or *abrasian platforms*) appear on promontories of embayed coasts where wave refraction is active, or along straighter coasts where the cliffs and benches may extend for several kilometers. Their evolution is initiated when surf abrades a horizontal notch into a sloping shore. As the notch deepens by erosion, the land is undermined. Lacking support, rock materials above the zone of surf erosion crumble to form a steep sea cliff that continues to retreat into

FIGURE 7.16

MINOR WAVE ERODED FEATURES. A *sea cave* is formed by the surf hollowing out more erodible bedrock. A sea cave cut through a headland becomes a *sea arch*. Isolated *stacks* stand where the surf has torn away parts of the bedrock from the mainland.

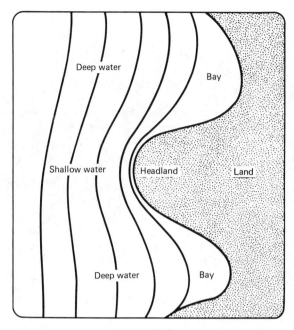

FIGURE 7.17

WAVE REFRACTION. This map shows a series of waves that become increasingly distorted as they approach the shore over a bottom that is deepest at a point opposite the bay. Because wave energy is concentrated along the headlands, they are eroded more rapidly than the bays; thus, in time the coast becomes smoother and less indented.

the land as horizontal erosion continues. Other erosional features found in association with wave cut cliffs are *sea caves, sea arches,* and *stacks* (Figs. 7.16 and 7.19). An abrasian platform is formed as the cliffs retreat. Some of these wave cut benches are bare or partially bare but, by far, most are covered with sediments that are gradually being transported from the shore to deeper water.

The sweeping action of waves constantly carries eroded debris across an expanding wave cut bench and deposits it at the platform's lower end, thereby producing a wave built terrace. When the terrace is narrow only scant beach deposits—usually stripped away with the passage of each storm—are present. As the bench widens, however, the waves reaching the land lose considerable energy through bottom drag, and permanent beaches become part of the shore profile.

The *beach* is thought of by most people as a place to lie on the sand and enjoy the sea and sun. Technically, a beach is a depositional landform consisting of sand, and sometimes wave-worn cobbles, lying along a coast in a position extending from the low-tide line to an inland limit demarcating the highest point reached by storm waves. Included in this category are a variety of sand reefs that extend from promontories into open water, such as spits, bars, and barrier islands. Spits and sand bars are essentially extensions of mainland beaches wherein the sand has been transported parallel to the coast by longshore currents and beach drifting. *Long-shore currents* are water flows within the surf zone that flow parallel to the shore. They are formed as oblique waves pile water against the shoreline. *Beach drift* results from the motion of water in the surf. The *swash* (forward motion) of the surf carries materials obliquely forward, while the *backwash* (related to gravity) carries sediments straight down the beach's slope. Thus, debris is carried forward in a zig-zag manner. Pebbles identified and timed have been recorded to move more than 800 m (2600 ft) per day by beach drift.

Barrier islands differ from spits and bars by

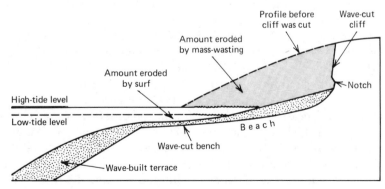

FIGURE 7.18
PRINCIPAL FEATURES OF THE SHORE PROFILE.

FIGURE 7.19
SEA STACKS AND A SEA ARCH AT THE CLIFFS OF ETRETAT (CHANNEL COAST), FRANCE.

being separated from the mainland by broad lagoons. Barrier islands are low sandy ridges initially built by wave action and subsequently increased in height by the growth of sand dunes.

From the preceding discussion, it might seem obvious that coastal areas are regions in which energy for erosion is at relatively high levels. This is true. It is also a fact that coastlines experience a cycle of erosion in which their landforms are being continuously transformed. A typical erosion sequence is shown in Figure 7.20. Remember, however, that the cycle shown is dependent upon relatively stable conditions. If sea-level should rise or fall, the shoreline would either retreat seaward or move landward. This would give rise to a whole series of adjustments, causing erosion in some places and deposition in others. It should also be kept in mind that the speed with which a coastline is altered is largely dependent upon the type of rock materials present. Resistant rocks change form much more slowly than do less resistant materials.

Coastal zones have been favored for human settlement, though the reasons for living in them have changed through time. Initially, the reasons were proximity to oceanic food resources and a reliable source of salt. Later, as civilizations began to flourish and trading became established, cities sprang up in coastal zones from which goods could be dispatched and received. Many ancient coastal settlements established for the purpose of supporting either a fishing industry or international trade still function today. In recent decades, settlement in the coastal zone of the United States has taken place at an unprecedented rate. The new motive is recreation, "going to the beach." The recent impact of these densely clustered populations on coastal zone resources has become a matter of grave concern. Water and land pollution, increased erosion of shorelines, diminished availability of groundwater supplies, sewage disposal limitations, and degradation of the coastline's aesthetic character through commercialization are but a few of the problems facing coastal populations.

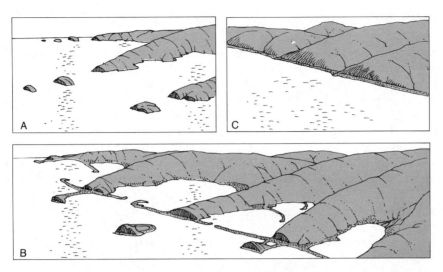

FIGURE 7.20

CYCLE OF COASTAL EROSION. (*A*) Coast at start, with headlands, islands, and deep bays. (*B*) Beaches and spits are built; cliffing reduces headlands to short stumps; spits join to form bay barriers. (*C*) Headlands and bays are eliminated; shoreline has become nearly straight.

Coastal zones are fragile environments subject to rapid and often catastrophic events. How to balance the demand for coastal land with wise land use in the coastal zone is a problem that can only be resolved through comprehensive planning, encompassing a thorough understanding of the physical characteristics of the coastal zone and the processes, physical and human, acting upon it.

WIND FORMED LANDSCAPES

Wind erodes rock and transports unconsolidated materials in areas where vegetation is sparse. This may suggest that wind formed landscapes are associated with desert regions. Although "windscapes" are most prevalent in arid climatic regimes, the majority of desert landforms are produced by running water. And, as with stream erosion, the work of wind in modifying land surfaces transcends climatic boundaries. True, wind is active in arid climates, but it also shifts the sand of beaches in humid regimes, moves finely-ground debris peripheral to glaciers, and erodes soil from the farmer's freshly plowed field.

Wind erodes the land through *deflation* and *abrasian*. Loose particles may be lifted or rolled along by the wind, a process called deflation. Since air's density and ability to lift matter is relatively low, only fine sediments such as clays and

silts are normally carried away by lifting. But, under the right circumstances, enormous quantities of surface matter may be redistributed over considerable distances (Fig. 7.21). During the drought of the 1930s, it is estimated that areas within the Dust Bowl were lowered in places by as much as 1 meter (3 feet) through deflation. Light soils originating in areas from Oklahoma to Colorado are

FIGURE 7.21

DUST STORM IN BACA COUNTY, COLORADO.

From Strahler *Modern Physical Geography* © 1978, John Wiley & Sons, New York.

known to have been transported as far as New England and deposited as a dirty film on snow. Heavier particles, such as sand grains, cannot be lifted very high. They move by either rolling or skipping across the surface, but only after winds attain moderately strong velocities (Fig. 7.22). Blowing sand travels close to the ground, seldom rising over 2 meters (6 feet) above the surface. When wind speeds are high, small pebbles may even roll along in the same fashion. Rarely, however, do they move very far before becoming lodged.

The most obvious landscape features produced by deflation are broad, shallow depressions, known as *blowouts* or *deflation hollows*. There are thousands of blowouts within the Great Plains region of the United States. They are normally less than 1 meter deep, with horizontal dimensions ranging from a few meters to several kilometers across. Most blowouts originate in small depressions in the surface of plains. Conditions for their formation are especially enhanced where the grass cover has been disturbed. Rains intermittently fill the depressions, creating temporary lakes or ponds. Later, as water evaporates, the muddy bottom of the depression dries out, cracks, and leaves

small scales of mud exposed. The small fragments are, in turn, removed by the wind in a continuous cycle that keeps enlarging the hollow.

In the majority of earth's desert regions the surface is covered by a layer of coarse pebbles and gravel. This layer, called *desert pavement*, or *reg*, forms when fine materials are deflated and coarse fragments remaining are jostled about by wind action until they fit into a tight mosaic. Although possibly taking hundreds of years to form, desert pavement, once established, effectively protects the surface from further extensive deflation.

When armed with sand, the lowest layer of air can erode through *abrasion*, a grinding or sandblasting action. Abrasion is most effective within 30 to 60 centimeters (1 to 2 feet) of the surface, where the impact of sand grains is at a maximum. In areas where lower-layer winds have an ample supply of sand, wind-carved notches are prominent in unprotected fence posts and telephone poles; rock formations are pitted, grooved, or hollowed out; hard rock surfaces are smoothed out and polished; and *ventifacts*—wind-blasted cobbles with faceted surfaces joined in sharp edges—are frequently present (Fig. 7.23).

Material removed from one location by ero-

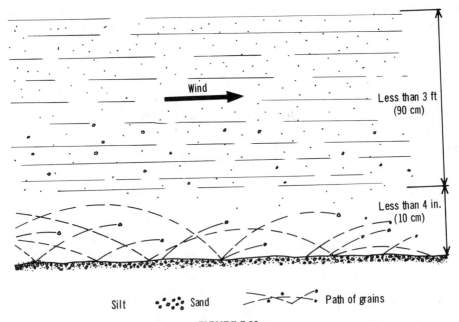

Silt   Sand   Path of grains

FIGURE 7.22

SAND MOVEMENT. Sand grains travel along the surface, near the ground, in a skipping motion.

FIGURE 7.23
A WIND ERODED ROCK FORMATION IN NATURAL BRIDGES NATIONAL MONUMENT, UTAH.

sional processes is deposited in another. The depositional landforms created by winds are far more impressive than those landforms which they erode. Based upon the size of material laid down, such deposits can be classified into two broad groups: (1) accumulations of sand, and (2) deposits of fine silt.

Dunes are sandy deposits that form when rapidly moving air, laden with sand particles, encounters an obstruction. The obstacle may be anything: a rock, a bush, or an old, deserted cabin. As wind passes about the obstacle, it sweeps over and around it, but leaves pockets of slower moving air both immediately behind and in front of the obstacle (Fig. 7.24). In these low velocity pockets, sand grains drop out to form mounds. The mounds themselves become obstacles to air flow, causing more sand to drop out about them. Even-

tually, the mounds may grow above the height of the obstacle and coalesce to form a dune.

A dune is not symmetrical; rather, it has a steep, lee slope and a gentler windward slope (Fig. 7.25). Many dunes are considered "live;" that is, they move. As illustrated in Figure 7.26, the processes that cause sand grains to roll and skip along with the wind also shift rearward dune sands forward. In places this forward movement may amount to as much as 20 meters (60 feet) per year. Such dune migrations, especially along sandy coastlines, have been known to bury cottages and threaten the existence of towns. Yet, when vegetation establishes itself on dune surfaces, forward motion ceases, since the plants anchor sand grains in place.

Dunes occur in a multitude of sizes and shapes. Many grow to heights between 30 to 100 meters (100 to 300 feet), with a few attaining heights of

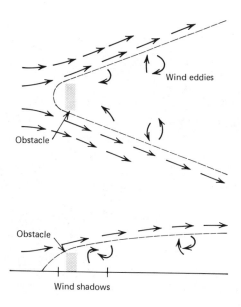

FIGURE 7.24
WIND SHADOW. The shaded area indicates where an obstacle produces a reduction in wind speed, or a wind shadow. Within wind shadows, wind velocity is low and air movements are marked by eddies. It is within this zone that sand grains accumulate.

200 meters (600 feet) or more. Their shapes can be simple or complex, depending largely upon the consistency of wind direction and strength. Figure 7.27 illustrates the most common forms of dunes.

Sand deserts are much less common than reg deserts. Nevertheless, sand deserts are expansive in certain regions of the earth. Probably the best-known are the sand deserts, referred to as *ergs*, that cover a little over 10 percent of the African Sahara. Others include the sand deserts of Arabia, central Asia, and Australia, along with a few small scattered sand deserts in the Americas.

Loess represents another form of wind deposited sediment (Fig. 7.28). Unlike the sand dunes,

FIGURE 7.25
BARCHAN SAND DUNES IN SAUDI ARABIA.

its chief component is silt, but it often includes some clay and fine sand particles as well. While dunes are located in close proximity to the source of their materials, loess is frequently deposited thousands of kilometers from the point where winds first lifted their airborne load. Loess lacks any layering associated with variable sediment sizes. Its soils tend to be fertile and, where water is available, well suited to the cultivation of grains. Deposits have a uniform buff color.

Loess originates in deserts and the sediments of streams formed in the melting of glaciers. The deep loess cover of western China covers an area more than 800,000 square kilometers (309,000 square miles). It is derived from the desert floors of central Asia, where an abundance of mechanically weathered sediments are present. Glacial

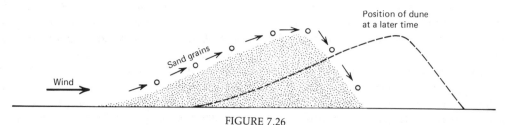

FIGURE 7.26
DUNE MIGRATION. Sand dunes migrate along the surface as sand is rolled up the windward side of the dune and then tumbles down the lee slope.

Kind	Definition and Occurrence	Illustration (Arrows indicate wind directions)
Beach dunes	Hummocks of various sizes bordering beaches. Inland part is generally covered with vegetation.	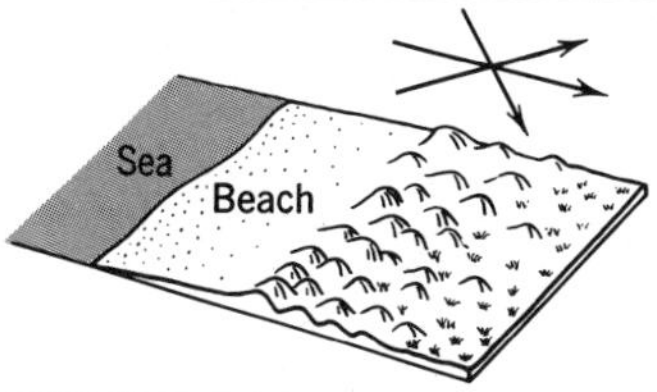
Barchan dune	*A crescent-shaped dune with horns pointing downwind.* Occurs on hard, flat floors in deserts; constant wind, limited sand supply. Height 1m to more than 30m.	
Transverse dune	*A dune forming a wavelike ridge transverse to wind direction.* Occurs in areas with abundant sand and little vegetation. In places grades into barchans.	
U-shaped dune	*A dune of U-shape with the open end of the U facing upwind.* Some form by piling of sand along leeward and lateral margins of a growing blowout in older dunes.	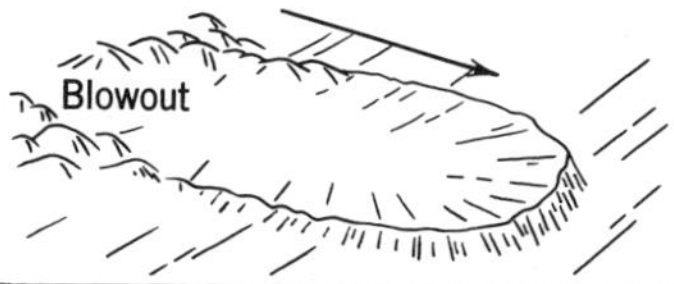 Blowout
Longitudinal dune	*A long, straight, ridge-shaped dune parallel with wind direction.* As much as 100m high and 100km long. Occurs in deserts with scanty sand supply and strong winds varying within one general direction. Slip faces vary as wind shifts direction.	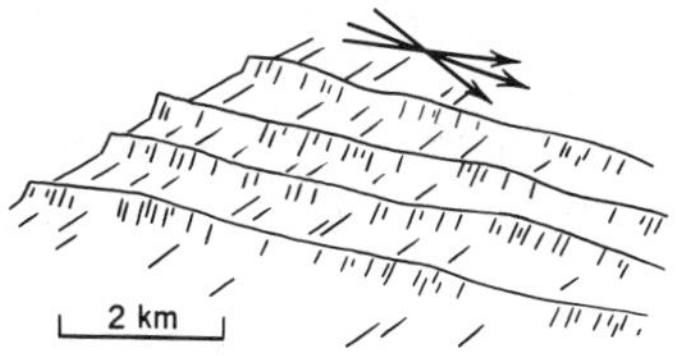 2 km

FIGURE 7.27
SAND DUNE TYPES BASED ON FORM.

loess is more prominant in North America and Europe. During the glacial ages, the climate near the margins of ice sheets was cold and windy. Rivers that formed from the melting ice were choked with debris and filled their valleys so rapidly that plants could not establish a foothold within them. Hence the floors of the valleys remained bare and were easily deflated. Their fine-sized particles were lifted by strong air currents and carried downwind, settling out in blankets 8 to 30 meters (26 to 100 feet) thick near source valleys, and thinning further downwind to thicknesses of 1 to 2 meters (3 to 6 feet). There they settled in more stable environments and were subsequently covered either by grass or forest that protected them from further deflation.

Landforms are the basic structural units of the earth's outer crust. Roughly fashioned by forces sufficiently intense and pervasive to move continents, their ultimate shape is sculpted by humble raindrops, snowflakes, the wind, and gravity. The processes effective in etching surface details also give rise to the development of soils and support their mantle of vegetation. The character of the soil can limit or enhance the growth of plants anchored within it. Vegetation, in turn, can affect soil properties and reduce the susceptibility of the ground to erosion. All the components of earth's physical environments are intimately interrelated in a complex web of process and form. Herein, man makes his home.

Although present for only a very small portion of earth's history, mankind has had an impact upon the physical landscape greater than that of

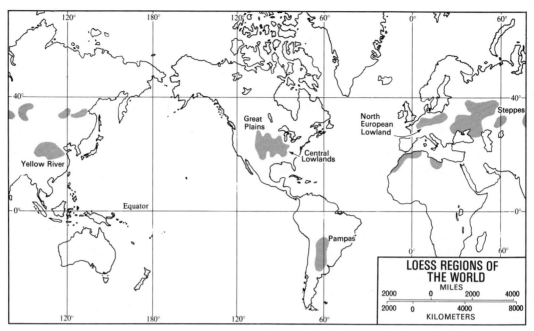

FIGURE 7.28
LOESS DEPOSITS OF THE EARTH.

any other organism. In accordance with his attitudes, objectives, and his degree of technological skills, man uses the natural resources available to him to meet his primary needs. At the same time he makes imprints upon his environment that are indicative of his existence and which provide his domain with a character that uniquely exhibits his presence. The geographic elements related specifically to man's culture and economic systems are the subject of the chapters that follow.

KEY TERMS AND CONCEPTS

glacier
snowfield
mountain glacier
glacial trough
glacial abrasion
glacial plucking

fiord (fjord)
hanging valley
cirque
arete
erratic
horn

till
stratified drift
moraine
outwash plain
drumlin
esker
iceberg
density current
surface currents
cold current
warm current
oceanic upwelling
tides
tidal bores
characteristics of ocean
    waves
surf
wave refraction
coastal landforms
wave cut cliffs and
    benches
sea caves

sea arches
sea stacks
beach
barrier island
spits
bars
longshore currents and
    beach drifting
coastal cycle of erosion
windformed
    landscapes
deflation
wind abrasion
blowout (deflation
    hollow)
desert pavement
reg
ventifact
dunes
windshadow
erg
loess

DISCUSSION QUESTIONS

1. What is a glacier?

2. Why has the earth experienced times of wide-spread glaciation? Discuss some of the ramifications of another ice age.

3. Describe landform features associated with mountain glaciers; with continental glaciers.

4. Discuss economic factors associated with glacial landforms.

5. Explain: density currents; oceanic upwelling; tides.

6. Describe the general circulation patterns of the world's oceans. Explain where cold and warm currents are normally found and what effect they have on global climates.

7. How do water molecules travel in deep oceanic waves?

8. What is wave refraction?

9. Define: sea stacks and arches; spits; bars; barrier islands.

10. Discuss the cycle of landform changes resulting from coastal erosion.

11. What is the difference between deflation and abrasion?

12. How do blowouts form? How can blowouts affect land use?

13. Describe the differences between a reg and erg desert.

14. Describe the evolution and development of sand dunes.

15. What is loess? Where are the most important regions of loess located in the world? How does loess originate?

REFERENCES FOR FURTHER STUDY

Bagnold, R. A., *The Physics of Blown Sand and Desert Dunes*, Methuen and Co., London, 1941.

Flint, Richard F., *Glacial and Quarternary Geology*, John Wiley and Sons, New York, 1971.

Gautier, E. F., *Sahara: The Great Desert Dunes*, Methuen and Co., London, 1941.

King, C. A. M., *Beaches and Coasts*, 2nd ed., St. Martin's Press, New York, 1972.

Paterson, W. S. B., *The Physics of Glaciers*, Pergamon Press, Oxford, England, 1969.

Weyl, Peter K., *Oceanography*, John Wiley and Sons, New York, 1970.

8
POPULATION

A SPACE TRAVELER VISITING THE EARTH FOR THE FIRST time would find considerable unevenness in the distribution of various phenomena over the earth's surface. Of the earth's 510 million square kilometers (197 million square miles) of surface area, only 148 million square kilometers (57 million square miles), or 29 percent, consist of exposed continental surfaces. Upon closer examination, our traveler would note that 80 percent of the land surface is located in the Northern Hemisphere. Just as the earth's land area is unevenly distributed, so too is its population. Man has congregated in certain areas while others remain virtually empty. An examination of a population map will show that most of the world's people live on plains areas along the margins of continents (Fig. 8.1). In fact, nearly 90 percent of the world's population lives on about 10 percent of the land. The area where people live has been called the *ecumene*. The *nonecumene* consists of those portions of the earth's surface that are largely uninhabited. It should be noted that advances in civilization and technology have permitted man to expand the ecumene. The spread of European peoples during the past five centuries fostered the intensive use of lands formerly used in an extensive manner by nomadic and subsistence folk. Irrigation, windmills, the steel plow, and other innovations have brought new lands into use and expanded the ecumene.

The ecumene is not expanding as rapidly as it once did. The obstacles to expansion are largely climatic; unoccupied lands are usually too cold, dry, hot and wet, and in some cases, too high or rugged in slope and relief.

THE NONECUMENE

High latitude and high altitude (cold) lands present many climatic barriers to support human life. Farming is virtually impossible because of a short growing season, permanently frozen subsoils, poor drainage, and unsuitable soils. With the exception of externally supplied mining camps, trading posts, and military installations, it is unlikely that the cold lands will experience widespread settlement in the future (Fig. 8.2).

Dry lands occupy vast areas in Africa, Asia, Australia and, to a lesser extent, North and South America. A lack of reliable precipitation explains why dry land settlement is to be found in scattered oases or along the banks of perennial streams where fresh water can be used for irrigation. Multipurpose irrigation schemes have permitted settlement of some dry land areas. The recently completed Aswan High Dam project in Egypt expanded that country's irrigated land. Not all dry lands, however, are accessible to vast amounts of fresh water, and not all nations with surplus water are willing to share it with less fortunate neighbors. Desalinized water is used in some municipalities,

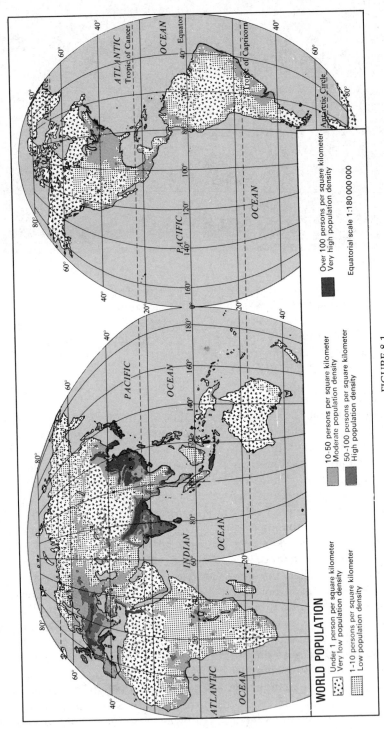

FIGURE 8.1
WORLD POPULATION.

WORLD POPULATION

Under 1 person per square kilometer
Very low population density

1-10 persons per square kilometer
Low population density

10-50 persons per square kilometer
Moderate population density

50-100 persons per square kilometer
High population density

Over 100 persons per square kilometer
Very high population density

Equatorial scale 1:180 000 000

FIGURE 8.2

SETTLEMENTS IN COLD LANDS OFTEN SURVIVE THROUGH EXTERNAL SUPPLIERS. This is a view of Hallet Station, Antarctica.

but is too expensive at this time for irrigating basic food crops. Costly dry land reclamation projects will continue to be pursued by some nations, but the impact of these on the world's settlement frontier will remain limited.

Hot and wet lands are found in the tropical areas of Latin America, central Africa, and southern Asia. One geographer has observed that the wet tropics, unlike the cold and dry lands, ". . . are affected with a superabundance of climatic energy in the forms of solar radiation, heat, and precipitation." It is this climatic energy that has retarded settlement in some areas. Excessive heat and rainfall are responsible for leached soils that deteriorate rapidly under cultivation. The hot, wet lands are something of a population anachronism because some areas, especially in Asia, are densely populated, while others (e.g., in Latin America) are sparsely inhabited. These differences suggest that some tropical areas can support larger populations provided that suitable land use systems are adopted. The opening of Brazil's Transamazonica Highway reflects that country's attempt to encourage settlement in its wet tropical lands.

THE ECUMENE

The world's more than 4 billion people are clustered in four concentrations, three of which are located on the Eurasian continent. The fourth and smallest concentration is found in east central Anglo-America. Three fourths of the world's people live on the periphery of Eurasia while the drier and rugged lands of the interior are sparsely populated. The largest human concentration is found in East Asia, consisting of China, North and South Korea, Japan, and Taiwan. The center of this cluster is located at 35° N. latitude in a climatic regime similar to that found along the east coast of North America. The East Asia cluster has about 25 percent of the world's population.

The second concentration is located in South Asia and is the only one that is tropical in location. Centering on India, it extends into Pakistan, Bangladesh, and Sri Lanka. The more than 900 million people in this area make up about 20 percent of the world's total. The two Asiatic clusters together represent about 45 percent of the world's inhabitants. The most outstanding characteristic of these two areas is that they consist of less developed countries (exclusive of Japan) that are overwhelmingly rural and whose farmers are engaged in subsistence agriculture. High birth rates, declining death rates, and large annual increases in population are typical of most countries in these areas.

Centered at 50° N. latitude, the European cluster extends from Great Britain in the west to the Ural Mountains in the Soviet Union. This population axis closely parallels Europe's principal coal fields—a resource that was crucial to the development of the Industrial Revolution. Early industries were located near the coal deposits; the locations of many European cities mirror this early dependence upon coal. The three Eurasian clusters collectively account for nearly 65 percent of the world's total population.

The fourth largest population cluster is in east central Anglo-America and includes *Megalopolis,* that large urbanized seaboard area that extends from Boston to Washington, D.C. The European and Anglo-American agglomerations are part of the developed world. Annual birth rates are lower than the world average, death rates are also low, and life expectancy is high—as are levels of technology. Most of the population is urbanized.

Seven of the 10 most populous countries of the world are found in these four major population concentrations (Table 8.1). Other populous nations are found outside of the four large clusters. One such nation is located in the southeast Asian country of Indonesia, especially on the Island of Djawa (Java). The most populous countries in Africa are Nigeria and Egypt. In Latin America, Brazil and Mexico have large populations; each is a less develped nation, with birth rates and economies characteristic of such areas.

POPULATION DYNAMICS

The characteristics of the world's population patterns are determined by a variety of demographic components, among which are distribution and density, fertility and growth rates, birth and death rates, and age-sex structure in a population.

Distribution and Density

We have seen that the world's population is distributed irregularly. This means that some areas have very high population densities (the number of people per square kilometer or mile) while others, particularly the nonecumene lands, have very

TABLE 8.1 ESTIMATED POPULATION OF THE WORLD'S TEN LARGEST COUNTRIES (millions)

1932		1980		2000	
1. China	425	China	975	China	1212
2. India (br)	360	India	676	India	976
3. USSR	160	USSR	266	USSR	311
4. United States	125	United States	223	United States	260
5. Japan	66	Indonesia	144	Indonesia	219
6. Germany	66	Brazil	122	Brazil	205
7. Indonesia (neth)	63	Japan	117	Bangladesh	157
8. United Kingdom	46	Bangladesh	91	Nigeria	149
9. France	42	Pakistan	87	Pakistan	152
10. Italy	42	Nigeria	77	Mexico	129

Source. United Nations *Demographic Yearbook,* 1948; Population Reference Bureau, *World Population Data Sheet, 1980.*

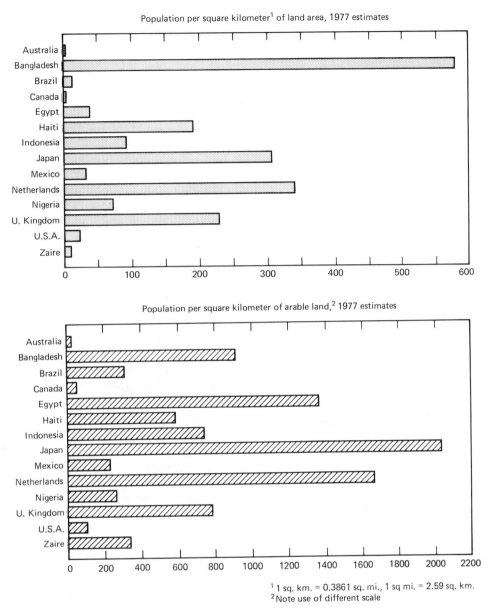

Population per square kilometer[1] of land area, 1977 estimates

Population per square kilometer of arable land,[2] 1977 estimates

[1] 1 sq. km. = 0.3861 sq. mi., 1 sq mi. = 2.59 sq. km.
[2] Note use of different scale

FIGURE 8.3

POPULATION DENSITY IN SELECTED COUNTRIES. Source: Population Reference Bureau, Inc., Washington, D.C.

low densities. The world's population density (or *arithmetic density*) is about 27 people per square kilometer (70 per square mile). The figure for the United States is about 23 per square kilometer (60 per square mile). Figures such as these though often used, can be misleading, because they include the world's uninhabited areas. A case in point is Egypt, where the population is concentrated along the Nile valley and delta. Egypt has

an arithmetic density of about 38 people per square kilometer (100 per square mile). This figure, by itself, tells us very little. It would be more meaningful to exclude unusable land, which in the case of Egypt is the desert that accounts for about 96 percent of total land area. Density ratio that includes only arable land is called the *nutritional* or *physiological* density. Egypt's nutritional density is about 1400 people per square kilometer

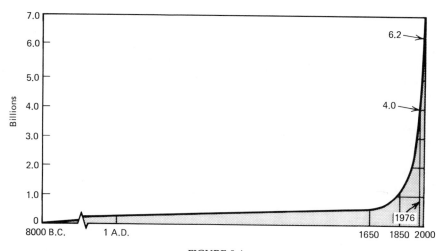

FIGURE 8.4
WORLD POPULATION GROWTH THROUGH TIME, 8000 B.C. TO 2000 A.D. Source: Population Reference Bureau, Inc., Washington, D.C.

(3600 per square mile). Even this ratio, however, can be misleading. It does not include productive forests or pasture, and does not take into account the productivity of various arable lands. In spite of these inadequacies, the nutritional density figure is more meaningful than a simple arithmetic density. Figure 8.3 illustrates the significance of nutritional densities.

World Population Growth

Figure 8.4 depicts population growth through time. For thousands of years human populations remained stable or expanded only slightly. Estimates indicate that there were about 250 million people living at the birth of Christ. It took 1850

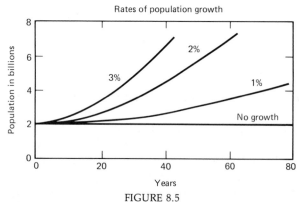

FIGURE 8.5
POPULATION GROWTH AT VARIOUS RATES. Source: Population Reference Bureau, Inc., Washington, D.C.

years for man's numbers to reach the first billion. Then in only 80 years this figure doubled to 2 billion in 1930. It took less than 50 years to double again to its present 4 billion. If the world's present annual growth rate of 1.7 percent is maintained, population will double again in nearly 40 years. Doubling time at various rates of growth are shown in Figure 8.5. The current estimate for the year 2000 is a world with 6.2 billion inhabitants. This figure assumes that the current annual growth rate will continue, which would result in an additional 135 people per minute or about 194,000 per day; an annual addition of 70 million people. According to the latest United Nations estimates, the growth rate of the world is projected to decline to about 1.5 percent by the year 2000. Is this a new trend? Evidence suggests that the annual world population growth rate is no longer increasing and, in fact, may be beginning a gradual decline from a high of 1.9 percent during the 1960s. The rate has recently declined to 1.7 percent, the 1950 level (Fig. 8.6).

A population that doubles periodically is increasing *exponentially* or *geometrically* as in the series of numbers 1, 2, 4, 8, 16, . . . Exponential growth can be illustrated by the familiar story of the boy who requested a monthly allowance that began with a penny on the first day and two cents on the second, four cents on the third, and so on. This arrangement would have netted the boy $10,737,418.24 on the 31st day of the month. According to the "law of 70," doubling time for a

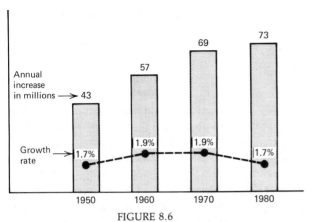

FIGURE 8.6
WORLD POPULATION GROWTH. Source: Population Reference Bureau, Inc., Washington, D.C.

population with an exponential growth is calculated by dividing the percentage of growth into 70. Therefore, a population growing at one percent would double in size in 70 years; at 2 percent in 35 years; at 3 percent in 23 years, and so forth. The United States in 1979 had a natural increase of 0.6 percent, which, if continued, would double its population in 116 years. The last United States doubling period, between 1921 and 1978, took just 57 years.

Not all regions of the world are growing at the same rate. Table 8.2 reveals that Latin America, Africa, and Asia have the highest percentage growth rates. These regions are dominated by the less developed nations, those that can least afford

enormous population increases. Most less developed countries are growing by at least two percent a year, while many have rates as high as 3.4 and 3.5 percent. Today 76 percent of the world's inhabitants live in these countries; if these rates continue until the year 2000, 82 percent of all mankind will be included.

The Soviet Union, North America, Europe, and Japan have distinctly lower growth rates. Most developed nations are found in these areas. These regions tend to be industrialized and urbanized. Their citizens enjoy most of the amenities associated with a high standard of living: high literacy rates, balanced diets, adequate housing, and high per capita incomes. Most developed countries are growing at an annual rate of less than one percent.

Our world, then, may be viewed as consisting of two types of countries: those with high standards of living and low fertility rates, and those with low standards of living and high fertility rates. Lester Brown, in *World Without Borders*, 1972, succinctly expresses this point: "Our world is in reality two worlds, one rich, one poor; one literate, one largely illiterate; one industrial and urban, one agrarian and rural; one overfed and overweight; one hungry and malnourished; one affluent and consumption-oriented; one poverty-stricken and survival-oriented."

In 1932, seven of the world's 10 most populous countries belonged to the developed world. By the year 2000, eight of the top 10 will be less developed nations (Table 8.1). How can this rank order shift so dramatically? The answer lies in different growth rates. We can illustrate the point by looking at recent population changes in Mexico and the United Kingdom (Table 8.3). Mexico has a very high natural increase rate of 41 per 1000 pop-

TABLE 8.2 AVERAGE ANNUAL GROWTH RATES
FOR WORLD REGIONS

(1980)	
Africa	2.9
Latin America	2.6
Asia	1.8
Oceania	1.1
USSR	0.8
North America	0.7
Europe	0.4
World	1.7

Source. Population Reference Bureau, *World Population Data Sheet,* 1980.

TABLE 8.3 POPULATION GROWTH OF MEXICO
AND THE UNITED KINGDOM (millions)

Year	Mexico	United Kingdom
1932	16.6	46.0
1979	67.7	55.8
2000[a]	132.0	56.6

[a] Assuming current rate of natural increase of 3.4 percent for Mexico and near zero percent for the United Kingdom.

ulation. The United Kingdom has almost reached a zero population growth rate. The difference in growth rates between the two countries accounts for the fact that Mexico's population, once one-third that of the United Kingdom, will probably be double the latter's by 2000.

Can the earth sustain its present growth rate? Can 5 billion people be adequately fed, educated, housed, and employed in 1989 and 6.2 billion in 2000? Can simple amenities be offered to billions more at a future date when billions are without them today? A declining growth rate is necessary. There is evidence that this has occurred over the past few years. All too often, however, much of this decline has taken place in the developed countries. More less developed nations are recognizing the need for family planning and population education. Some have implemented programs designed to eliminate large families. In other countries enforced sterilization programs have been suggested for family members with more than three children. A wide range of programs are needed in order to reduce growth rates, particularly in the less developed countries.

Birth and Death Rates

Whether a country's population increases or decreases is determined by the interaction of three factors: birth rate, death rate, and migration. Both birth and death rates are the ratio of birth and deaths to the total population, in a specific year, per 1000 population. The ratios are shown as follows:

$$\text{Birth rate} = \frac{\text{births per year}}{\text{population}} \times 1000$$

$$\text{Death rate} = \frac{\text{deaths per year}}{\text{population}} \times 1000$$

The rate of natural increase is the difference between birth and death rates and measures the degree to which a population is growing due to reproduction. This ratio is shown below:

$$\text{Rate of natural increase} = \frac{(\text{birth rate} - \text{death rate})}{10}$$

Annual growth rates are computed by adding natural increase and net migration and dividing by

10 to get a percentage. This ratio is shown as follows:

$$\text{Annual growth rate} = \frac{(\text{natural increase} \pm \text{net migration})}{10}$$

What do we mean when we speak of high birth and death rates as opposed to low birth and death rates? What is a moderate rate of natural increase? For our purposes birth rates greater than 35 are considered high, those between 25 and 35 are moderate, and birth rates below 25 are considered low. The world birth rate of 28 is, therefore, considered a moderate rate (Table 8.4). Death rates greater than 25 are high, those between 15 and 25 are moderate, and those below 15 are low. In this case, the world death rate of 11 is considered to be low. A rate of natural increase greater than 3.0 percent is high; a moderate rate lies between 1 and 3.0 percent; and a natural increase rate of less than 1.0 percent is considered low. What is the world's current rate of natural increase?

Table 8.4 depicts birth and death rates and the rate of natural increase for the world, its six major regions, and selected countries. Again, it will be noted that the developed regions and countries have the lowest birth rates and the lowest rates of natural increase. The high birth rates and the declining death rates in the less developed nations account for their high natural increase rates.

Throughout most of human history both birth and death rates were high. Hunger and disease meant that many infants did not reach parenthood. Europe's population, for example, was reduced by 25 percent in the 1350s by an outbreak of bubonic plague. Societies needed high fertility rates in order to maintain their numbers. Shortly after the Industrial Revolution, however, improvements in medicine and sanitation lowered death rates. Following the decline in death rates (and particularly in infant mortality), birth rates began to decline. During the Industrial Revolution many persons moved from rural to urban areas and the need for larger families was lessened. The shift from high birth and death rates to lower rates has been called the *demographic transition*.

Figure 8.7 depicts the transition's four distinct stages. During the first stage, both high birth and death rates produce a period of low growth. This stage is typical of preindustrial agrarian countries

TABLE 8.4 DEMOGRAPHIC CHARACTERISTICS OF REGIONS AND SELECTED NATIONS

	Birth rate	Death rate	Natural increase
World	28	11	1.7
Developed regions	16	9	0.6
North America	16	8	0.7
Europe	14	10	0.4
Soviet Union	18	10	0.8
Less developed regions	32	12	2.0
Africa	46	17	2.9
Asia	28	11	1.8
Latin America	34	8	2.6
Developed countries			
United States	16	9	0.7
Japan	15	6	0.9
France	14	10	0.3
Less developed countries			
Nigeria	50	18	3.2
India	34	15	1.9
Brazil	36	8	2.8

Source. Population Reference Bureau, Inc., *World Population Data Sheet,* 1980.

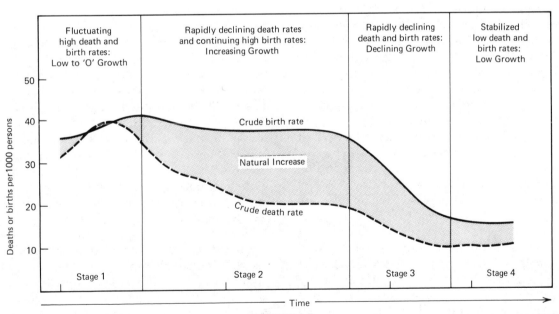

FIGURE 8.7
STAGES IN THE DEMOGRAPHIC TRANSITION MODEL.

where birth and death rates balance each other. The second stage is one of high birth rates, declining death rates, and a rapid increase in population. The falling death rate reflects the spread of scientific and medical knowledge. Better sanitation facilities also contributed to this downward trend. (All societies welcome changes that reduce deaths, but changes that reduce birth rates tend to be accepted more slowly.) The third stage occurs when the fertility rates drop to decrease the large gap between birth and death rates. In this stage people make private decisions to deliberately control the number of births, and there is a slowing in the rate of population growth. A slowdown of this kind characterizes most of today's developed nations. In the fourth stage, birth rates and death rates stabilize at a low level. Growth rates are slow, and population totals tend to be stationary.

The demographic transition has not been duplicated in the less developed countries of Latin America, Africa, and Asia. After World War II, western technology for death control was increasingly made available to underdeveloped nations. Improved public health measures brought about a drop in mortality but not in fertility. In other words, the death rate has been dropping precipitously while birth rates have remained high. The result is a rapid increase in the population of the world's less developed countries. This unprecedented decline in the death rate has caused the world's population explosion. It has been suggested that all societies will pass through a European-type demographic transition as they modernize. But a question remains as to whether or not the world can afford to wait out the natural sequence of the demographic cycle.

Figure 8.8 shows that fertility levels appear to be closely associated with economic development. The world pattern of fertility suggests that attitudes toward family size change slowly. Control over death rates has been easier for populations to accept than control over birth rates. In some societies birth control may conflict with treasured value systems. In others, children may be viewed as economic assets and may also serve as a form of old-age insurance. Times, however, are changing. About 25 years ago no less developed country had a population policy. Since the mid-1960s, many have adopted policies and programs designed to reduce high birth rates. By 1980 the majority of

the world's people were living in countries with positive family planning programs. Evidence from many sources indicates a decline in fertility in many parts of the less developed world. These declines are the rule rather than the exception in countries with strong family planning programs. Indeed, all countries having vigorous family planning programs have experienced considerable decreases in fertility. The Peoples Republic of China, for example, has recently been hailed as "family planning's greatest success story." Among the less developed regions, Africa remains the major new frontier for family planning programs and fertility reduction. The 1965–1975 decade may be thought of as a period of world awakening to the problems of rapid population growth. The new attitudes will undoubtedly encourage smaller families. Nevertheless, the youthful age structure of many developing countries will guarantee continued high growth rates for many years to come.

Age-Sex Structure of Populations

The age-sex structure of a population is the proportion by sex of people at each age. Fertility and mortality rates determine a population's age structure. It is important that nations and political leaders have information about age structure in their populations. A country with a large number of young people will certainly be required to allocate more of its resources to the building of schools and the employment of teachers, for example.

The age-sex structure of a country is represented by a population pyramid that provides an easily understood visual image (Fig. 8.9). The vertical axis represents the different age groups, the horizontal axis and bars show the percentage of people in each age group by sex. Male populations appear on the left side, female on the right. Countries with high birth and death rates have age-sex pyramids that look like an ancient Egyptian pyramid. Countries with low birth and death rates have elongated population pyramids.

Mexico's rapid population growth and declining mortality result in a large number of young people. The small number of elderly reflect the high death rates of the past. This pattern is typical of many less developed nations. Mexico's young population places a severe strain upon the nation's economy. First, this sizable age group (15 years and under), representing 46 percent of Mex-

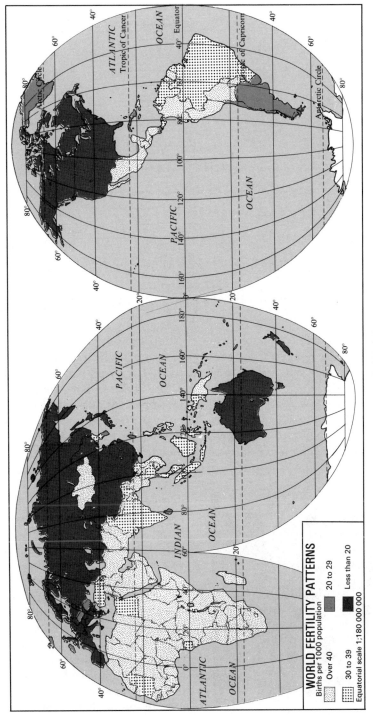

FIGURE 8.8
WORLD FERTILITY PATTERNS.

WORLD FERTILITY PATTERNS
Births per 1000 population
Over 40   20 to 29
30 to 39   Less than 20
Equatorial scale 1:180 000 000

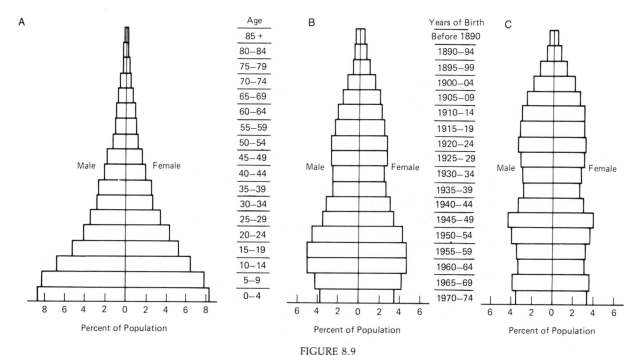

FIGURE 8.9

AGE-SEX POPULATION PYRAMIDS. (*A*) Rapid growth (Mexico). (*B*) Slow growth (United States). (*C*) No growth (Sweden).

ico's total population, is dependent upon the smaller economically active segment of society. Second, expensive educational and social programs must be provided for this age group (Fig. 8.10).

A fairly even distribution among age groups, as found in Sweden, is representative of an industrialized society that is approaching a zero popu-

FIGURE 8.10

LARGE NUMBERS OF DEPENDENT CHILDREN CAN PLACE A SEVERE STRAIN ON THE ECONOMICALLY ACTIVE SEGMENT OF A SOCIETY. These are Guatemalan school children.

lation growth rate. Following an extended period of low birth and death rates, Sweden has a relatively large number of older people. That is, most of the children born reach old age. Whereas only three percent of Mexico's population is age 65 and over, the figure for Sweden is 15 percent. Compare the young and old populations of Mexico and Sweden.

The United States has a slow population growth rate. Although it once had high birth rates, recent decades have witnessed a downward trend. This trend was interrupted briefly by the postwar baby boom that continued until 1961. This event significantly altered the age structure of the United States. The bulge near the base of the pyramid depicts this boom. The dependent young (under 15 years) in the United States represent 26 percent of the total population while the dependent elderly (65 years and over) equal 10 percent.

A number of factors besides birth and death rates can affect the shape of a pyramid. Heavy war losses will increase the death rate of young men; the bars representing that age group will consequently be smaller. Heavy migration patterns will also alter the age-sex ratios.

Several measurements are used to determine the "age" of a population. One is the median age; that is, half the population is older than the median age and half is younger. In 1975 the median age for Mexico was 16.9 years, obviously a very "young" country. In Sweden it was 35 years and in the United States it was 28.6 years. For the world as a whole the figure is 22.4 years, a figure reflecting the large number of young people in the less developed countries.

Another useful indicator of age structure is the *dependency ratio*. This ratio indicates the number of people who are dependent upon the economically active members of a society: people aged 15 through 64 years. The dependency ratio is shown as follows:

$$\text{Dependency ratio} = \frac{\text{Ages 0 to 14 plus 65 and over}}{\text{Ages 15 to 64}} \times 100$$

Obviously those countries with large numbers of young people have high dependency ratios. These are usually the developing countries who are least able to provide adequate education and social services to their citizens.

THE DEMOGRAPHER'S LANGUAGE

Age-dependency ratio. The ratio of persons in the ages defined as dependent (under 15 and over 64 years) to those in the ages defined as economically productive (15–64 years) in a population.

Antinatalist policy. Policy of government, society, social, or religious group aimed at slowing population growth.

Baby boom. The period following World War II from 1947–1961, marked by a dramatic increase in fertility rates and in the absolute number of births in the United States, Canada, Australia, and New Zealand.

Birth control. Measures of individuals, countries, or societies aimed at influencing—decreasing or increasing—numbers of births. The expression is often used synonymously with family planning, fertility control, and responsible parenthood.

Birth rate. The number of births per 1000 population in a given year. (Also known as crude birth rate.)

Growth rate. The rate at which a population is increasing (or decreasing) in a given year due to natural increase and net migration, ex-

pressed as a percentage of the base population.

Infant mortality rate. The number of deaths to infants under one year of age in a given year per 1000 live births in that year.

Life expectancy. The average number of additional years a person would live if current mortality trends were to continue. Most commonly cited as life expectancy at birth.

Natural increase. The surplus (or deficit) of births over deaths in a population in a given time period.

Pronatalist policy. Policy of a government, society, social, or religious group aimed at increasing population growth.

Rate of Population Growth. The annual rate of natural increase (birth rate less death rate) plus or minus net migration.

World Population Year. The year 1974 was proclaimed "World Population Year" by the United Nations to focus world attention on population problems and specifically on the UN World Population Conference, Bucharest, August, 1974.

Zero population growth. A population in equilibrium, with a growth rate of zero, achieved when births plus immigration equal deaths plus emigration.

Sex ratio is yet another measure of population structure. It is the number of males per 100 females. Most countries have sex ratios falling between 95 to 102 males to each 100 females. Sex ratios greater than 110 or less than 90 are considered imbalanced. War losses and migration can affect this ratio. Because more male babies are born than female babies, the sex ratio at birth ranges between 104 and 107. This difference is reflected in the bottom bar of the age-sex pyramids presented earlier. This trend quickly reverses itself since at all ages males are more likely to die than females.

POPULATION AND FOOD

At the global level population growth is the dominant cause of increased demands for food. If a nation can increase its food production and hold its population constant, then the amount of food available for each person is increased. However, in most less developed nations, recent gains in food production have been cancelled by increased gains in population. As a result, per capita grain consumption has declined since 1973. The late 1960s witnessed an improvement in the food situation of many less developed nations. This reflected the success of the Green Revolution whose high yielding varieties of wheat and rice raised hopes that the world's food problem could be solved. Beginning in the early 1970s, several seasons of poor weather in various countries led to a reversal of the world food situation. Large surpluses and low prices are no longer available. Prices for grains have risen, as have prices for the petroleum and fertilizers needed for mechanized farming. This dilemma is particularly acute for those less developed nations that were counting on the benefits of the Green Revolution. Also, much of the world is becoming increasingly dependent upon North America for grain. Grain exports from this area doubled between 1970 and 1976 and no new country has emerged as a cereal exporting nation during this period. Indeed, the United States now exports 65 percent of all grain that enters the world's export market. Europe, a developed region, is the principal recipient of these exports.

Many nations have begun to look to the sea as a source of protein. Vigorous competition, however, has resulted in overfishing that may have exceeded the regenerative capacity of table-grade fish. Since 1970 the world fish catch per capita has declined. In a protein-hungry world, competition for this resource will probably intensify.

In addition to population growth, rising affluence in developed nations adds to the demand for food. High income levels in the developed world create a strong demand for livestock products. Therefore, grains, legumes, and fish that could feed people directly are used instead to feed livestock. Because of this inefficient conversion system, the developed countries consume more food per capita than developing countries. In less developed countries, for example, annual grain consumption averages 182 kg (400 lbs) per capita—almost all of it consumed directly. In the United States, by contrast, there is an annual average consumption of 905 kg (2000 lb) per capita. However, less than 91 kg (200 lb) of this is consumed directly (as bread, pasta, pastries, and breakfast cereals),

the rest is consumed in the form of meat and such animal products as milk, butter, and eggs.

The population-food balance is potentially most critical in the south Asian countries of India, Bangladesh, and Sri Lanka. The situation remains precarious in the Sahel area of Africa. In the Western Hemisphere, the situation is potentially dangerous in Guyana, Honduras, Paraguay, and several Caribbean nations. Although actual famine may not be endemic in many less developed nations, hunger and malnutrition are. It has been estimated that more than 450 million of the world's people suffer from severe malnutrition. Malnutrition not only saps a person's vitality and productivity, it also makes one more susceptible to disease and illness. But the greatest tragedy is found in the effect on children (Fig. 8.11). Each year more than fifteen million children under five years of age die of the combined effects of malnutrition and infection. For those that survive the outlook remains grim. There is increasing evidence that

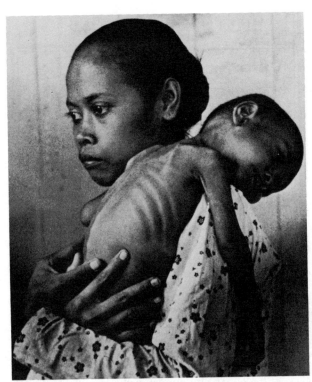

FIGURE 8.11
MILLIONS OF PEOPLE SUFFER FROM SEVERE MALNUTRITION, A CONDITION THAT MAKES ONE MORE SUSCEPTABLE TO DISEASE AND ILLNESS. An Indonesian mother and child.

a lack of adequate protein retards the development of the human brain, which reaches 90 percent of its structural development within the first two years of life. In this critical period of growth, the brain is susceptible to nutritional deficiencies that can cause a 10 to 25 percent impairment of normal mental ability. These impairments are permanent. Equally tragic is that the sequence is likely to be repeated when these children become parents—not because of genetic inheritance, but because they will be poorly equipped to avoid the nutritional deficiencies in their own children that they themselves suffered.

THOMAS R. MALTHUS

Perhaps the most influential exposition of the population dilemma came from the pen of Thomas R. Malthus (1766–1834). An English clergyman and economist, Malthus published his classic work *An Essay on the Principle of Population,* in 1798. This work focused on the relationship between population growth and food production. He knew that the population of England had doubled in the eighteenth century, and he feared it was outstripping the food supply.

Malthus believed that an unchecked population increases geometrically while food production increases arithmetically. He argued that population was always limited by food production (which he called means of subsistence), and that population would always increase where the means of subsistence increased. He believed that if diseases, epidemics, and wars did not stop population growth, then hunger and famine would.

Writing in the late eighteenth century, Malthus could not have foreseen the impact of modern agricultural technology, the opening to cultivation of vast new lands, and the development of transportation systems that could move food from one part of the world to another. From Malthus' time to the present, food production has increased enormously. He was correct, however, in his broad contention that population growth presses against food supplies. Today's neo-Malthusians make this same point as reports of famine and malnutrition come in from around the world.

Agricultural production has been increasing in all regions of the world during the past decade and a half. This has been accomplished by expanding areas under cultivation and by intensifying farming operations. Unfortunately, population increases in some areas and affluence in others have nullified any gains that might have been made. The population-food equation can be solved only if several measures are successfully undertaken. First, each country must increase the production of grains, legumes, and other staple food crops. Efforts are needed to make available protein-rich foods such as meat, fish, eggs, and dairy products. At the same time national population policies are needed that will permit and/or support vigorous population related programs, including family planning. It is necessary to realize and accept the fact that population control will come about either through intelligent planning by society or through the more painful alternative of overcrowding and disaster.

International Migration

Migration, either forced or voluntary, has traditionally been an available alternative for crowded, hungry people with limited opportunities. People may migrate from areas that suffer from natural catastrophies such as earthquakes, floods, famine, or from areas that experience manmade disasters such as war and revolution. Lack of economic and sociocultural opportunities also can induce large-scale movements. The world's greatest international migration occurred between 1840–1915 when more than 50 million Europeans emigrated. Thirty million came to the United States while the others moved to Canada, Australia, South Africa, New Zealand, Argentina, and Brazil. This movement filled in many of the world's empty spaces and led to a better balance between population and resources.

Both forced and voluntary migrations are still part of the world population picture. In 1972 Uganda expelled 40,000 Asians in its effort to Africanize the nation. Thousands of Portuguese-speaking people have fled the civil war of the 1970s in Angola. Today more than 500,000 Cuban and 120,000 Vietnamese refugees reside in the United States. In southeast Asia thousands of starving Cambodians fled their homeland in 1979

FIGURE 8.12
THE INDUSTRIAL AREA ALONG THE CUYAHOGA CANAL, CLEVELAND, OHIO.

to seek relief in refugee camps in neighboring Thailand. Voluntary migrations have occurred in Europe since the establishment of the European Economic Community (Common Market). Millions of workers from southern and eastern Europe and North Africa have migrated to Western Europe because of job opportunities. Other important migration streams include movement from Europe to the United States, Canada, Australia, and South Africa; from Latin America and Asia to the United States; and from Latin America and Asia to Canada.

Is migration a solution to today's population problem? Probably not. The days of free international migration are over. Most countries now have restrictions against permanent immigration and those countries that admit immigrants screen them carefully. Countries may welcome scientists and trained technicians but no country wants masses of illiterate and unskilled people. If migration were the answer to the world's population

dilemma, recipient nations would have to absorb scores of millions of people each year. The logistics of moving so many people would create impossible transportation, housing, health, and education problems.

THE OTHER DIMENSIONS

To the average citizen, the food aspect of the population dilemma remains paramount. There are, however, other dimensions to the problem of an expanding population. Some of these reflect ecological, social, economic, and political facets of the situation.

The record-breaking cold of the winters of 1976–1978 suggest that man may still be a pawn in the hands of the elements. Some studies suggest, however, that man is inadvertently modifying the climate of large regions. Airborne dust, created by urban and rural pollution, is a common man-made pollutant whose particles often act as cloud seeders. It is possible for large amounts of atmospheric dust to reflect the sun's rays away from the earth. Carbon dioxide may exert warming trends through the "greenhouse effect." Man's impact on climate is not fully understood; we do know, however, that larger populations will generate more dust, carbon dioxide, and thermal pollution.

Pollution, like population, is increasing at an exponential rate (Fig. 8.12). The expansion of population and consequent air and water pollution have increased so rapidly over the past several decades that nature can no longer absorb the contamination. Non-biodegradable wastes now reach our seas and oceans. A growing population will generate a demand for food, goods, and services whose production will increasingly pollute our atmosphere and hydrosphere. Airborne and waterborne pollutants can induce illness and death. People who live in heavily polluted urban areas, for example, have a higher incidence of heart disease and lung cancer than those who live in unpolluted rural areas.

For centuries man has been clearing forests to make room for agriculture and to provide wood for fuel and shelter. In many developing countries, wood is scarce and expensive. The two main causes of deforestation is man's ever increasing need for agricultural land and wood for fuel. Population growth multiplies the need for cropland, grazing land (now often overgrazed), and firewood. It has been estimated that nine-tenths of the people in many of the poorest countries depend upon firewood for fuel. Deforestation and overgrazing can also lead to the expansion of deserts and soil erosion (Fig. 8.13).

Minerals and mineral fuels are non-renewable resources whose consumption is growing at a faster rate than the world's population. Rising affluence, particularly in the developed nations, compounds the problem because those countries consume more than their share of the world's resources. In so doing they both exhaust their indigenous reserves and create problems of waste disposal. Recycling may prove helpful in both aspects of this situation.

The economic development of western nations is closely related to energy consumption. Energy supports their industrial economies. High productivity in these nations supports a high standard of living, which in turn generates a demand

for consumer goods such as cars and appliances that require additional energy for their use. Fossil fuel energy sources are finite and are being consumed at increasingly rapid rates. Thus, in the developed countries per capita energy consumption is 40 to 50 times higher than in the less developed countries. The industrialized world, for example, has 30 percent of the world's population but consumes 80 percent of total world energy. The United States, with about 5 percent of the world's population, consumes 30 percent of the world's annual energy consumption—20 percent of the world's coal; 49 percent of its natural gas; and 30 percent of its oil. And population growth plus higher standards of living lead to greater and greater total demand for energy, in a process to which there seems to be no end.

Every person needs water, but the amount consumed directly by an individual is only a small portion of society's daily needs. Vast quantities of water are needed to produce food. Hundreds of gallons are needed to produce a bushel of grain and thousands of gallons are needed to produce a few pounds of meat. This includes water for fodder crops, water consumed directly by the animals, and water needed for meat processing. Water is also needed to generate energy and for

FIGURE 8.13
THE PROCESS OF DESERTIFICATION IN AFRICA IS CAUSING THAT CONTINENT TO LOSE A SIZABLE SHARE OF ITS FOOD-PRODUCING CAPACITY. The skull of a long-dead cow is all that remains of what was once a pasture full of grazing cattle in the Sahel.

THE OTHER DIMENSIONS

irrigation purposes. Water for irrigation represents an important future demand, and here the needs of industry in developing nations will create conflict. Although water is a renewable resource it can be depleted; water tables, for example, are dropping in many areas dependent upon fresh underground water. The earth's supply of water is finite. Most of it, 97 percent, is seawater, usable only for fisheries and navigation. Three-fourths of the earth's fresh water is locked

in polar ice caps and glaciers, while much groundwater is too deep to tap. Surface water, of rivers, streams, lakes, and swamps, is man's most important source of this important resource, yet surface water amounts to less than one percent of total supply.

The less developed nations are faced with the overwhelming task of providing basic social services. Educating millions of children is a difficult task for a poor nation. Countries with the highest

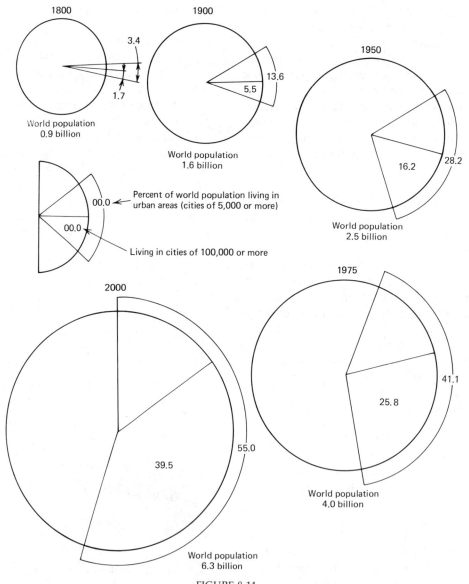

FIGURE 8.14
WORLD URBANIZATION, 1800 TO 2000. Source: Population Reference Bureau, Inc., Washington, D.C.

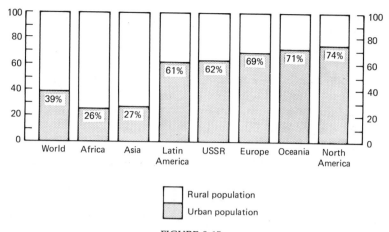

FIGURE 8.15

PERCENT URBANIZATION OF WORLD REGIONS, 1980. Source: Population Reference Bureau, Inc., *1980 World Population Data Sheet.*

birth rates are those that can least afford to build and staff schools. The problem is not confined to that of educating youngsters; indeed, one-third of the world's adult population is illiterate.

The desire for housing is a cultural universal. All people desire homes in which to eat, sleep, and raise children. Yet housing requires space, building materials, capital, and energy—commodities that are in short supply in much of the world. Shantytowns are becoming permanent features of almost every city in less developed countries. Housing the world's 4 billion and the billions to come presents a monumental challenge to rich and poor nations alike. As populations soar, the number of persons living in each dwelling increases, and there is considerable evidence that high density living leads to tension, anxiety, neurosis, and crime. This is especially the case in crowded cities where traditional patterns of social behavior tend to break down.

For most of human history man lived in rural villages. The proportion of city dwellers did not increase significantly until about 100 years ago. The twentieth century is witnessing, however, a rapid transition from rural to urban life. In 1900 only 5.5 percent of the world's population lived in cities with populations of 100,000 or more (Fig. 8.14). In 1975 the figure was about 41 percent. More than 1 billion people lived in these cities in 1975; an estimated 2½ billion will reside in them by the year 2000. If present trends continue, the twentieth century will be that in which the world converted from a rural society to an urban one.

Most industrialized nations are highly urbanized. Figure 8.15 depicts the degrees of urbanization found in various regions of the world. North America, for example, is about 75 percent urban. Asia and Africa are predominantly rural, while Latin America, a less developed region, is more urban than rural. The growth of cities in Latin America is a manifestation of the region's modernization.

Industrialization in the nineteenth and twentieth centuries attracted rural labor to the cities of today's developed nations. At the same time increasing agricultural mechanization meant that fewer rural folk were needed to work the land. City growth, accordingly, resulted from "urban pull." In the developing countries, however, the rural to urban migration is one of "rural push"— a situation that reflects destitute and overcrowded rural areas. Most migrants end up living on the outskirts of cities in slums and shantytowns where they must contend with overcrowding, poor housing conditions, lack of sanitation facilities, and in some cases scarcity of potable water. Even space is at a premium and competition for it forces many families to live in one-room dwellings. Many cities in poor nations are growing at more than seven percent a year; some of these will double in size within the next 15 years.

Social and economic problems often lead to political stress. Inflation has affected the whole world. Prices of petroleum, grain, fish, and even firewood, have soared. To the rich inflation may be a nuisance; to the poor it can be catastrophic.

THE OTHER DIMENSIONS

Reduced incomes and unemployment can lead to political conflict within national borders and between countries. Population growth can complicate resource allocation decisions, and conflicts over the distribution of food have toppled a number of governments. Resource scarcity can create international strife such as the "cod war" that took place between the United Kingdom and Iceland when the latter country extended its offshore fishing limits to 200 miles. Resource competition shows signs of spreading to Morocco and Algeria with their rival claims to phosphate-rich Spanish Sahara.

As resources become increasingly scarce and as more people inhabit the earth, it seems likely that more rules and regulations will be required to supervise the use of resources and space. Pollution guidelines are needed to protect the environment. Speed limits for motor vehicles are established to conserve energy. Zoning laws regulate the use of privately owned land. Enforced sterilization after three children has been considered in some of India's states. A few decades ago these rules would have been unimaginable. However, as the world becomes more crowded there will be, inevitably, increasing limitations on personal freedom.

The population problem is a many faceted dilemma. It reflects a greatly expanding world population, rising standards of living in many parts of the world, and increasing scarcity of natural resources. The rapid population growth occurring in the less developed nations can place a severe strain on efforts to improve agriculture, education, employment, and health care. Pressure on agricultural systems can result in overgrazing, deforestation, and soil erosion which, in turn, can hinder agricultural programs. The developed nations are not immune. Pollution, inflation, climatic change, unemployment, and housing shortages affect all. The population problem goes beyond merely producing enough food. Perhaps the most encouraging sign is that political leaders no longer ignore the population issue. Successful international conferences on food supply, water supply, and population, represent efforts to coordinate programs on a worldwide basis. The most encouraging development over the past few years is the increase in family planning programs now available in many countries.

The earth's human family is composed of a mo-saic of culture groups. Members of each group share common languages, ideologies, technologies, and perceptions of the physical environment they inhabit. These culture traits, characteristic of all people, are examined in the following chapter on cultural geography.

KEY TERMS AND CONCEPTS

ecumene
nonecumene
population density
physiological density
exponential growth
birth rate

death rate
rate of natural increase
annual growth rate
demographic transition
dependency ratio

DISCUSSION QUESTIONS

1. What are the four most heavily populated areas of the world's ecumene?
2. How might one characterize the world's nonecumene areas?
3. Who was Thomas Malthus and what were his views of population and food supply?
4. Why is the physiological density of an area a more significant statistic than an arithmetic density?
5. Discuss the "law of 70."
6. Discuss the significance of each of the four stages of the demographic transition.
7. What is the significance of the dependency ratio? How might this ratio vary from a developed country to a less developed one?
8. In what ways does rising affluence contribute to population problems?
9. What is the possible role of international migration as a means of solving population problems?
10. Discuss some of the ecological dimensions of an expanding population.

REFERENCES FOR FURTHER STUDY

Borgstrom, G., *The Food and People Dilemma*, Duxbury Press, North Seituate, Mass., 1973.

Peters, G., and R. Larkin, *Population Geography: Problems, Concepts, and Prospects*, Kendall-Hunt Publishing Co., Dubuque, Iowa, 1979.

Stanford, Q., ed., *The World's Population*, Oxford University Press, New York, 1972.

Tapinos, G., and P. Piotrow, *Six Billion People*, McGraw-Hill Book Co., New York, 1978.

Zelinsky, W., *A Prologue to Population Geography*, Prentice-Hall, Englewood Cliffs, N.J., 1966.

Zelinsky, W., and L. Kosinski, R. Prothero, eds., *Geography and a Crowding World*, Oxford University Press, New York, 1970.

CULTURAL GEOGRAPHY

CULTURE MAY BE DEFINED AS THE LEARNED PATTERNS OF behavior of the members of any society at a particular time and place. Geographers, concerned as they are with observing and explaining the spatial distributions and interaction of phenomena, view cultural geography as the application of the concept of culture to geographic problems. Cultural geography deals with the spatial variation of human creation that gives character to an area. The distribution of people, their use and modification of the natural environment, and their culture traits are the essence of cultural geographic studies.

Man differs from other forms of life in his ability to think abstractly. His position in the animal world is designated by the words *Homo sapiens,* which mean "wise man." Man's outstanding characteristic is his intelligence—his ability to think, learn, reason, give meaning to things, and develop a culture. Humans have to satisfy certain basic needs (food, water, clothing, shelter) in order to survive. How these needs are met is determined by the culture of individual groups. One person may wear a coat and tie while another person may wear only a loin cloth. One may prefer steak and another fried octopus. One may live in a grass hut while another lives in a large apartment complex (Fig. 9.1). Thus, culture, may be defined as learned and shared behavior transmitted from generation to generation by means of abstract communication (that is, by the use of

symbols, of which speech and writing are the most important).

To satisfy basic needs, all humans have established modes of communication, means of satisfying hunger and thirst, methods of housing and dress, a religion (or an ideology), laws, and economic systems. These institutions are known as *cultural universals.* It must be kept in mind, however, that the culturally derived methods of satisfying basic needs vary greatly from time to time and from one culture group to another. Economic and political systems, patterns of land use, religious and social practices, and levels of technology will vary from group to group, and are the key elements of the varying cultural landscapes found throughout the world.

As a society uses its natural resources, it develops unique patterns of land use in keeping with its particular technology. Technology is one manifestation of culture. Given the technology that a society has developed, man may hunt, fish, farm, mine, or produce manufactured goods from raw materials that have been shipped thousands of miles. Technical systems range from the simple technologies of hunting-gathering folk to the complex technologies of the urbanized-industrialized societies of the developed world. Simple societies tend to exploit a limited range of accessible environmental resources, while advanced societies exploit a broad array of resources collected from all over the world.

217

FIGURE 9.1
HOUSING IS A BASIC HUMAN NEED. The method of satisfying this need varies from time to time and from culture group to culture group.

PERCEPTION AND LANDSCAPES

Each culture uses the resources of the earth to satisfy its needs. But each will analyze and interpret its natural resources within the framework of its own cultural structure. Not all societies perceive and interpret the natural environment in the same way. A society views its habitat through the lens of its own culture and uses, or misuses, its resources accordingly. The notion that every person and culture has a particular image and awareness of the natural environment is known as *environmental perception*.

Usage determines what constitutes a natural resource. Interpretations and patterns of usage will vary according to place and time. The eastern American Indians of the seventeenth century considered game, fish, corn, and squash as their principal resources. They did not, for example, consider coal a resource. Europeans perceived the forests of the area as a source of timber that could be used for the construction of ships, homes, furniture, fences, and tools. Our industrial society values minerals and mineral fuels, commodities that were unappreciated until relatively recently. Not all cultures, however, are attempting to find new and intensive uses for nature's raw materials. There are many examples of different cultures inhabiting similar environments but using different natural resources. The Bedouin nomads of the Middle East and North Africa use their arid lands in a manner quite unlike that of the Bushmen of the Kalahari Desert or the irrigation farmers of the American southwest. Culture is a dynamic factor that releases mankind from the bonds of nature and determines how, or whether, any given element of the physical environment will be perceived and utilized. The physical environment may, to some degree, influence what man will not do within a given environment, although this negative influence often changes as human technology advances.

When man uses the resources of the earth, he modifies the natural environment and adds his own features. The interplay between man-made forms of human communities and the altered physical environment produces a *cultural landscape*, a term used by geographers to describe the earth as modified by the activities of man.

A cultural landscape is not fixed and immutable; rather, it is subject to constant change. Human activities, by which man imposes his imprint upon the land, are an ongoing process. Landscapes undergo change perpetually, and alteration in them may be pronounced and rapid. New skills, economies, and technologies have made man ecologically dominant in many areas of the world. Studies have been undertaken that start with a *natural landscape* and trace its evolution into a cultural landscape. A problem, however, is that few areas of the world remain untouched by man. Almost all parts of the earth's surface have been modified, at least to some extent, by humans. When Europeans arrived in the New World they found "wild" or "primitive" landscapes of which some original features had been altered by the Indians. Unfortunately, the landscapes of the distant past are poorly known.

Landscape studies focus upon the occupation and use of land. Here the imprints of culture may be seen in such features as types of houses and buildings, field patterns, roads, cultivated plants, engineering works, and architecture. Landscape studies not only help diagnose particular cultures but provide documentation for regional classification and culture change.

CULTURE CHANGE

Culture is not a static phenomenon. Although some culture groups are more conservative than others, most tend to be dynamic. Cultures may change as a result of invention, discovery, or innovation by individuals. Or, inventions may reflect the cumulative effort of many people over extended periods. It is doubtful that writing, coinage, paper, and the wheel were the inventions of individuals.

Change may come to a society from the outside. *Diffusion* refers to the spread of cultural elements from one society or person to another. New customs, ideas, art forms, and technological advances may penetrate the established patterns of a society. Diffusion may be restricted if an area is inaccessible, and change may come slowly to remote areas. On the other hand, ideas and customs may penetrate an accessible society suddenly and violently, as occurs at times of military invasion. The ancient Hyksos invaders penetrated Egypt in 1730 B.C., bringing with them an invention new to the Nile Valley—the horse–drawn chariot. It took the Egyptians a little more than a century to learn to make and use the weapons of their masters. In a patriotic fervor they launched a war of liberation against the Hyksos and drove them out of Egypt using the weapons introduced by their foes. Similarly, Alexander the Great spread the culture of Greece to the Middle East and India in

the fourth century B.C. Diffusion, however, usually takes place gradually as a result of peaceful activities involving transportation and communication. Trade, tourism, films, cultural exchanges, and television are all important factors in promoting diffusion.

Diffusion can lead to *acculturation*, a process of interaction between two societies whereby one (recipient) group adopts culture traits of another (donor) group. When two culture groups come into contact, and especially when the invading group has a more complex socioeconomic structure, one of three conditions usually develops. First, the native culture may be totally destroyed; second, the native culture may be absorbed; or third, it may be transformed in some manner by selectively adopting aspects of the dominant culture. When the latter happens, acculturation occurs.

THE 100% AMERICAN

Our solid American citizen awakens in a bed built on a pattern which originated in the Near East but which was modified in Northern Europe before it was transmitted to America. He throws back covers made from cotton, domesticated in India, or linen, domesticated in the Near East, or wool from sheep, also domesticated in the Near East, or silk, the use of which was discovered in China. All of these materials have been spun and woven by processes invented in the Near East. He slips into his moccasins, invented by the Indians of the Eastern woodlands, and goes to the bathroom, whose fixtures are a mixture of European and American inventions, both of recent date. He takes off his pajamas, a garment invented in India, and washes with soap invented by the ancient Gauls. He then shaves, a masochistic rite which seems to have been derived from either Sumer or ancient Egypt.

Returning to the bedroom, he removes his clothes from a chair of southern European type and proceeds to dress. He put on garments whose form originally derived from the skin clothing of the nomads of the Asiatic steppes, puts on shoes made from skins tanned by a process invented in ancient Egypt and cut to a pattern derived from the classical civilizations of the Mediterranean, and ties around his neck a strip of bright-colored cloth which is a vestigial survival of the shoulder shawls worn by the seventeenth-century Croatians. Before going out for breakfast he glances through the window, made of glass invented in Egypt, and if it is raining puts on overshoes made of rubber discovered by the Central American Indians and takes an umbrella, invented in southeastern Asia. Upon his head he puts a hat made of felt, a material invented in the Asiatic steppes.

On his way to breakfast he stops to buy a paper, paying for it with coins, an ancient Lydian invention. At the restaurant a whole new series of borrowed elements confronts him. His plate is made of a form of pottery invented in China. His knife is of steel, an alloy first made in southern India, his fork a medieval Italian invention, and his spoon a derivative of a Roman original. He begins breakfast with an orange, from the eastern Mediterranean, a canteloupe from Persia, or perhaps a piece of African watermelon. With this he has coffee, an Abyssinian plant, with cream and sugar. Both the domestication of cows and the idea of milking them originated in the Near East, while sugar was first made in India. After his fruit and first coffee he goes on to waffles, cakes made by a Scandinavian technique from wheat domesticated in Asia Minor. Over these he pours maple syrup, invented by the Indians of the Eastern Woodlands. As a side dish he may have the egg of a species of bird domesticated in Indo-China, or thin strips of the flesh of an animal domesticated in Eastern Asia which have been salted and smoked by a process developed in northern Europe.

When our friend has finished eating he settles back to smoke, an American Indian habit, consuming a plant domesticated in Brazil, in either a pipe, derived from the Indians of Virginia, or a cigarette, derived from Mexico. If he is hardy enough he may even attempt a cigar, transmitted to us from the Antilles by way of Spain. While smoking he reads the news of the day, imprinted in characters invented by the ancient Semites upon a material invented in China by a process invented in Germany. As he absorbs the accounts of foreign troubles he will, if he is a good conservative citizen, thank a Hebrew deity in an Indo-European language that he is 100 percent American.

Source. Ralph Linton, *The Study of Man* (Englewood Cliffs, N.J.: Prentice-Hall, Inc., 1936, pp. 326–327. Reprinted by permission.

Most Amerindian tribes underwent varying degrees of acculturation after initial contact with early European settlers. Cherokee Indians living in the southeastern United States are a case in point. Prior to their first contacts with Europeans, the Cherokees were primarily agriculturalists whose numerous towns and villages were usually located along streams scattered throughout the valleys of the southern Appalachians. Their subsistence economy, in addition to agriculture, included hunting, fishing, and gathering. The extensive woodlands surrounding their settlements were used for hunting; buffalo, deer, and bear were the principal large game in the area while animal domestication was limited to the dog. Fields were prepared for cultivation by girdling and subsequent burning of trees and undergrowth. Corn was the chief cultivated crop, with beans and squash almost as important. In addition, pumpkins, strawberries, and tobacco were raised. A limited number of hand tools, such as the hoe and pointed stick, were used for cultivation. The hoe blade usually consisted of animal bone, wood, stone, or shell.

Generally speaking, seventeenth-century fur traders were the first Europeans to penetrate the wilderness and act as agents of land use and cultural change. In exchange for pelts, traders gave the Indians firearms, hatchets, knives, traps, and other manufactured goods. The introduction of these items not only increased the Indians' hunting efficiency but also led to a more rapid depletion of an area's game supply. Some traders took Indian wives and established permanent residence in the Cherokee country. They raised European crops and livestock and, in so doing, became agents of agricultural diffusion. The trader's methods were viewed favorably by some Cherokees and adopted as alternatives to the traditional subsistence approach.

After the Revolutionary War the Federal government sent agents to live among the Cherokees. It was the task of the agents to encourage Indians to raise livestock and poultry and use the more efficient plow to cultivate their fields. They also taught the women how to raise cotton and spin and weave the fibers. It was hoped that such changes would result in the Indians being content with less land. A number of religious societies established missions in the Cherokee territory. In addition to building schools, these societies played an important role among the adults in demonstrating cropping procedures, care of livestock, and the building and maintenance of grist mills and dams. The government agents and missionaries lived and worked closely with the Cherokees. Through gifts, annuities, advice, and education they contributed to the substantial acculturation of the tribe.

During the eighteenth century most Cherokees continued to adhere to their time-honored subsistence economy. But by the last decade of that period and the early decades of the nineteenth century, the Cherokees had become a nation of farmers and had progressed far toward developing an agrarian economy. The Cherokees, the most acculturated tribe at that time, made great strides in adopting European agricultural techniques. As a result, their cultural landscape was beginning to change. Success in establishing farms, making improvements, and overall progress in "civilization" gave the nation a look of permanence that was resented by proponents of Indian removal. Forced to abandon their homes in 1838, the Cherokees embarked upon the "Trail of Tears" that took them to lands west of the Mississippi River.

CULTURE HEARTHS AND THE DIFFUSION OF CIVILIZATION

A *culture hearth* is a source area where important ideas and innovations developed and from which they diffused to surrounding areas. Four important hearths developed in the Old World and two developed in the New World. As we shall see, each contributed significantly to civilized life. Figure 9.2 depicts the world's early culture hearth areas.

Old World Culture Hearths

The world's oldest culture hearth developed in the Fertile Crescent—"fertile" because of the richness of its soil, "crescent" because of its shape. In the western part of the Fertile Crescent, bordering on the Mediterranean Sea, were Palestine, Phoenicia, and Syria; in the eastern part was Mesopotamia—the land between the Tigris and Euphrates rivers. About 10,000 years ago early plant and animal domestication occurred here and diffused to surrounding areas. Within a short period of time

(short relative to the millions of years humans have been on earth), mankind's nomadic hunting-gathering existence changed to a sedentary village life of tending crops and herds. But this new lifestyle created new needs. Containers were needed to store grain, articles were needed for cooking and eating, and better tools such as iron scythes, sickles, and hoes were needed to harvest crops and improve fields. Food surpluses released some individuals from agricultural activities. These people became artisans, merchants, religious functionaries, and other specialists. Gradually, these specialists left their agricultural communities for larger towns and cities. With the rise of cities great civilizations began to appear in Mesopotamia. This highly accessible area eventually became the crossroads of the ancient world and the center of innovation and diffusion.

The Fertile Crescent witnessed many wars, invasions, and conquests. At times, Mesopotamian armies conquered neighboring groups and brought them back as slaves. The "Babylonian Captivity," for example, marked the deportation of thousands of Jews to Mesopotamia when Jerusalem fell to conquering armies in 516 B.C. But the importance of these people lies not in their wars and conquests but in what their cultures contributed to civilization. Mention should also be made of the contribution of merchants and traders to cultural diffusion. With the development of specialized crafts came a great expansion of trade. A flourishing trade exchanged the agricultural and industrial produce of Mesopotamia for goods, especially metals, not produced in the area. The following excerpt from *The History of Herodotus* is indicative of many aspects of the areas's commerce at that time:

The boats which come down to Babylon are circular, and made of skins. The frames, which are of willow, are cut in the country of the Armenians above Assyria, and on these, which serve for hulls, a covering of skins is stretched outside, and thus the boats are made, without either stem or stern, quite round like a shield. They are then entirely filled with straw, and their cargo is put on board. . . . Their chief freight is wine, stored in casks made of the wood of the palm-tree. They are managed by the men who stand upright in them, each plying an oar. . . . Each vessel has a live donkey on board; those of larger size have more than one. When they reach Babylon, the cargo is landed and offered for sale; after which the men break up their boats, sell the straw and the frames, and loading their donkeys with the skins, set off on their way back to Armenia.

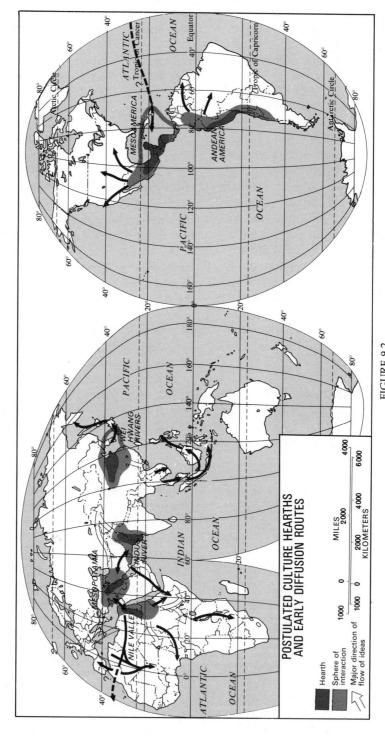

FIGURE 9.2
CULTURE HEARTHS AND EARLY DIFFUSION ROUTES.

Within the figure:

POSTULATED CULTURE HEARTHS
AND EARLY DIFFUSION ROUTES

Hearth

Sphere of
interaction

Major direction of
flow of ideas

MILES
1000   0   2000   4000
1000   0   2000   4000   6000
KILOMETERS

The contributions of this culture hearth to western civilization are numerous. Systems of writing, for example, were developed in areas having more advanced technologies. Writing eased the burden of keeping records (important to the merchant and tax collector), and helped to transmit laws, rules, and accomplishments from one generation to another. With writing came mathematics, a development of tremendous advantage to all people. The Phoenicians, who lived along the Mediterranean coast north of Palestine, became the great merchants and sailors of the ancient world. They have been called the "Missionaries of Civilization" because they carried to other people the culture of the eastern Mediterranean. Needing a simple method of keeping records, the Phoenicians borrowed from the Sumerians who had developed *cuneiform* writing, and from the Egyptians who had developed *hieroglyphic* writing, and created an alphabet of 22 consonants. This was a phonetic system that the Greeks later improved by adding vowel sounds. From the Greeks, the alphabet was transmitted to the Romans who made some slight changes in the form of the letters. It is the Roman, or Latin, alphabet that we use today.

In addition to cuneiform, the Sumerians developed a system of numbers based on 60 as a unit. This unit is still used in our method of telling time. The Sumerians are credited also with the development of irrigation and the invention of wheeled vehicles. Other peoples of the Middle East made substantial contributions to the sum of human knowledge. Hebrews were the first people to accept monotheism; Hittites of Asia Minor were the first to use iron for tools and weapons; Lydians, finding barter inconvenient, invented coinage; Babylonians were interested in astrology (the forerunner of astronomy) and learned to recognize planets, foretell eclipses, and determine the length of the year. These and other innovations suggest that scientific achievements were probably the Fertile Crescent's greatest contribution to western civilization. Without the pioneering work of its people, the lives of all later peoples would have been substantially different. The influence of this area is surpassed by few civilizations and the honor of being the "Cradle of Western Cvilization" must go, as far as we know now, to Mesopotamia.

Ancient Egypt was another early culture hearth.

This was a relatively homogeneous and stable civilization whose boundaries were sharply demarcated by the fertile Nile Valley and delta. Protected by desert barriers, Egypt did not experience the range of invasions and cultural diffusion to which the Fertile Crescent was exposed. This long-lived civilization made its greatest discoveries during the first few centuries of its existence. These achievements were in the fields of writing, mathematics, medicine, and architecture. The Egyptians had three forms of writing: hieroglyphics was used for sacred texts; hieratic, a pen and ink form of hieroglyphic writing, was the official script; and demotic, a shorthand of hieratic, was the common Egyptian writing of everyday use. Papyrus, the writing material, was made from a reed found near the Nile. Ink came from various gums and pens were made of pointed reeds.

The Egyptians were efficient surveyors, as demonstrated by such precisely calculated structures as the pyramids. The Great Pyramid of Khufu, for example, stands on a square base facing the four points of the compass. Egyptian documents suggest that medical knowledge was surprisingly advanced for the time. And we know how successful the ancient Egyptians were in preserving the human body after death. In the field of architecture the Egyptians were the first to successfully use mass in working with stone. Their pyramids, obelisks, tombs, and temples reflect the solid mass and eternity of the surrounding cliffs where their building materials originated (Fig. 9.3). It is not surprising that the Greeks, whose classical architecture reflects Egyptian influences, considered Egypt a repository of ancient wisdom.

The Indus Valley was the site of a third culture hearth. Until recently, historians believed that civilization in the Indus Valley began in the second millennium B.C. with the Aryan invasion. As late as 1921, however, important discoveries of ancient cities were made at the sites of Mohenjo-Daro (mound of the dead) in southern present-day Pakistan, and at Harappa, 400 miles to the north, in the Punjab area. It appears to have been a civilization as advanced as those found in Mesopotamia, the Nile Valley, and along the Huang He (Huang-Ho) in China. Achievements consisted of writing, subsistence agriculture, urban life, a bronze and copper technology, the use of fired bricks for buildings, and a complex governmental

FIGURE 9.3
MASSIVE TEMPLES, OFTEN WITH HIEROGLYPHIC WRITING, WERE CHARACTERISTIC STRUCTURES OF THE EGYPTIAN CULTURE HEARTH.

organization. The life span of this culture hearth is estimated to have run from 4000 to 2500 B.C.

The cities at Mohego-Daro and Harappa were large and well built with planned rectilinear streets. Houses were several stories high; there were impressive drainage systems. The most imposing city edifice was a great municipal bath. The arts were well advanced in the Indus culture hearth. Smiths worked with copper, bronze, silver, and lead. Potters used the wheel and cotton was raised and woven into cloth. Various elements of this culture have been retained in the present pattern of Indian life. One of its deities, for example, appears to be a prototype of the Hindu god, Shiva.

The fourth Old World culture hearth was located along the banks of the Huang He (Huang-Ho) of North China. This relatively dry area is famous for its fertile wind-blown loess soils. These fine and porous soils can be worked easily and this undoubtedly contributed to the success of prehistoric Chinese agriculture in this area. Although there are numerous Neolithic sites within the Huang Basin, Chinese history doesn't begin until the time of the Shang Dynasty (1766–1122 B.C.), when the Egyptian civilization was already old. This highly developed civilization had a written language, the forerunner of modern Chinese. Religion was well developed, with numerous deities attached to particular places and elements of nature. Well developed legal and administrative systems existed, as well as that characteristic Chinese material, silk. This culture also built a great city, Anyang (An-Yang), the capital of the Shang Dynasty. The site has been excavated and has yielded evidence of an advanced metallurgy that enabled artisans of the time to craft intricate bronze vessels. The presence of artifacts that came from afar suggests that long-distance trade also existed. The ancient Chinese were a highly innovative people; many of their contributions eventually found their way to the western world. The Chinese invented a highly durable paper, use of which was passed to the Arabs in the eighth century and to Europe in the thirteenth. Playing cards were in use in China in the tenth century and were introduced to Europe near the end of the fourteenth century. Paper money was used in China in the tenth century, but it was not until 1656 that paper currency was issued in Europe. Other Chinese inventions were porcelain, gunpowder (restricted to fireworks), the compass, and the use of coal for fuel.

New World Culture Hearths

The Maya culture flourished in the Yucatan and northern Guatemala from about 2000 B.C. to 1000 A.D. Here were built hundreds of temple-cities, many with immense pyramids, that were used primarily as ceremonial centers. These sites had been abandoned by the time the Spaniards arrived. The Maya lived in widely scattered farm hamlets where the population raised corn and beans in forest clearings prepared by slash-and-burn techniques. The Maya have been characterized as the "intellectuals of the New World" because they repeated the ancient Egyptian inventions of hieroglyphic writing and the calendar and because of the complexity of their architecture. Although they had neither metals nor the wheel nor draft animals, they did manage remarkable achievements in sculpture, pottery, and murals. Maya sculpture is massive. The most impressive are the carved stone monoliths called *stelae*.

Of all new world culture groups, the Mayas had the greatest inclination to trade and travel by sea as well as land. The beans of the cacao tree, from which we derive chocolate, were the currency of Mesoamerica in pre-Columbian times.

A number of culture groups occupied central Mexico before the Aztecs appeared about 1200 A.D. The great temple-city of Teotihaucán, a relic of an earlier civilization, is the largest and most impressive urban site of ancient America. Located in the Valley of Mexico, it covered 20 square kilometers (8 square miles) and may have housed as many as 100,000 inhabitants at the peak of its power (Fig. 9.4). It remained an important trade and religious center from 100 B.C. to 750 A.D. Its religious monuments, especially the Pyramids of the Sun and Moon, remained sacred to the Aztecs who continued to worship at the site until Hispanic times.

Centuries after the decline of Teotihaucán, the Aztecs established their capital, Tenochtitlán, 41 kilometers (25 mi) to the south on an island in the western shallows of Lake Texcoco (the site of today's Mexico City). Planning for this island city included fountains, dikes, causeways, sluice gates, and removable bridges. The Aztecs also built pyramids, but sculpture (much of it magnificent by today's standards) was their greatest contribution to art. Although metal-working arrived late in Mexico, the Spaniards found magnificent examples of jewelry made from gold and precious stones, especially jade. Roads facilitated a flourishing trade in spite of the fact that wheeled vehicles and pack animals did not exist. All of the advanced cultures of Mexico had some form of

FIGURE 9.4
TEOTIHAUCÁN, AN ANCIENT URBAN SITE OF EARLY AMERICA, MAY HAVE HOUSED 100,000 INHABITANTS. The Pyramid of the Sun is shown in the background.

CULTURE HEARTHS AND THE DIFFUSION OF CIVILIZATION

paper and writing, but the Aztecs perfected these inventions. Indeed, the conquistadors were amazed to find that the Indians had books.

Corn was the basic Aztec crop. Diets were supplemented with beans, squash, pumpkins, yams, tomatoes, hot peppers, avocados, and cacao. The potato, so important to the prehistoric population of South America, was unknown in Mexico until it was brought in from the Andean Highlands by the Spanish. The Aztecs also are known to have developed the unique *chinampa*, or "floating garden," type of agriculture on their lakes. When agricultural land was limited, floating reed-woven baskets eight feet in diameter were filled with earth and anchored in shallow water with staves. Roots from planted trees and crops would eventually penetrate the baskets and firmly fix them to the lake bottom. These artificial islands were intensively cultivated and produced high yields.

Like the Aztecs, the Incas arrived late in an area that had witnessed cultural growth for thousands of years. The Incas established themselves about 1200 A.D. in the valley of Cuzco, Peru, and from there expanded by conquest until they controlled an empire that extended along the Andes Mountains from the Pacific coast in the west to the headwaters of the upper Amazon River in the east (Fig. 9.5). This area became an important center of plant domestication, and much of the food that the world eats today was developed by these Andean farmer-soldiers. Potatoes and corn were the vegetable base of the Inca realm. These staples were supplemented with beans, manioc (from which comes tapioca), peanuts, pineapples, chocolate, tomatoes, strawberries, papaya, and vanilla. Flat agricultural land was so scarce that much Inca skill was directed to the terracing and irrigation of hillsides. Many of these were so well made as to be still in use. The Incas had neither the plow nor draft animals, although the llama was used as a carrier of trade goods. The related alpaca was chiefly reared for its wool.

The Incas are most famous for their engineering skills. Aquaducts and canals brought water from distant sources to population centers. Although the Incas had no wheeled vehicles, they managed to hold their realm together by constructing a complex road system that was comparable to the ancient Roman roads. The longest, the Andean royal road, was 5200 km (3250 mi) in length and extended through Ecuador, Peru, Bolivia, and into

FIGURE 9.5
THE RUINS OF MACHU PICCHU, AN EARLY INCA COMPLEX, CONSIST OF TERRACES, PALACES, TEMPLES, AND RESIDENTIAL COMPOUNDS.

Argentina. Bridges, an integral part of road systems in a land of mountains and chasms, were another Inca achievement. The rope suspension bridge was an outstanding example of Inca engineering feats.

The Incas were masters of organization and conquest; this in spite of the fact that they lacked a system of writing. The *quipu*, a device consisting of strings and knots, was the closest thing to writing that the Incas developed. The quipu, however, was not a form of writing but a decimal system of counting that enabled the Incas to keep accurate records of what taxes and tribute were to be paid.

Unlike the Old World culture hearths, those of the New World did not arise in river valleys. Moreover, the hearths of the New World appeared much later and developed without the technical advantage of the plow, wheel, and draft animals. Nevertheless, each hearth evolved after an agricultural revolution had produced abundant food and a sedentary way of life that provided a foundation on which a high civilization could be built.

Western civilization, the dominant civilization

on earth today, was not a "foundation" hearth such as Mesopotamia or the Indus Valley. Rather it was a channel into which contributions flowed from many areas. Aided and transmitted by the Greeks, Romans, Arabs and others, the cultural streams converged and were carried forward in Europe to be later transmitted to the New World and beyond. The world has essentially become Europeanized. Today every region where an earlier civilization once flourished is permeated by Western ideas. Some communities are using the ideas and technology learned from the West to challenge the dominance of Western culture.

RACE, LANGUAGE, AND RELIGION

Although the human family belongs to the same subspecies, people are by no means alike. They possess at birth different physical characteristics that determine skin color, average height, eye color, hair texture, head and nose shape, and many other attributes. After birth they acquire, through learning processes, cultural behavior that induces them to speak a given language or worship in a certain faith. Language and religion are learned cultural universals; race does not in any way determine or influence a group's learned behavior. Cultural processes are the main factors in producing linguistic, religious, and all other differences among the world's population.

Racial Groups

Race is a biological concept that focuses upon physical characteristics genetically transmitted from parents to children. The term *race* refers to a large group of people who tend to have certain physical characteristics that set them apart from other humans. However, the distinctive characteristics used to denote races, such as head and eye structure, account for only about five percent of the total physical characteristics of man. The traits that human beings have in common are far more numerous and important than those differences that separate them in biological classification.

Mankind is probably descended from one homogeneous population. But different physical en-

vironments, and isolation, natural selection, mutation, and genetic drift fostered the emergence of distinct racial groups thousands of years ago. There are many classification schemes. The most commonly recognized divides humankind into three races: Caucasoid, Mongoloid, and Negroid.

Caucasoids are generally characterized by a high percentage of light colored skin, light eyes, thin lips, straight and dark hair, and narrow noses, often with a high bridge. The Caucasoids first appeared in East Asia and eventually migrated to India, the Middle East, North Africa, Europe, and various parts of the world settled by Europeans. Mongoloids have a high percentage of brown to yellowish skin, brown eyes with *epicanthic* eye-folds, broad noses with a low bridge, high cheekbones, and straight and dark hair. Mongoloids evolved in China and spread throughout East Asia, Polynesia, and America. The Negroid race in general terms exhibits brown to brown-black skin, dark tightly-curled hair, brown to brown-black eyes, broad noses, and prominent lips. The Negroid race developed in the savanna and steppe areas of sub-Saharan Africa where it remains the dominant group. Figure 9.6 depicts the world distribution of racial groups.

Scholars today agree that there is no such thing as a "pure race." People have been wandering and interbreeding over the earth for too many millenia for any group to be considered "pure." Indeed, no race has a physical characteristic so unique to itself that it never occurs in another race. The dark skin color typical of African Negroids, for example, is found in some Caucasoids as well as in the Melanesian and Australian races of the Pacific, while some people classified as Negroid have light skin. Some Mongoloid people are darker skinned than some Negroid people. Only Caucasoids have blue eyes but most Caucasoids are not blue-eyed. Only Negroids (both African and Oceanic) have tightly curled hair, but this trait is not characteristic of all Negroid people. Adult males of the Watusi tribe of East Africa are very tall and average about seven feet in height while the Pygmies of Central Africa are very short and average about four feet in height. The range of stature within the Negroid race is equivalent to the range within the entire human race.

What significance can be given to racial characteristics that vary so widely within each race? The concept of race is obviously imprecise and

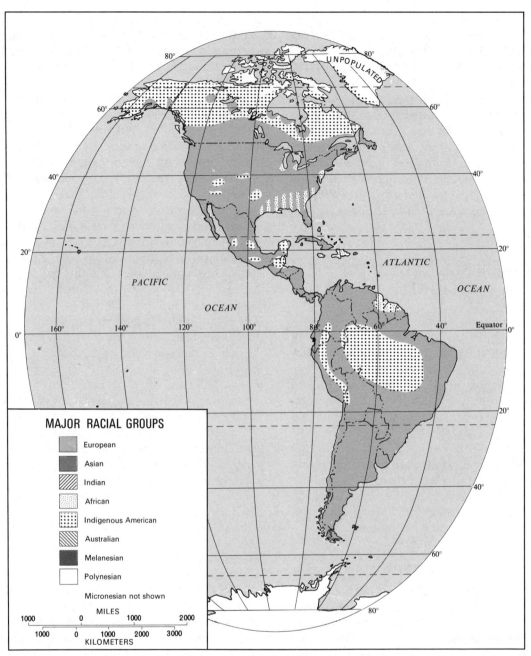

FIGURE 9.6
MAJOR RACIAL GROUPS.

limited. Some scientists may find the concept helpful in classifying certain physical features of the human population, while others will find it useful in studying the origins, evolution, and migration of groups. But as a tool for determining social, cultural, or intellectual classifications, the concept is useless. There is no known cause and effect relationship between race (which is inherited) and culture (which is learned) and no conclusive evidence indicating that any race possesses innate intellectual advantages or disadvantages.

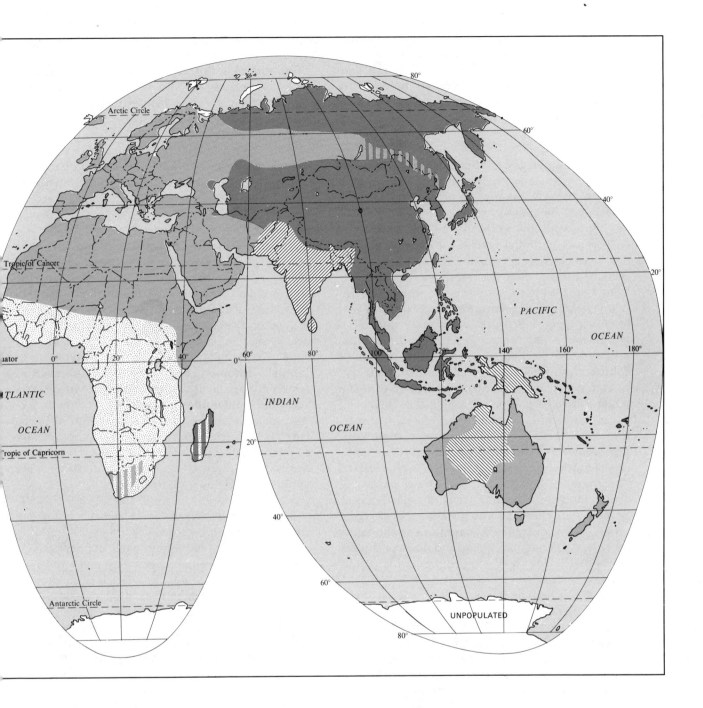

Language Groups

Within the animal kingdom man is unique in terms of the complexity of the linguistic systems over which he has command. *Speech* is the physical faculty by which language is expressed through sounds or words. *Language* is the learned system of verbal communication by which human beings express ideas and concepts by the use of symbols. It is the medium through which accumulated information can be preserved and transferred and through which man can speculate intelligently about the future. Languages may or may not have a written form. A *dead language* is one that is no longer spoken as a mother tongue although, like Latin, it may be used for religious or scholarly purposes. Hebrew, once a dead language, was revived and adopted as the official language of independent Israel.

Living Languages. Cultures create languages that are adequate for their own needs. With time, however, new words may be needed to deal with new ideas, inventions, or situations. These may be "borrowed" from other languages, or may be created by combining native words, or by making new uses of existing words. English has been enriched by borrowing thousands of words, many from non-European sources: *moccasin* (Amerindian), *tattoo* (Polynesian), *bamboo* (Malayan), *sofa* (Arabic), *typhoon* (Chinese), and *karate* (Japanese) are examples. At the same time English words, especially "Americanisms," have invaded other culture groups. The American expression "okay" is commonly used in many countries of Asia, while a host of baseball (pronounced "beisbol") vocabulary words are used in several Latin American countries where that sport has become popular. A number of English words used in continental Europe are "jazz," "gangster," "blue jeans," "hamburger," and "party." Sometimes old words are given new meanings when the need arises. Several words used in a new sense include computer, program, feedback, countdown, and space capsule. As situations change and new vocabulary needs arise, languages acquire the words people need.

Variant speech patterns, called *dialects*, differ from the generally approved vocabulary, pronunciation, and grammatical structures of a language. Dialects vary from the cockney form of English heard in the east end of London to the type of French heard in remote areas of Quebec. Regional dialects within the United States range from the "Southern accent" to the "Bostonian accent": where a broad "a" is used in words or where an "r" sound is used at the end of words, so that Indiana becomes "Indianer," or Cuba becomes "Cuber."

Language Families. Approximately 3000 languages are spoken in the world today. Philologists have classified them into about 40 to 50 linguistic families. Although the languages within each family are related to each other, they may exhibit different degrees of closeness. Indo-European languages are spoken by nearly half the world's people. This family includes the languages of Europe and the areas of the world settled by European colonizers. The western region of this language family includes the Latin-Romance languages of French, Spanish, Portuguese, Italian, and Romanian; the Germanic languages that include English, German, Dutch-Flemish, and the Scandinavian tongues; the Balto-Slavic, including Russian, Ukrainian, Polish, Czech, Slovak, Serbo-Croatian, Bulgarian, Slovenian, Lithuanian, and Latvian; the Celtic languages of the British Isles; Greek, Albanian; Armenian; and the Indo-Iranian languages of Hindustani, Bengali, Persian, and Pashto.

A second language family, Uralic and Altaic, is also spoken in parts of northern and southern Europe. Its languages include Magyar (Hungary), Finnish, Estonian, and Turkic. For the sake of simplicity a general classification of language families and subfamilies is depicted in Figure 9.7. As can be seen, the spatial distribution of world languages is complex. Moreover, this complexity is intensified when we realize that literally thousands of additional languages and dialects are spoken in their own distinct areas.

Multilingualism. A common language is a strong unifying force that greatly facilitates the building of nation states. Numerous well defined groups speaking different languages may make the pursuit of common national goals difficult or impossible. This is especially true if antagonistic culture groups speak different languages.

There are a number of countries where two or more languages have official status. In India, for instance, 15 major languages are officially recognized. Hindi, in a variety of dialects, is spoken by about half of India's people. Although Hindi has been declared the national language, this ruling has met with violent resistance in states where other languages are dominant.

Quebec is a Roman Catholic, French-speaking province in an otherwise Protestant, English-speaking Canada. In 1969, in response to grievances, the Canadian Parliament made French equal to English in all operations of the federal government. Cultural and linguistic differences

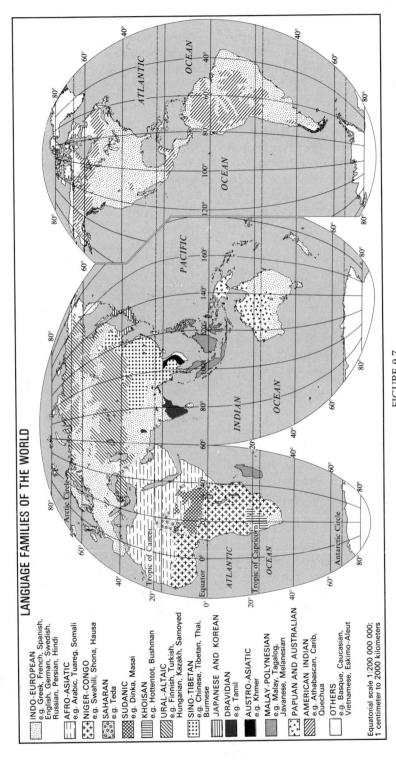

LANGUAGE FAMILIES OF THE WORLD

INDO-EUROPEAN
e.g. Greek, French, Spanish, English, German, Swedish, Russian, Persian, Hindi

AFRO-ASIATIC
e.g. Arabic, Tuareg, Somali

NIGER-CONGO
e.g. Swahili, Shona, Hausa

SAHARAN
e.g. Teda

SUDANIC
e.g. Dinka, Masai

KHOISAN
e.g. Hottentot, Bushman

URAL-ALTAIC
e.g. Finnish, Turkish, Hungarian, Kazakh, Samoyed

SINO-TIBETAN
e.g. Chinese, Tibetan, Thai, Burmese

JAPANESE AND KOREAN

DRAVIDIAN
e.g. Tamil

AUSTRO-ASIATIC
e.g. Khmer

MALAY-POLYNESIAN
e.g. Malay, Tagalog, Javanese, Melanesian

PAPUAN AND AUSTRALIAN

AMERICAN INDIAN
e.g. Athabascan, Carib, Quechua

OTHERS
e.g. Basque, Caucasian, Vietnamese, Eskimo-Aleut

Equatorial scale 1:200 000 000;
1 centimeter to 2000 kilometers

FIGURE 9.7
LANGUAGE FAMILIES OF THE WORLD.

RACE, LANGUAGE, AND RELIGION

231

and continued claims of political and economic discrimination have nourished the "Free Quebec" movement.

In Belguim, Flemish (a dialect of Dutch) is spoken in the northern half of the country while Walloon (a dialect of French) is spoken in the south. Walloon is the language of Brussels, the nation's capital, making the city a language enclave in the northern Flemish-speaking area.

English and Afrikaans (a dialect of Dutch) are spoken among the whites of the Republic of South Africa while several dozen different indigenous languages are spoken by blacks of that country. More than a thousand indigenous languages are spoken in Africa and most countries are multilingual. The national boundaries of many African nations generally conform to the colonial territories carved out by European powers in the past. These boundary lines were drawn rather arbitrarily, and as a result people with different languages and customs now find themselves citizens of the same nation. These groups are often antagonistic toward each other, a condition which has contributed to considerable civil disturbances in a number of African countries. Nigeria, Sudan, and Zaire are examples.

Lingua Franca. A *lingua franca* is a hybrid "trade" language used among peoples of mutually unintelligible languages. These hybrids developed to facilitate trade and commerce in areas where a multitude of languages made business transactions difficult. Swahili, a modified Niger-Congo language of the Bantu group, is spoken throughout East Africa while the Hausa language serves the same purpose in West Africa.

Pidgin English, a simplified lingua franca, is spoken throughout the southwest Pacific and parts of the Caribbean. The word pidgin is derived from the Chinese corruption of the English word business.

Religion

Religion, like language, is a cultural universal. Whether a particular religious interpretation helps to overcome fear of the unknown, to deal with the mysterious factors that control man's destinies, or to find ultimate meaning and purpose in life, all societies have a religion and all religions show a basic similarity. In one way or another religions

say that man does not, and cannot, stand alone. While each faith has an ethical, or invisible side, each also has ceremonial, or visible, aspects. These may consist of sacred structures, burial grounds, bodies of water, sacred groves and mountains, and sacred animals. Visible aspects are expressed geographically and find expression on the landscape.

Religion and Landscapes. Religion will affect the cultural landscape in either a positive or negative sense. Sacred structures may blanket an area and be conspicuous landscape features associated with a given religious system, while religious prohibitions in the form of taboos may account for the absence of an expected condition.

Churches, synagogues, mosques, and temples are familiar sacred structures that may house a congregation and/or a deity. Christianity, with its numerous denominations and sects, has made a great imprint on the landscape with its sanctified houses. Roman Catholic churches, especially cathedrals, are typically large and conspicuous religious structures while Protestant churches tend to be smaller and less ornate (Fig. 9.8). This partly reflects the Catholic belief that the church is the house of God. To Protestants, the church building is a place that God visits. Protestant churches tend

FIGURE 9.8
CHRISTIANITY HAS MADE A GREAT IMPRINT ON THE LANDSCAPE WITH ITS SANCTIFIED HOUSES. The Basilica of St. Peter in the Vatican.

to be more numerous because of the multiplicity of denominations and their relatively small congregations. Sanctified houses of worship are less important in Islam and Judaism. Only an assembly of the faithful (usually ten in number) can give sanctity to a mosque or synagogue; accordingly, the majority of these buildings tend to be less imposing structures. Hinduism and other religions have produced an array of visually impressive temples. Most of these do not have a large interior space for the public. The holy of holies is often a dimly lit shrine, entrance to which is often restricted to the temple priests.

Cemeteries, like sacred structures, may be conspicuous features of the cultural landscape in areas where the institution of burial prevails. Christian, Moslem, Jewish, and traditional Chinese societies require space for internment and for the erection of monuments. In some areas cemeteries and ancestral shrines have kept badly-needed land out of agricultural production. In many of our larger cities vast acreages once set aside for cemeteries are now surrounded by residential and sometimes incongruous land uses (Fig. 9.9). Buddists, Hindus, and Shintoist Japanese cremate their dead; cemeteries as cultural landscape features are not typical of these societies.

Many religions have conferred sanctity upon water, plants, and animals. The Ganges River is sacred to Hindus who, at death, may seek cremation along its banks. Water from the grotto of Lourdes, a small town located in southwestern

FIGURE 9.9
CEMETARIES MAY BE CONSPICUOUS FEATURES OF THE CULTURAL LANDSCAPE IN AREAS WHERE THE INSTITUTION OF BURIAL PREVAILS.

France, is considered sacred to millions of Roman Catholic pilgrims who seek miraculous cures at the shrine. Grape cultivation among Jews and many Christians reflects in part the sacred value of wine in meeting ritual needs. The bo tree is sacred to Buddhists and has been dispersed from its native habitat in the Himalayan Mountains to distant Sri Lanka and Japan. The evergreen spruce, once sacred to the ancient Greeks and Germans, lives on in a quasi-religious form as our Christmas tree.

A *taboo* is a rule established by a community to neutralize unfavorable supernatural forces. A religious taboo regulating food and work habits can not only have a negative effect upon the landscape but can also profoundly effect the economy of an area. Until recently, Catholics were obliged not be eat meat on Fridays. This rule greatly stimulated the fishing industry in Catholic areas. Religious taboos on killing and eating certain animals produces some opposite effects. Orthodox Jews and Moslems are forbidden to eat pork; pigs, therefore, are not raised throughout the Middle East and South Asia. In contrast to the pig taboo, the beef taboo of Hindu India has produced such an abundance of cattle that India ranks first in the world in cattle population. This enormous cattle population reflects the Hindu veneration of life and the sanctity of the cow (Fig. 9.10).

FIGURE 9.10
A STREET SCENE IN AGRA, INDIA.

INDIA'S SACRED COWS

In India, for many centuries, the cow has been the object of reverence or something bordering thereon. Cows, some old and gaunt, wander about at will.

A typical American attitude is that the cows compete with human beings for food and that if India would get rid of the cows, it would greatly ease the food problem. This reasoning comes from applying the relationship of cows to people in the United States. Here, cattle eat a lot of grain that might otherwise go to human consumers. Cows reduce the volume of food, even while contributing to it.

But in India, circumstances are vastly different. Most Indian cows go through life without ever consuming a mouthful of grain. They graze at the roadsides; they eat the straw left from threshing and scavenge whatever they can. True enough, they graze on some land that might otherwise be used to produce crops, but this does not cut deeply into the tilled acreage.

Consider the contribution of the Indian cow to food production. The bullock is the main draft power of Indian agriculture; without him crop production would come to a near halt. The cow produces milk, almost the only source of animal protein for most families. Cow dung is often the only source of fuel and of fertilizer. Even after the animal finally dies, it goes on serving man; its hide becomes a valuable item of commerce.

Hungry people do not deify an animal that basically competes with them for food.

There is, of course, some justification of the American's critical attitude toward the Indian cow. Some of these animals are kept alive, aged and decrepit, long after they are capable of productivity. But Americans might remember that they have a "sacred cow" of their own, the automobile. The automobile, like the cow of India, has its combination of good and bad points and is deeply built into the culture. It would be well for Americans to extend to other cultures the charity they desire for their own.

Source. Don Paarlberg, "For Those Who Would Criticize India's Sacred Cow . . . Consider the Automobile, *Agenda*, Agency for International Development, Vol. 2, No. 10, December 1979, p.24.

Religious taboos may prohibit work on certain days and thereby alter normal economic activities.

Sunday is a holy day in the Christian world while the Jewish sabbath is observed on Saturdays. Friday is the sacred day to Moslems. Added to these weekly holy days (hence, holidays) are numerous feast and saints days associated with religous calendars. In medieval France there were more than forty such feast days in addition to Sundays. This meant that during the course of a year economic activities were suspended on one out of every four days for institutionalized religious observations.

In some societies, religious taboos restrict various crafts and economic activities to selected kin groups or castes; to the nonmember, the occupation is ritually taboo. The most elaborate work taboos are associated with the caste system of Hinduism, where such enterprises as fishing, tanning, leather working (animal by-products), and seafaring are considered defiling. These occupations are not only penalized on religious grounds but also are removed from sources of investment capital that would permit stimulation and expansion.

Characteristics of Religious Systems. The world's religions may be divided into two broad categories. The first group includes *ethnic* religions; those characteristic of particular ethnic or tribal groups. These include the indigenous religions of native American tribes; the Shinto faith of Japan; the traditional Chinese religions; Hinduism; and Judaism, probably the most dispersed ethnic religion. Ethnic religions do not actively seek converts and rarely acquire new members in larger numbers. The second category consists of *universalizing* religions, faiths characterized by their efforts to gain converts because they believe their religion is suitable for all people. Christianity, Islam, and Buddhism are the three main universalizing religions. These have spread rapidly from their source areas, partially through the efforts of missionaries. Figure 9.11 depicts the origin and dispersal of the world's major religions.

Religions are divided into two other categories depending upon whether the faithful pay homage to one or more dieties. The ancient Hebrews and the Persian religious leader Zoroaster were the first to reject the notion of numerous lesser deities and assert that there is only one supreme divine being, God, the creator of the universe and the ultimate source of all reality. Later, the Christians and Moslems followed this basic belief. Today Judaism, Christianity, and Islam are the leading *monotheis-*

tic faiths. Belief in many gods is called *polytheism.* Mankind has worshipped higher spiritual beings, usually associated with forces of nature, for thousands of years. Some gods look after the living, others look after the dead, while others are personal gods who look after an individual. Hinduism, Buddhism, and the Chinese religions are the leading polytheistic faiths.

Monotheistic Religions

Judaism. Judaism, the oldest of the great monotheistic faiths, evolved among the Hebrews, a small Semitic tribe that wandered through the deserts of southwest Asia nearly four thousand years ago. Under the leadership of Abraham, the tribe left the city of Ur in southern Mesopotamia and eventually settled in Palestine and later in Egypt. During the Exodus, Moses led his people out of Egypt (celebrated during Passover) and back to Palestine where a monarchy was established. But the small kingdom could not withstand the might of its enemies. In 586 B.C. Jerusalem was destroyed by invading Babylonians and the Jews taken into captivity. In 538 B.C. Cyrus the Great conquered Babylon and granted permission for the Jews to return to their homeland. Here they remained under the control of various powers until 70 A.D., when they revolted against Rome. Again their holy city Jerusalem was destroyed and the Jews dispersed throughout the Roman Empire.

During the nearly two thousand years that followed, the Jews continued their existence in the face of tremendous odds. In the agrarian Middle Ages of Europe they could not own land or practice many occupations. It is not surprising that many became merchants and bankers in a landowning society that disdained these occupations. The economic and professional successes of Jews in late nineteenth century Europe stirred a new wave of anti-Semitism. Convinced that their only security lay in a national homeland in Palestine, the Zionist movement arose. During the following decades thousands of Jews migrated to Palestine. The Nazi persecution of Jews accentuated the pressures for the formation of an independent Jewish state. Despite Arab opposition, the state of Israel was proclaimed on May 14, 1948. The Jews, who number only about fourteen million, have contributed to civilization out of all proportion to their number. Whether one agrees or dis-

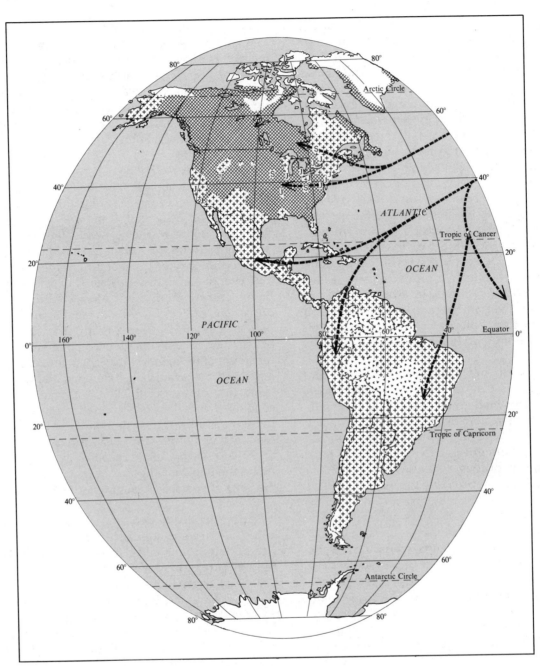

FIGURE 9.11
RELIGIONS OF THE WORLD.

agrees with the views of Marx, Freud, and Einstein, the influences these men have had on the last century cannot be denied.

Judaism's basic contribution to religious thought was the belief in one omnipotent God, Yaweh (Johovah), and in his law, the *Torah*. Unlike the amoral and indifferent native gods of the Greeks and Romans, the God of the Jews was caring, righteous, and loving. The Ten Commandments, which according to the Old Testament were given by God to Moses, established man's social order. They were taken over by Christianity and Islam

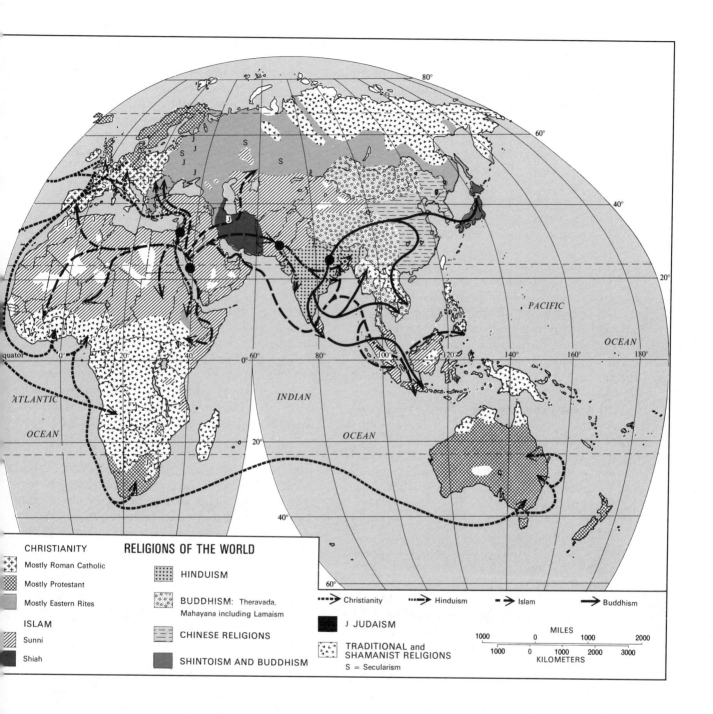

RELIGIONS OF THE WORLD

CHRISTIANITY
- Mostly Roman Catholic
- Mostly Protestant
- Mostly Eastern Rites

ISLAM
- Sunni
- Shiah

HINDUISM

BUDDHISM: Theravada, Mahayana including Lamaism

CHINESE RELIGIONS

SHINTOISM AND BUDDHISM

J JUDAISM

TRADITIONAL and SHAMANIST RELIGIONS
S = Secularism

- ▸▸▸ Christianity
- ▸▸▸ Hinduism
- ▸ Islam
- → Buddhism

MILES
1000   0   1000   2000

1000   0   1000   2000   3000
KILOMETERS

and today constitute the moral foundation of half the world's population. Like other religions, Judaism, has its sects. Reformed Judaism began as an eighteenth century movement. Unlike orthodox Jews, the reformed Jews rejected many restrictions of the law and much of the ritual. The conservatives, who occupy a median position between orthodoxy and reform, retain some orthodox customs.

The massacre of millions of Jews during World War II and subsequent migrations have drastically changed the distribution of Jewish populations.

A century and a half ago more than 80 percent of the world's Jews lived in Europe. At present the figure is less than 30 percent. Fifty percent live in the Americas and approximately 40 percent live in the United States alone. Only two and a half million, or 18 percent of the world's Jews, now live in Israel.

Christianity. Of all the religions of man, Christianity is the largest. From humble origins in Palestine, it has grown into a great universalizing faith that now includes one-third of the world's population. Based upon the teachings of Jesus Christ, disciples preached the gospel (good news) throughout the Graeco-Roman world. The apostle Paul, frequently called "the second founder of Christianity," travelled widely throughout the Mediterranean world on missionary journeys to the Gentiles. The early Christians fought an uphill struggle in the face of Roman persecution until 313 A.D. when Christianity gained equality with other religions of the Roman Empire. Nearly a century later it was declared the official state religion.

With the exception of a few minor splinterings, the Church remained intact until 1054, when a major schism developed between the Orthodox Church in the East and the Roman Catholic Church in the West. Differences of dogma and ritual continue to separate the two churches today. More importantly, however, the Eastern churches do not recognize the primacy of the Pope and his infallibility in matters of faith and morals. The Eastern Church split into a number of local groups corresponding more or less to the national states in which they exist.

The Roman Catholic Church reached the pinnacle of its power and influence during the Middle Ages only to be challenged by demands for reform. The Protestant Reformation of the sixteenth century split western Christianity into two irreconcilable groups. Scholarly studies have suggested that issues other than religious concerns played a major role in the growth and expansion of Protestantism. These include the rise of a new middle class, increased commerce and trade, and the ascendency of towns and cities. While the Catholic Church claimed to maintain an elaborated and developed version of early doctrines and rites, Protestant churches claimed to restore pristine doctrines and forms, cleansing them of false additions and developments. The early Protestant rally cry became, "justification by faith alone." Today more than 250 Protestant denominations exist in the United States alone. Protestant diversity, however, is not as great as these hundreds of denominations would suggest; indeed, 85 percent of all Protestants belong to only 12 denominations.

Roman Catholicism is the world's largest Christian denomination. On the whole, it is the dominant faith in central and southern Europe, in Poland, Ireland, and Latin America. Eastern Orthodoxy is the major religion in Greece and the Slavic countries. Protestantism is dominant in northern Europe, Anglo-America, and the Pacific Islands.

Islam. Islam (submission to God), the youngest of the world's major religions, is derived from the teachings of the prophet Muhammed. He was born in Mecca in 571 A.D. and eventually entered the caravan business. His religious call came through a series of revelations.

At a time when the Arabs were polytheists and idol worshippers, Muhammed began to preach what was to become the basic tenet of Islam and the greatest phrase of the Arabic language: *La ilaha illa Allah* (There is no God but *Allah*). His teachings were not well-received in Mecca, a city that for centuries had already been a holy place because of a black meteorite stone housed in the cube-shaped *Kaaba* and worshipped there. His life threatened in Mecca, Muhammed fled to the city of Yathrib, whose name was changed in his honor to Medina (City of the Prophet) in 622 A.D. The flight of Muhammed, known in Arabic as the *Hegira*, is considered by Moslems to be the turning point in world history and is the year from which they date their calendar. In 630 A.D. he returned to Mecca as a conquerer of the city that had once rejected his revolutionary ideas. Making his way to the Kaaba stone, he rededicated it to Allah. Two years later, in 632 A.D., Muhammed died; the Arab tribes, however, had become united in a theocracy under his control. A century later his followers had conquered most of southwest Asia, North Africa, and Spain, and had crossed the Pyrenees into France, where they were defeated by Charles Martel in the Battle of Tours in 732 A.D. Were it not for this defeat, the entire Western world might be Moslem today.

The *Koran* is the holy book of Islam. Moslem doctrine asserts it is the undistorted and final word of Allah as revealed to Muhammed. Many theological concepts of Islam are identical to those of Judaism and Christianity. Scriptures from these religions are believed to be good but not final and complete like the Koran. Islam also honors Hebrew prophets and honors Jesus as a true prophet of God, even accepting the doctrine of virgin birth but not the doctrines of incarnation and the trinity.

The Five Pillars of Islam are the principles that regulate the private lives of Moslems in their relationships to God. The first pillar is Islam's creed: "There is no God but Allah, and Muhammed is his Prophet." Practicing Moslems repeat this creed several times a day. The second pillar is prayer five times a day—upon rising, at noon, in mid-afternoon, after sunset, and before retiring—facing Mecca. Friday is a holy day when Moslems gather at noon to pray, usually in mosques (Fig. 9.12). The third pillar is charity and Moslems are admonished to give alms to the poor. The fourth pillar is the observance of *Ramadan*, Islam's holy month that commemorates Muhammed receiving his initial commission as a prophet and the Hegira. Being a month in a lunar calendar, Ramadan

rotates throughout the year. An orthodox Moslem will take no food or drink from sunrise to sunset during this month of fasting. The fifth pillar is to make a *hajj* (pilgrimage) to Mecca at least once in a lifetime if finances permit. This tenet serves a useful purpose in international relations. It brings together Moslems from around the world. In Mecca they can learn about their brothers from distant lands and return home with a better understanding of each other. Traditionally, Moslems also have abstained from gambling, drinking, and eating pork.

Today there is a religious resurgence in the Islamic world. Islam extends from Morocco on the Atlantic coast of Africa eastward through the entire Middle East and to India, Bangladesh, Indonesia, and parts of the Philippines. Claims have been made that it is the fastest growing religion in the world.

Polytheistic Religions

Hinduism. Hinduism is the world's oldest extant religion. It began to evolve about 1500 B.C. with the arrival of invading Aryan tribes in northwest India near the Ganges River. It had no founder and no rigid dogma. It has been said that Hinduism "means all things to all men." To the western mind its diversity is mystifying. Hinduism is recognized as a polytheistic religion, yet a Hindu may choose from among a wide selection of beliefs and practices. He may be a monotheist, pantheist, atheist, or agnostic. He may or may not worship in a temple. His only real obligation is to abide by the rules of the caste into which he is born.

The caste system, the most distinguishing aspect of Hinduism, evolved into its final form about 500 B.C. The order of rank is as follows: *Brahmins* or priests; *Kshatriyas* or nobles; *Vaisyas* or vassals (peasants or artisans); and last the *Shudras* or servants. Each of these main castes is subdivided into hundreds of subcastes. Marriage outside one's caste is forbidden, as are social intimacies such as drinking and eating together.

Outside the caste system are the "untouchables." These outcastes perform "defiling" tasks such as fishing, street sweeping, and leather cutting. Technically, there are more than 90 million untouchables in India today. In the past their lives were pitiful. They were despised by all; their touch, or even shadow, was believed to pollute a

FIGURE 9.12
TWIN MINARETS AND DOME OF THE MASJID JAMI (THE FRIDAY MOSQUE), ISFAHAN, IRAN.

person of high caste. In some areas of India they had to loudly announce their approach while walking down a street so others could move out of reach. They could not visit certain temples, use certain roads, or draw water from public wells. Under the leadership of Mahatma Gandhi, who championed the cause of the outcastes and referred to them as *Harijans* (Children of God), India's constitution of 1949 abolished untouchability and forbade its practice. The lot of the Harijans has improved somewhat, especially in larger cities, where common eating establishments, factory jobs, public schools, public transportation, and the overall crowding and density of population make segregation more difficult. Yet untouchability is a deeply rooted tradition and old ways die slowly.

Hinduism is characterized by yet another doctrine, the transmigration of souls, or reincarnation. Although this belief is not unique to India, the Indo-Aryan form runs as follows: The soul of a person at death does not pass into a heaven or hell; rather, it is repeatedly reborn into another existence. Successive births may be on any plane of existence—vegetable, animal, or human—and may be higher or lower than the present existence. An individual of low social status may be reborn as a Brahmin, an untouchable, a vegetable, or a worm. What determines the level of rebirth? The answer is determined by the unique Hindu doctrine of *Karma*, the law of "deeds" and "works." One's future existence is determined by one's thoughts, words, and deeds; that is, a man reaps what he sows. The caste system fits perfectly into this doctrine, for the inequalities of life have a simple explanation. A man born a Shudra must have sinned in a previous existence while a Brahmin has every right to enjoy his high station because of good deeds in previous existences. This attitude suggests that any attempt to eliminate caste inequalities and establish social justice must be considered morally wrong.

Hinduism is essentially a religion of India. While it has absorbed dissident groups over the years, it has not expanded beyond its native land to any great extent.

Sikhism. Sikhism is an offshoot of Hinduism. Dating from the fifteenth century A.D., it is one of the world's youngest faiths. Its adherents, numbering about 10 million, are concentrated in northwestern India's Punjab region. Sikhism, an example of syncretism, grew out of the confrontation between Islam, which reached India in the tenth century A.D., and Hinduism. Monotheism, its basic conviction, is drawn from Islam. Nanak, the founder of the religion of the Sikhs (literally, disciples), was born a Hindu but later renounced the traditional Hindu rites of pilgrimages, asceticism, and idolatry. Later, the caste system was abolished. Wars against the Moslems developed a long-lasting military heritage among the sect. British colonial administrators found that the meat-eating Sikhs were of fine physique, courageous, and made ideal police and military recruits; they became Britain's favorite native soldiers throughout the East. Sikhs are distinguished by their beards and long hair carefully cloaked in a turban.

Jainism. Jainism was a dissident religion that originated in the sixth century B.C. and challenged the Hindu notions of Karma and caste. It won adherents in India only and survives today as a small group with several million followers. Essentially nontheistic, Jains are noted for their extreme asceticism and the practice of *ahimsa*, noninjury to any and all living things, as a way to personal enlightenment.

Buddhism. Like Jainism, Buddhism was formed as a protest movement against traditional Hinduism. It also is an essentially nontheistic philosophy, although its popular manifestations may be polytheistic in form. It was founded in the sixth century B.C. by the Hindu Prince Siddhartha Gautama who was later called the Buddha, the "enlightened one" (Fig. 9.13). Rejecting the severe asceticism of Jainism, he sought deliverance in a moderate "middle-way" that he preached for 45 years until his death at age 80. Although the Buddha believed that the universe abounded in gods and goddesses, he did not believe that praying to them could direct man's destiny. He believed in the Hindu Law of Karma and the transmigration of souls, but he modified both doctrines. The cause of rebirth could be eliminated, he believed, by following the four noble truths. The first is recognizing that life is full of suffering; the second is the understanding of the causes of suffering, mainly desire; the third is that suffering ends through a cessation of desire; the fourth is understanding that suffering can end by following the noble eightfold path, the path that leads to no desire. The path consists of right belief, right resolve

FIGURE 9.13
FOUNDED IN INDIA BY A HINDU PRINCE, BUDDHISM
DATES FROM THE SIXTH CENTURY B.C. It is the dominant
religion in a number of southeast Asian countries.

(renounce carnal pleasure and harm no living thing), right speech, right conduct, right occupation or living, right effort, right mindedness, right meditation. Finally, through intense meditation one can achieve sainthood and the assurance at death of an end of rebirth forever and entrance into a state of blissful nonexistence—*nirvana.*

Buddhism reached great heights in India in the third century B.C., when Emperor Asoka became a convert. Asoka might be considered the religion's second founder, for he helped turn an Indian sect into a world religion. He sent missionaries to Sri Lanka, Thailand, and Cambodia. Eventually the religion spread to China and became popular in the fifth century A.D. Later it spread to Korea, Japan, Burma, and Tibet. Buddhism enjoyed great prestige in India for 800 years but later died out in its native land where it was absorbed by a rejuvenated Hinduism.

There are two main Buddhist sects, each calling itself a *yana*, a raft, ferry, or vehicle. One group claimed to be larger and called themselves *Mahayana*, the Great Raft. The second group, *Theravada* (Way of the Elders) came to be known as *Hinayana*, or Little Raft. Mahayana Buddhism accepts a number of divine beings, accepts Buddha as a savior, is more liberal, and has incorporated many aspects of local religions. Most of its followers are in East Asia. The Hinayana group is simpler and more conservative. Prayer is confined to meditation, and religion is considered a full-time job, primarily for monks. Buddha is regarded as a saint. Most of its followers are in Sri Lanka and Southeast Asia.

East Asian Religions. Traditional Chinese religious ideology consists of an intermingling of Confucianism, Taoism (pronounced Dowism), and Buddhism. Both Confucius and Lao-tzu, the legendary founder of Taoism, lived in the sixth century B.C. In the midst of tyranny and corruption, Confucius preached an ethical behavior based upon a social system of good personal conduct and high moral quality. The basic moral principle of the system was the maintenance of *jen* (humaneness) between men: Treat those who are subordinate to you as you would be treated by those in positions superior to yours. This is the Confucian Golden Rule. Confucian thought became the basis of Chinese education and the basic principle for the training of government officials. For more than 2000 years his teachings affected a quarter of the population of the world.

Tao means "path" or "way;" in general, "the way to go." To the Taoists, the secret of life was to be found in nature and her laws. Taoism, with something of a modern-day ecological approach, sought to be in tune with nature. Man, to follow the Tao, gives up striving and escapes desire by mystical contemplation. Later Taoism adopted many gods. It was, however, a philosophy too subtle for the average mind. Seeking more emotion than that offered by Confucianism, aspects of sorcery and magic entered popular Taoism at various times.

Buddhism, when it came to China, was considered too other worldly by orthodox Confucians. They did not like the attention given to transmigration, birth, death, and self-salvation, for they believed these doctrines diverted men from service to society. Nevertheless, the novelty and freshness of Buddhism negated their protests, and this faith intermingled with Confucianism and Taoism to form the Chinese religious ideology.

Shintoism. Shintoism (The Way of the Gods), the native religion of Japan, was originally a cult of nature worship loosely tied to ancestor worship. Syncretism occurred, and Shintoism later was modified by Buddhism and Confucianism. Japanese emperors were held to be descendents of the sun goddess, but the belief of divine ancestry was disavowed by Emperor Hirohito in 1946.

East Asian religions are characterized by ancestor worship and by their deep aesthetic appreciation of nature. Unlike the otherworldliness of Hinduism, man and nature are viewed as organically related. This philosophy is reflected in oriental art, particularly landscape scenes.

Secularism. There is a difference between having a religion and being religious. Today, in many parts of the world, traditional organized religions are declining in membership. This noninvolvement may reflect individual personal choice or it can result from a government's active support of a secular ideology. Communism, for example, as a universalizing ideology, is viewed by some as constituting a widespread quasi-religion with a worldwide mission.

Mankind has evolved into a seemingly endless variety of human communities. Within each there developed taboos, languages, religions, perceptions, and technologies. The emergence of agriculture was also a part of humanity's cultural evolution. Perhaps the latter step was inevitable as people came to understand the ecological relationships between themselves and the plants and animals they used for food. Out of these relationships evolved a number of agricultural systems that are examined more fully in the following chapter.

KEY TERMS AND CONCEPTS

Homo sapiens
cultural universals
environmental perception
cultural landscapes
diffusion
acculturation
culture hearth
language

dialects
multilingualism
lingua franca
taboo
ethnic religions
universalizing religions
secularism

DISCUSSION QUESTIONS

1. What is culture? What is cultural geography?
2. What is meant by "environmental perception?"
3. Discuss the term "cultural landscape." Why may a cultural landscape vary from one area to another?
4. Distinguish between "diffusion" and "acculturation." List several examples of each.
5. Define "culture hearth" and list the world's principal hearth areas.
6. Distinguish between speech, language, and dialect. What is a lingua franca?
7. How might the religion of an area be reflected in its cultural landscape?
8. Distinguish between universalizing and ethnic religions. Give examples of each.
9. Discuss several characteristics common to the world's principal monotheistic religions.
10. What is secularism?

REFERENCES FOR FURTHER STUDY

Dohrs, F., and L. Sommers, eds., *Cultural Geography: Selected Readings*, Thomas T. Crowell Co., New York, 1967.

Jordan, T., and L. Rowntree, *The Human Mosaic: A Thematic Introduction to Cultural Geography*, Harper & Row, New York, 1979.

Sopher, D., *Geography of Religions*, Prentice-Hall, Englewood Cliffs, N.J., 1967.

Tuan, Yi-Fu, *Topophilia: A Study of Environmental Perception, Attitudes, and Values*, Prentice-Hall, Englewood Cliffs, N.J., 1974.

Wagner, P., and M. Mikesell, eds., *Readings in Cultural Geography*, University of Chicago Press, Chicago, 1962.

Zelinsky, W., *The Cultural Geography of the United States*, Prentice-Hall, Englewood Cliffs, N.J., 1973.

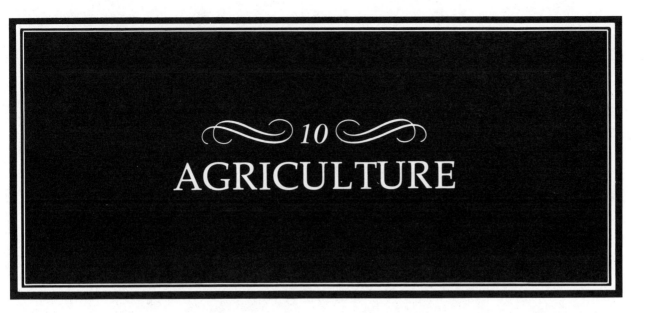

10
AGRICULTURE

AGRICULTURE IS THE FOUNDATION OF CIVILIZED LIFE. Mankind's transition from a hunting and gathering to a herding and farming economy has been called the Agricultural Revolution, an event that apparently occurred independently and simultaneously between 6000 and 10,000 years ago in Asia, Africa, and America. With the advent of agriculture, people learned to domesticate plants and animals; the food supply became more dependable, and life more certain. It was the European explorers of the Renaissance, however, who fostered the great exchange of crops, domesticated animals, and farming practices between the different regions of the world, thereby greatly increasing the earth's capacity for food production. The white potato, native to Peru and Boliva, was taken to Europe, where it became the staple food of millions of people. Wheat made its way from Southwest Asia to Europe and thence to North America, today the world's greatest area of surplus grain production. Corn, domesticated by the Indians, was brought to Europe by Columbus. Soybeans, introduced from China, are now a major crop in the United States. Rice, indigenous to the Orient, is grown in Europe, Africa, and North and South America. Sugar cane, a major crop in Middle America, came from Southeast Asia. And the Europeans, with their plows and draft animals, opened new lands to agricultural production in the Americas.

The agricultural patterns that have emerged throughout the world are a result of a number of ecological factors—a combination of climate, soil, and slope—as well as of a number of cultural and economic factors. Wherever farming is practiced, agricultural systems are characterized by certain dominant crops (Fig. 10.1). Yet, considerable variation in farming practices occurs within each broad agricultural region. Farms vary from large to small, and from privately owned to communally shared. The intensity of managment varies as do the secondary crops and the types of livestock raised. Notwithstanding the great diversity in agricultural patterns, food production may be classified as either *subsistence* or *commercial*.

There are three forms of subsistence economies: gathering, herding, and farming. In subsistence cultures there is little surplus food produced and little or no exchange. The subsistence farmer, for example, grows food only for himself and his family and consumes only the produce of the land he occupies. Subsistence farmers rarely participate in market activities, either as buyers or sellers. In an economic sense they live cocoon-like existences. The commercial farmer, on the other hand, grows a crop for sale in the market, takes the money so earned, and buys the goods he wants to consume.

In reality, of course, many subsistence farmers do have some limited interaction with the market. Thus, a subsistence farmer in the Amazon Basin may carry a basket of papaya into town on market day and exchange it for rum, salt, or flashlight

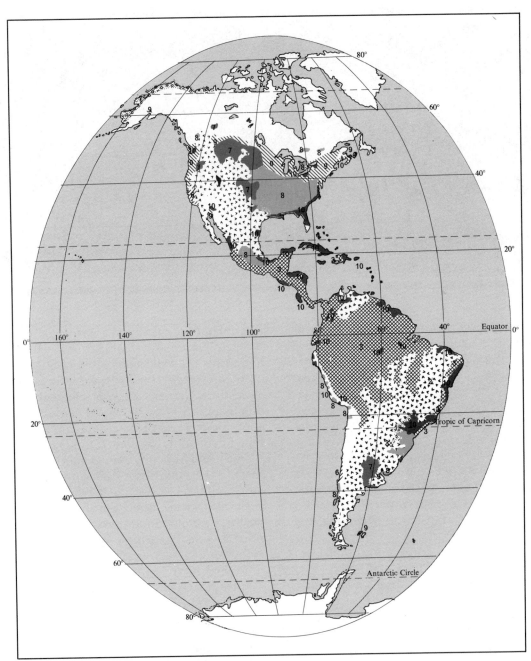

FIGURE 10.1
WORLD AGRICULTURE.

batteries. To that extent, he is something less than a pure subsistence farmer. By the same token, most commercial farmers produce at least some food for their own tables by tilling a vegetable patch, or by keeping a few chickens or a cow, and to that extent they also are something less than purely commercial farmers.

SUBSISTENCE GATHERING

The most ancient of agrarian lifestyles, *subsistence gathering*, was common to all peoples prior to the development of agriculture. The technique still survives, but subsistence gatherers are now found only in limited areas of the world. Examples

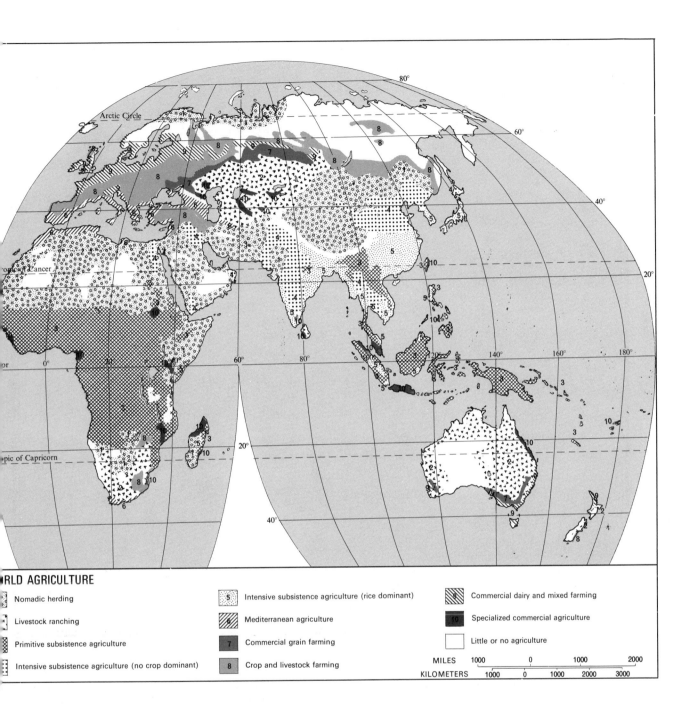

are: the Eskimos in the Arctic, the Auca in the Amazon Basin, and the Semang in Malaysia. Subsistence gathering is a vestigial way of life that persists primarily because of the isolation of the peoples involved.

Subsistence gatherers live by collecting berries, nuts, fruit, seeds, and roots; they may also gather insects and supplement their meager diets by fishing and hunting. These activities represent an inefficient land use; returns per hectare (2.5 acres) are low. Accordingly, it takes a large area to support a small group of people. Population densities tend to be low in those areas devoted to subsistence gathering, and the people exist in small

SUBSISTANCE GATHERING

REGIONS OF PLANT DOMESTICATION

Mediterranean Basin

Barley	Oats
Celery	Peas
Grapes	Olives
Dates	Asparagus
Lettuce	Carrots
Garlic	Sugar beets

Southwest Asia

Wheat	Beans
Poppies	Rye
Onions	Beets
Spinach	Apples
Peaches	Cherries
Plums	Walnuts

Southeast Asia and Indonesia

Bananas	Bamboo
Sugar Cane	Yams
Cucumbers	Coconuts
Taro	Breadfruit
Durian	Ginger
Nutmeg	Cloves

Southeast China

Cabbage	Tea
Citrus fruit	Ramie
Water chestnuts	Litchi

West Africa Hill Lands

Oil seeds	Oil palms
Tamarinds	Kola nuts

East India and Burma

Rice	Mangoes
Indigo	Sorghum
Kapok	Jute
Eggplant	Millet
Safflower	

Northern China

Buckwheat	Rhubarb
Jujubes	Mustard
Persimmons	Mulberries
Radishes	Soybeans
Apricots	

East Africa Highlands

Oil seeds	Melons
Coffee	Okra
Gourds	Castor beans

Mexico and Central America

Sweet potatoes	Tomatoes
Cotton	Chili peppers
Muskmelons	Avocados
Beans	Squash
Maize (corn)	

Andrean Highlands

White potatoes	Strawberries
Papayas	Cubio
Pumpkins	Quinoa

Eastern Brazil

Peanuts	Cacao
Tobacco	Cotton
Brazil nuts	

bands that rarely exceed a few dozen in number.

The Bushmen of the Kalahari Desert in southwestern Africa are an example of the subsistence gathering life style (Fig. 10.2). They live in clans of extended families that typically include a few adult males, their wives, and children. They are nomads, each clan having recognized authority over a specific territory through which they wander in the course of the seasons. Only occasionally do they construct shelters; more usually they spend the night lying in the open.

The Bushmen have developed important skills that enable them to cope with the rigors of life in the Kalahari Desert. Thus, they know where to find water; they know when and where certain fruit, seeds, and roots can be found; they know where to find succulent grubs valued as delicacies; they know how to make simple tools and weapons; and they are skilled in tracking and killing game. Hence they survive. However, the Bushmen produce no surpluses. Infant mortality rates are high and life expectancy is short. Bushmen enjoy few comforts and live most of their lives on the verge of starvation. When times are harsh, as

FIGURE 10.2
SUBSISTENCE GATHERING REMAINS A VIABLE WAY OF
LIFE FOR A FEW GROUPS OF PEOPLE WHO LIVE IN BAL-
ANCE WITH NATURE. A woman of the Bushman group dig-
ging for edible roots in Botswana.

hair, and wool. Animal leather is used for sad-
dlery, harnessing, belts, and straps; teeth and
bones are used for making tools and artifacts,
while animal dung is used as fuel. The camel,
horse, and burro are used for transport.

Like subsistence gathering, subsistence herding
is a vanishing way of life. It still persists, however, in the dry regions of North Africa, the Middle East, and inner Asia. Another subordinate region of herding is found in the Arctic, extending
from northern Scandinavia eastward across the
northern reaches of the Soviet Union into Alaska
and Canada. These Arctic herders are dependent
on caribou or reindeer for their needs.

Subsistence herders follow a nomadic existence
on lands that are marginal for other types of use.
Their flocks graze on the meager forage available
to them in areas where climatic limitations inhibit
the rate of plant growth. Pastures are sparse. Large
flocks cannot be permanently supported. It is necessary, therefore, to rotate the herds between
areas of pasture, to move with the seasons, and
to use relatively large areas to support a small
number of animals and people. The term *transhumance* is used to denote this seasonal migration
of herds.

Nomadic herders tend to be organized along
tribal lines where loyalties are strong. They follow
a way of life that has changed little in thousands
of years. These people are characterized as fierce,
proud, independent, and austere in their habits.
Because they inhabit areas that are uninviting to
others, because of their mobility, and because
their life-style requires them to be hardy, many
nomadic tribes have been successful in retaining
a considerable measure of autonomy. Inevitably,
efforts are made to induce the tribes to settle down
in one place under conditions making for more
easy control over them. Thus, in recent decades
the Soviet Union has attempted to settle the tribes
of Turkestan. The Kurds who recently wandered
between Iran, Iraq, Turkey, and Soviet Armenia,
have been settled, although they still engage in
occasional warfare with the authorities in Iraq.
Bakhtiari still wander in Iran, as do Somalis in
East Africa, Tuareg in the Sahara, Chukchi in Siberia, and Eskimos in Canada.

Efforts leading to detribalization have sometimes caused ecological disasters, as nomads have
been persuaded to settle down and take up farming in areas that are marginal for that purpose.

in a drought or period of natural disaster, weaker
members of the tribe may perish. Still, subsistence gathering remains a viable way of life in
which man can live in balance with nature, and
2000 Bushmen still follow this ancient pattern in
the inner reaches of the Kalahari Desert.

SUBSISTENCE HERDING

Subsistence herding represents a fundamental
step in technological advancement over subsistence gathering, because, in herding, man intervenes in nature to increase the productivity of the
environment. Thus, the herder differs from the
gatherer, who collects the bounty of nature but
does nothing to increase that bounty. The herder
intervenes by domesticating animals and then
protecting and encouraging them to flourish and
increase. The dependency of man on animals can
be complete, as in the case of the Bedouin nomads
of the Sahara Desert whose flocks provide them
with food (meat, milk, butter, and cheese), clothing, tentage, carpets made from camel and goat

For example, in the Sahel region along the southern edge of the Sahara, numbers of Tuareg were settled and established as commercial farmers and ranchers. As rainfall was inadequate to sustain this activity, wells were sunk and water deficits were made up by tapping underground sources. A dry cycle began in 1967, and for six years the rains failed. The water table fell and underground sources ran dry. Since underground sources represent rainfall accumulated over decades or centuries, the water supply system collapsed. As the herds died off from lack of food and water, the situation of the people grew desperate. An aid program brought in foodstuffs from outside, but the effort was hampered by inadequate transport facilities and large numbers of people died of starvation.

In the case of the Sahel, disaster was invited because the land was pushed beyond its capacity. The Tuareg would have been better advised to follow the dictates of their ancient tribal ways, rather than adapt to the short-run solution offered to them by modern technology.

SHIFTING CULTIVATION

Shifting cultivation, or *slash-and-burn farming*, represents the simplest level at which mankind works with domesticated plants. As a result of tending crops, the food supply is increased and becomes more reliable, and mankind is motivated to live a sedentary existence so as to protect crops and harvest them. Settlements come into existence. Although these settlements remain small, they nevertheless constitute a fundamental step in the evolution of human society and in mankind's efforts to gain control over the environment.

Shifting cultivation is dominant in large areas of the humid tropics where it is particularly suited to cope with the environmental conditions found in the rainforest (Fig. 10.3). It prevails, therefore, in a broad belt that circles the earth near the equator and that includes most of the Amazon Basin and considerable areas of Middle America, Africa, Indonesia, New Guinea, and the Philippines. It occurs also in the interior of the Indo-Chinese peninsula.

In shifting cultivation the farmer, using an axe or machete, makes a clearing in the forest. This is done by cutting down small trees and bushes; the

FIGURE 10.3
OPENING OF NEW AGRICULTURAL LAND BY USE OF FIRE IN CENTRAL SUMATRA, INDONESIA.

larger trees may be felled or girdled. On a dry day the cuttings are piled and burned. The mineral content of the ashes contributes to the fertility of the soil. Crops are then planted, sometimes separated according to species, more often together. For example, in the Americas beans are often planted between rows of corn. Local usages differ, but there tends to be a heavy dependence on root crops such as cassava (manioc). Other crops planted are corn, beans, upland rice, millet, sorghum, sugarcane, yams, squash, tomatoes, melons, and bananas. Although most of these crops occur in all continents where shifting cultivation is practiced, in America corn and cassava are the principal crops, in southeast Asia, rice, and in Africa, millet and sorghum. Agricultural implements consist of a dibble stick or hoe. Draft animals are absent. Soil preparation is minimal; a small hole is scratched in the soft soil containing a layer of wood ash, and a seed is planted and covered.

Because the soils are fragile and quickly lose their fertility when stripped of forest cover, yields of successive plantings diminish rapidly. Destructive erosion may also set in as the soils become exposed to falling rain and running water. After two or three plantings, returns diminish to the point where the farmer must change to another location. There, he repeats the cycle of clearing, burning, and planting. Attention is then gradually shifted to the new locale and the old field is abandoned. Eventually the old field reestablishes itself with a forest of second-growth trees. Under these conditions the field's fertility will be restored by natural processes in a period of from 10 to 20 years, at which time the land may be cleared again. Cropping a clearing for shorter periods of time than it is allowed to lie fallow is a distinguishing characteristic of shifting cultivation. The ratio between the productive period of a field (2 to 3 years) and the fallow period (10 to 20 years) means that only a small fraction of an area devoted to this system can be productive at any given time. This factor limits the number of people that can be supported by such a system; accordingly, population densities remain low. In the Amazon Basin, for example, the average population density is approximately the same as that found in the interior of the Sahara Desert.

Shifting cultivation is a vanishing way of life in some areas, but it still supports more than 100 million people in tropical lands. It is known as *milpa* in Latin America, *fang* in Africa, and *ladang* in southeastern Asia. Geographers usually refer to it as *swidden*, on the grounds that the term lacks regional association. It is expected that shifting cultivation will decline as regions where it now flourishes are penetrated by transport routes, which tend to bring commercial farming in their wake.

INTENSIVE SUBSISTENCE AGRICULTURE

Intensive subsistence agriculture supports one-third of the world's population. A crescent-shaped area of intensive subsistence farming skirts the Orient, and is broken only by a population of shifting cultivators in the interior of the Indo-Chinese Peninsula. Other regions devoted to this activity include the Nile Valley, Mesopotamia, and some of the intermontane valleys of the Andes and Middle America (Fig. 10.1).

Two systems of intensive subsistence agriculture are recognized. The first, *sawah agriculture* (wet rice cultivation), is dominant in humid areas; the second, found throughout cooler and dryer areas, depends on cereals raised in accordance with dry-farming techniques. Of the two, high-yielding sawah agriculture is more important. It covers a wider area and supports more people. In the sawah regions, such irrigation facilities as dikes have been constructed, permitting precise control of water levels. In some areas where level land is limited, terraces have been carved into hills and mountainsides (Fig. 10.4). During the warm season, the fields are flooded and paddy rice is cultivated. If the growing season is long enough, and if water is available, two successive rice crops can be produced annually. In other areas, however, the fields may be drained to permit the raising of a second "dry" crop that may include grains, beans, or sweet potatoes during the cool or dry months. As another alternative, flooded paddy fields may be stocked with fish to provide an additional harvest.

Whatever the local system, it should be noted that man/land ratios are high throughout the region of sawah agriculture and, in consequence, the farming population is forced to maximize returns from the land. All possible acreage is pressed into service, including mountainsides that are terraced and brought under cultivation. There is heavy use of fertilizers, including human and an-

FIGURE 10.4
THE IFUGAO PEOPLE OF THE PHILIPPINES HAVE BEEN CULTIVATING THEIR SPECTACULAR RICE TERRACES FOR 3000 YEARS.

imal wastes and other waste material from households. The farmers, heirs to generations of accumulated folk wisdom, exhibit great skill in crop selection and soil science. They have learned, for example, that a mixture of soil particles overlying an impervious clay layer two feet below the surface is ideal for rice; the loam topsoil permits root development while the impervious layer maintains saturated conditions at the surface. Implements may be crude and techniques may appear to be outmoded, but these appearances are deceptive because the intensive subsistence farmer is closely attuned to the environment. Admittedly, sawah farmers tend to be undercapitalized and have little modern equipment, but they compensate by large investments in human labor. As regards technique, the sawah farmer exhibits a high degree of ingenuity and sophistication. Yields per hectare are high.

Sawah farms tend to be small; the average size ranges from less than two hectares (five acres) in the lowlands of Japan to about four hectares (10 acres) in India. Sawah farms in China averaged about three hectares (seven acres) before the Chinese Revolution of 1949.

Animal husbandry plays only a minor role in intensive subsistence farming. Farmers in these densely populated areas cannot devote large acreages to raising beef and dairy animals. Goats, hogs, and poultry are their principal sources of meat. These animals also serve as scavengers, consuming products that might otherwise go to waste. Sawah farmers, because they operate small farms, tend to be poor. Their diet consists primarily of inexpensive foods, mostly vegetable carbohydrates. They may occasionally enjoy the luxury of an egg, or chicken, fish, or pork. Chinese cooking, for instance, exhibits great skill in making a little meat go a long way.

THE IMPACT OF THE GREEN REVOLUTION

The term "Green Revolution" was coined in the late 1960s to describe the impact of the new high-yielding wheat and rice strains, and of their associated agricultural technologies, on the food-production capacities of several developing countries. Since that time the Green Revolution has been both praised and damned. In the late 1960s, as use of the new strains spread rapidly in Asia, optimism was the rule, and claims were made that many underdeveloped countries would soon be self-sufficient in, and even exporters of, cereal grains. But in the early 1970s, poor weather and global economic constraints slowed the progress of the revolution considerably. The possibility that population growth would soon outstrip the world's capacity to produce food was once again apparent. Pessimists proclaimed that the green revolution had been oversold by the developed nations and over-bought by the developing nations, and that mass starvation was inevitable. As in many scientific controversies, the truth lies somewhere between the extremes of optimism and despair.

Source. Maarten J. Chrispeels and David Sadava, *Plants, Food, and People* (San Francisco: W. H. Freeman and Co., 1977), p. 208.

Rice yields more food per acre than any other grain and more people depend on it than any other single foodstuff. Approximately 90 percent of the world's rice crop is produced in the paddy regions of the Far East, while another five percent is grown in the adjoining dryland regions of the interior (Fig. 10.5). This rice is not mobile in an economic sense, but is consumed close to where it is grown. Only a small percentage of it moves in international trade. Transport facilities are not well developed within sawah areas; thus, most Oriental rice farming remains at least a quasi-subsistence activity.

Northern China is too cool and western India too dry for paddy farming, so emphasis in these locations is shifted to other crops such as upland (or dryland) rice, wheat, millet, sorghum, soybeans, and a wide variety of vegetables. Man/land ratios are lower here, farm size is larger, and yields per hectare are lower than in the sawah regions.

Outlying regions of intensive subsistence agriculture show local characteristics in keeping with the peculiarities of the environment and the heritage of the people. Thus, Mesopotamia and the Nile Valley are hydraulic civilizations with farming techniques dependent on river water. Farmers raise a broad array of crops including rice, corn, wheat, and vegetables. On the other hand, Amerindian farmers in the high valleys of Peru and Ecuador depend upon the potato and corn, while in Guatemala they live mainly on corn and beans.

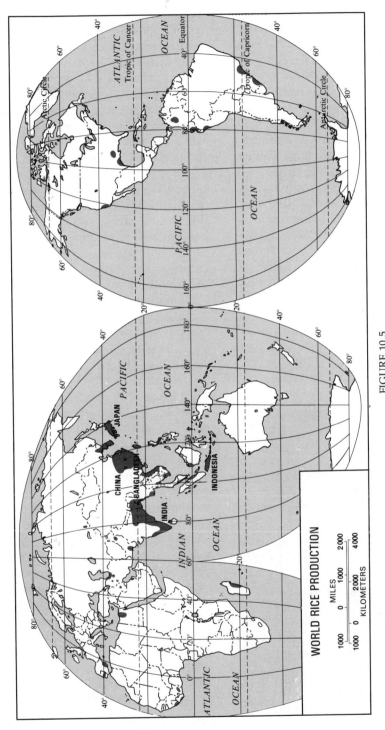

FIGURE 10.5
MAJOR WORLD AREAS OF RICE CULTIVATION.

WORLD RICE PRODUCTION

MILES
1000   0   1000   2000
1000   0   2000   4000
KILOMETERS

INTENSIVE SUBSISTANCE AGRICULTURE

253

Thus, though the crop mix and the environment may vary, the life style and techniques of these peoples permit us to classify them as intensive subsistence farmers.

PLANTATION AGRICULTURE

Thus far we have examined predominantly subsistence types of agriculture. We now turn to types that are classified as commercial agricultural systems. We will first look at plantation agriculture.

A plantation may be defined as a large landed estate that specializes in growing some single crop for export. Traditionally, plantation products have included sugar, rubber, palm oil, cotton, bananas, coffee, tea, cacao, copra, and jute. Plantations tend to be concentrated in humid coastal areas of the tropics (Fig. 10.6). Coastal locations are preferred because proximity to seaports reduces transport costs. Humid areas are selected because plantation crops tend to need a lot of water. For these reasons, plantations are concentrated in the following regions: the West Indies and tropical coastal areas of Latin America, Africa, southeast Asia, and Australia.

Most plantations are controlled by foreign in-terests, frequently under a corporate system of ownership. Management is exercised by a small cadre of professionals who supervise the field hands. And while the trend is toward mechanization, plantations are responsive to cost relationships; because they exist in areas where machines are expensive and wage rates low, plantations still employ much field labor.

Well established commercial linkages connect plantation areas and market areas. The general pattern of this trade moves plantation commodities from less developed tropical nations to the industrialized regions of the middle latitudes. There is a counterflow of manufactured goods. A certain amount of regional specialization exists. Southeast Asia produces more than 90 percent of the world's natural rubber, while 90 percent of the banana exports originate in Latin America, which also dominates in coffee. West Africa leads in cacao production, while most copra comes from the islands of southeast Asia.

Plantations often have facilities for initial processing, although the degree will differ with the crop. Sugar cane, for example, is crushed soon after cutting to extract the juice, which is then processed into raw brown sugar. Because processing involves considerable weight loss, crushing takes place near the plantations. Accordingly, sugar mills dot the landscape in areas where cane is grown, and those plantations that are large enough have their own mills.

Plantation agriculture exists within a setting that sometimes give rise to conflict. To some, the system may be viewed as an enterprize organized by and for people of high-technology societies and works to their advantage while it ignores the interests of the host country. Also, the charge is made that plantations have traditionally been exploitive and are frequently accompanied by a stratified social order and inequality of income. It is further charged that the multinational corporations that own many plantations are simply too big and too powerful and are thus an unhealthy presence in a small, poor country.* Another dilemma is that the host country may find itself saddled with an unstable economy if its export sector is dominated by one commodity, sales of which

FIGURE 10.6
CUTTERS AT WORK ON A BANANA PLANTATION IN ECUADOR.

* For instance, W. R. Grace & Co., which has extensive interests in Central America, has a larger annual income than any of the local republics.

may be highly vulnerable to world market conditions. Questions of land tenure can be another irritant, particularly when local farmers are pushed onto the hillsides because the rich bottom lands are owned by plantations. This particular condition may inhibit the growth of food for the local market, while job opportunities offered by plantations may be limited and seasonal.

Because the profits earned by plantations tend to be exported, the almost inevitable result is to generate local support for *economic nationalism:* a feeling that every country should retain control over its own productive facilities. Even though the plantation may be able to demonstrate that it uses what was formerly idle land; that it built roads, railroads, and ports; that it supports schools and hospitals; that it introduced new crops and scientific methods of cultivation; that it pays wages that exceed the regional norm; that it pays heavy taxes; and that the crops it sells provide much of the foreign exchange needed to purchase manufactured goods; still the plantation system exhibits social defects and is often viewed as foreign economic interference. Thus, current trends are moving in the direction of land reform and confiscation. Some countries have nationalized all landholdings and former plantations are now op-

erated as state farms. The Cuban experience provides an extreme example of this trend, but similar pressures can be identified in virtually every tropical country in Latin America, Africa, and Asia where plantations are found.

MIXED CROP AND LIVESTOCK FARMING

Mixed farming, also known as *general farming*, is an agricultural system in which the farmer raises both crops and livestock. The system is dominant in two large areas, one of which stretches across Europe from the British Isles to northern France, Germany, Poland, the northern Balkans, and across the Soviet Union into Siberia; the other lying in the American Midwest, most intensively in the Corn Belt that stretches westward from central Ohio for 1600 km (1000 mi) to eastern Nebraska and includes adjacent areas of Michigan, Wisconsin, Minnesota, and Missouri (Fig. 10.7). Mixed farming has also spread throughout much of the southeastern part of the United States, but less intensively. Subordinate areas of mixed farming are found in southern Brazil, eastern Argentina, and South Africa.

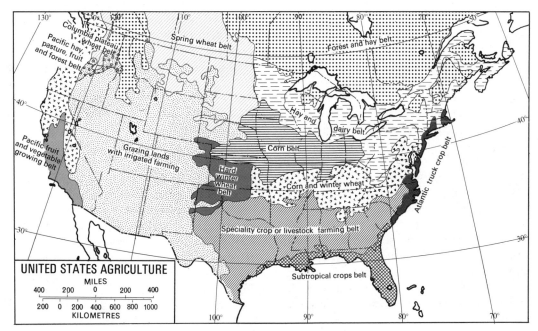

FIGURE 10.7
UNITED STATES AGRICULTURE

The mixed farmer operates by raising crops and livestock. The crops can be sold directly, or fed to livestock also raised for sale. The mixed farmer thus enjoys a certain flexibility and can shift his production in response to changes in market prices. The crop-mix and technology will vary from one region to another according to local circumstances. For illustration, we will compare conditions in the American Midwest with those in Europe.

In the United States, the Corn Belt is not only the agricultural heartland of the country but is also the most productive single farming area in the world. The area, devoted to mixed farming, supports the densest population of cattle and hogs in Anglo-America. The leading crop is corn, for which the hot, humid summer weather, and fertile, well drained, loamy soils of the area are ideal. Corn requires a frost-free period of at least 140 days and from 60 to 100 cm (25 to 40 in.) of rain. In the Corn Belt, most summer rains fall in the form of local showers, thus limiting cloud cover and creating ideal conditions in terms of sunshine, temperature, and moisture. As a result, the United States produces about 45 percent of the world's corn. Three-quarters of this comes from the Corn Belt.

The popular term "Corn Belt" suggests a system of monoculture, but corn, although the most widely planted crop within that region, occupies only about 40 percent of the harvested cropland. Soybean production has become increasingly important, and many farmers now follow a two-crop rotation regime of corn and soybeans. The soybean is a shallow-rooted legume whose climate and soil requirements are about the same as for corn. The area of greatest production is coextensive with the Corn Belt. No crop in the United States has experienced such a rapid expansion in production (Fig. 10.8). Today soybeans are one of the most valuable crops raised, and although production is concentrated in the Corn Belt states, it flourishes also in the lower Mississippi Valley and in the coastal regions of the southeast. Soybean oil is used as a component in margarine, soap, paint, and plastics; the meal is used as a source of protein for human consumption and livestock feed. The United States is the world's leading producer (65 percent) and exporter of soybeans. A variety of other crops are raised in the Corn Belt

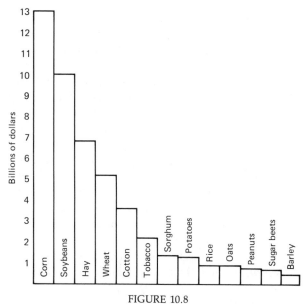

FIGURE 10.8
MOST VALUABLE CROPS IN THE UNITED STATES. Source: USDA, 1977.

including wheat, oats, hay, sorghum, and vegetables.

About 90 percent of the corn grown in America is fed to livestock, and the Corn Belt, a major producer of animal products, concentrates on growing this feed. Some yearling cattle are brought in from ranching areas of the West, while others are bred within the Corn Belt itself. Here they are fattened until they reach market weight. In recent years farmers of the Midwest have tended to specialize, many becoming either cash-grain farmers or feed-lot operators. Others still follow the traditional practices of mixed farming. Within the Corn Belt, the kind of livestock fattened usually depends upon the size of the farm. The small farms of Indiana and Ohio produce mainly hogs—animals that require less space and more attention than cattle. Cattle and hogs are fattened on the medium-size farms of Iowa, while the large farms of central Illinois sell corn and soybeans as cash grain. The hog population of the United States is largely concentrated in the Corn Belt, whose farms sustain about three-quarters of the national total. Of all domesticated animals, hogs are the most efficient in converting corn into meat. Hogs are less mobile than cattle; they tend to be born and live out their lives on the same farm. Iowa leads the nation in pork production.

The average size farm of the Corn Belt varies from 70 hectares (170 acres) in Ohio to over 110 hectares (250 acres) father west in Iowa and Missouri. The farms are well capitalized and represent a considerable investment in land, buildings, and equipment. Operations are highly mechanized, and returns per man/hour high. The average Iowa farmer raises enough food to feed himself and 45 other people. The farmers are well trained, many of them being graduates of the agricultural colleges that dot the region. They are thus prepared to take advantage of the most recent improvements in agricultural technology and employ the most advanced techniques with regard to fertilization, weed and pest control, high yielding seed strains, and cost control. Mixed farming in the American Corn Belt is modern, efficient, productive, and profitable. It is also found in other parts of the United States, particularly in the mid-Atlantic states and the South, but there it tends to be scattered and makes a smaller contribution to national food supplies.

The environment of northwest Europe's mixed farming areas differs in several important ways from the American Midwest. Europe is located farther north, which results in a shorter growing season with lower summer temperatures. Corn does not do as well under these conditions, so the farmers of this area use a different crop-mix than the Americans. Root crops are more important in European than in American agriculture. East and West Germany, Poland, and the Soviet Union, for example, rely heavily on root crops, as evidenced by the regional diet which features potatoes, beets, and turnips. Potatoes, tolerant of poor soils and a cool, damp climate, are raised in enormous quantities; they are used for human food, stock feed, and the production of industrial alcohol and potable spirits. Also grown are sugar beets, cabbages, and rutabagas. Sugar beets, grown in Europe since the 1870s, are raised on the better loamy soils throughout the region. The beet tops are fed to livestock, as is the beet pulp after the juice has been extracted for sugar production. Wheat is the leading cereal and is grown to some degree in most European farming areas, often in rotation with sugar beets. Rye is another important grain of the area; it is grown on the sandy plains of northern Europe. Rye is an important bread grain for many people in Germany, Poland,

and the Soviet Union, and accounts for the dark breads of the area. Barley and oats are raised in various parts of Europe; most is used for livestock feed. Corn, with its need for long, hot, moist summers, is raised in the Danubian plains of eastern Europe, in the Po Valley of northern Italy, and in southwest France (Fig. 10.9). A variety of fruits are also grown including apples, grapes, cherries, pears, and plums.

Livestock are important in Europe's mixed farming economy. Throughout much of the area, grazing land is limited and livestock are penned in courts or yards and fed farm by-products, coarse grains, and root crops raised for them on mixed farms. In the United Kingdom cattle and sheep are important, with the latter dominant in the uplands of Wales and Scotland. Beef and pork are produced throughout France, but especially in the cooler and wetter coastal areas and the more rugged Massif Central. More cattle are raised in France than in any other European country. Livestock production plays an important role in the agricultural economy of all parts of Germany. Emphasis is placed on the raising of cattle and hogs, and West Germany produces more hogs than any other European country.

The average mixed farm in Europe is smaller than its American counterpart. Its operations tend to be less mechanized and there is greater use of draft animals and manual labor; the average income per farm is lower. Although mixed farming in western Europe is commercialized, considerable remnants of subsistence activities still exist towards the east. Traditional farming methods and low yields are found in many of the overpopulated rural areas of East Central Europe. Here state authorities have introduced programs for the purpose of increasing agricultural productivity. The process of collectivization is complete and in most countries nearly all farmland is now in collectives or state farms. Collectivization, however, was strongly resisted by the farmers of Poland and Yugoslavia. Today approximately 85 percent of the cultivated land in these two countries is privately owned; the remaining 15 percent is held by cooperatives, collectives, and large state farms.

On a worldwide basis, the regions devoted to mixed farming are located in the humid mid latitudes where adequate growing seasons, abundant moisture, and fertile soils are available. The

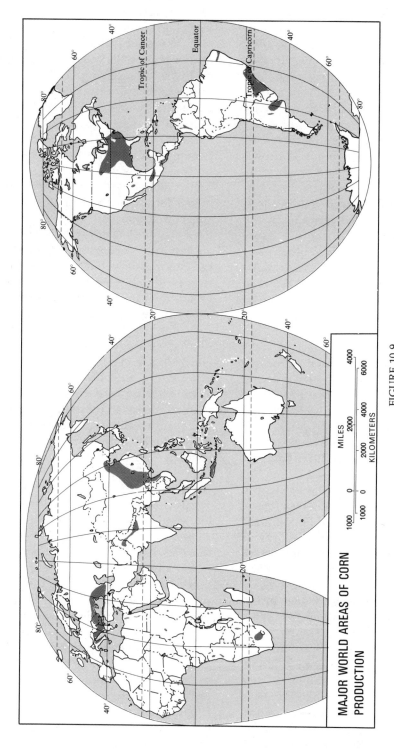

MAJOR WORLD AREAS OF CORN
PRODUCTION

MILES

KILOMETERS

FIGURE 10.9
MAJOR WORLD AREAS OF CORN PRODUCTION.

mixed farmers have access to an advanced agricultural technology that produces hybrid seeds, pesticides, fertilizers, equipment, and improved breeds of livestock, items that are all designed to help the farmer produce one of the most expensive farm products available—meat.

DAIRYING

Mankind has kept livestock, and produced and consumed milk, butter, and cheese, since prehistoric times. Only within the last century, however, has commercial dairying emerged as a specialized type of farming, oriented in particular to urban markets. *Commercial dairying* of this type is concentrated today in three major areas: northeastern United States, northern Europe, and Australia-New Zealand.

The so-called "dairy belt" in Anglo-America refers to those areas where the majority of farmers derive most of their income from the sale of milk. This is true of a belt extending from Minnesota to New England (Fig. 10.7). Adjoining areas of eastern Canada form part of the same zone. Some valley areas along the Pacific Coast also specialize in dairying. Wisconsin is the leading American state in milk production, followed by New York, California, and Pennsylvania.

In the dairy belt moisture supply is good throughout the year. However, the eastern area was subjected to glaciation and the soils that have evolved are acid, thin, stony, and occur in a terrain that is hilly to rolling. The western areas of the dairy belt are too wet for tillage but adequate for pasture. The area has a short growing season with cool temperatures characteristic of the humid continental warm-summer climatic zone. Under these conditions corn does not ordinarily produce abundant grain, but can be cut green for silage. Other crops grow well in this environment. In addition to corn for silage, hay and oats are raised to provide feed for dairy herds. These crops also form a continuous protective covering on hillside fields that might be badly eroded if planted in row crops. It should be noted that of all major types of farming, dairying is the one that is most conducive to good soil conservation practices (Fig. 10.10). This combination of environmental conditions, coupled with the economic advantages of

proximity to large urban markets, have encouraged dairying and discouraged competing types of farming in the dairy belt.

The American dairy belt has an excellent market orientation; it extends across an area that is densely inhabited and highly urbanized. Many large cities lie within the region, and dairy products must be made available every day to these market areas. However, fluid milk is 87 percent water and the most expensive dairy product to transport. Not only is it bulky per unit price of weight, it is also perishable and must be transported in refrigerated tanks. Therefore, there are limits to the distance that milk can be hauled to market.

Every city in the United States constitutes a market for fresh milk, and thus calls into existance a "milk shed" of its own, made up of dairy farms that exist for the purpose of supplying that city. For example, Miami, although far removed from the dairy belt, is surrounded by a number of dairy farms. The same is true of Atlanta, Shreveport, Phoenix, and every other American town.

Butter and cheese are easier to transport than milk; they are less perishable, of higher price per unit weight, and can absorb higher transportation costs. It follows that dairy production forms a pattern in terms of distance from market. Those dairy farms that are closest to urban areas will find it advantageous to specialize in producing fresh milk to be consumed in its fluid state. Those farther out will produce milk that is converted to butter, cheese, or other manufactured items such as condensed and powdered milk, or casein. This pattern of distribution can be seen in the dairy belt where the most easterly sections, such as New England, that are closest to the enormous urban markets of the Northeast, are devoted almost exclusively to the production of fresh milk. The western areas of the dairy belt, such as in Wisconsin and Minnesota, are farther removed from market and produce more cheese and butter.

Dairy farms vary in size, becoming progressively larger as one goes farther west. This pattern conforms to economic pressures; land in the densely settled East is high priced and must be used intensively. Thus, in New England the average dairy farm is not always self-sufficient in terms of forage and depends on supplemental feed grains brought in from the Midwest. This practice

FIGURE 10.10
THE LAND NEAR ROARING SPRING, PENNSYLVANIA HAS BEEN TURNED INTO FAMILY DAIRIES, ORCHARDS, AND
FIELDS OF HAY AND CORN.

is justified by the advantages enjoyed by the New England farmers with regard to access to market, but is also indicative of the small size of the average New England farm. The average dairy farm ranges from 50 hectares (120 acres) in Massachusetts, to 57 hectares (142 acres) in Pennsylvania, to 74 hectares (183 acres) in Wisconsin.

In Europe dairying is dominant in a zone that sweeps across the British Isles, northern France, Benelux, southern Scandinavia, northern Germany, Poland, and into the Soviet Union. In these areas a number of factors favor the dairying industry: a mild, moist climate; productive clay and sandy loam soils, an excellent growth of hay, feed grains, and root crops suitable for feed; and a dense rural and urban population. Man/land ratios are high in Europe and land use is intensive. The average European dairy farm is smaller than its Anglo-American counterpart. Much of the cropland is in pasture and supplemental feed grains are often purchased from abroad.

In Europe the farms become progressively larger toward the east where environmental limitations are more severe in terms of climate and where population densities are less.

Modern dairy farming represents an activity that generates maximum returns in areas environmentally limited for other types of farming. It is labor intensive and land intensive. It produces relatively expensive foodstuffs that tend to be consumed by higher-income groups. Cultural traits also affect the pattern of dairy product consumption; a dietary preference primarily characteristic of people of European heritage.

COMMERCIAL GRAIN FARMING

In *commercial grain farming,* a farm specializes in producing dry grain for sale at market. The largest crop is wheat; corn, barley, rye, oats, and sorghum comprise a modest percentage of the total grain tonnage produced. Wheat enjoys an overwhelming domination.

Wheat can be classified as either *winter wheat* or *spring wheat*. Winter wheat is planted in late summer or fall and establishes its roots before the winter freeze. It lies dormant through the cold months and begins to grow in spring, making use of the winter moisture supply. It is ready for harvesting by late May or early June. In areas of severe climate, however, winter wheat would be killed by frost, so spring wheat is produced. The seeds are sown as early as possible in the spring, grow through the long days of summer, and the crop is ready for harvest in August or September.

Wheat does well with 24 to 62 cm (10–25 in.) of rain and a growing season of 90 days. Minimum temperatures of about 15.5° C (60° F) are needed for ripening. These conditions prevail in the mid-latitude steppes, and it is in these regions that we see concentrations of wheat culture. The largest single region, which measures 3200 km (2000 miles) from east to west, lies within the Soviet Union (Fig. 10.11). It includes the Ukraine and extends eastward into Siberia where it forms a narrowing ribbon that ends between the taiga to the north and the desert to the south. Winter wheat is grown in the Ukraine, but spring wheat is dominant eastward in the more rigorous environment of Siberia. Another wheat region, distinctly subordinate, lies farther south in the semi-arid parts of Turkestan. Because of the enormous area covered by these two regions, the Soviet Union leads the world in wheat production.

North America has two major areas of wheat production. The winter wheat belt centers on Kansas and areas in adjacent states. The spring wheat belt extends from the Dakotas across Montana into the prairie provinces of Canada, reaching 57° N latitude in the Peace River Valley of Alberta. Between the two wheat-growing areas is a belt (in southern South Dakota and northern Nebraska) where little wheat is raised. This is the rolling Sand Hill country, an area ill-suited to cultivation. A smaller, but important wheat-growing area dominates the Columbia Plateau in eastern Washington and north-central Oregon (Fig. 10.7).

One of the advantages of wheat is its ability to grow in areas of limited rainfall. Most of the wheat raised in the Great Plains grows under natural rainfall conditions, amounting to as little as 11 inches annually in some areas. Commercial wheat farming can be a risky business in this climatically marginal area. Not surprisingly, there is a "boom or bust" character associated with wheat farming in parts of the Great Plains. This has encouraged moisture-saving "dry-farming" techniques such as strip-cropping, where rows of wheat alternate with fallow rows (the fallow rows serving to conserve this year's moisture for next year's strip of wheat). Although wheat is the dominant crop, considerable acreages are devoted to pasture, hay, and sorghum for cattle feeding—an increasingly important aspect of agriculture in the area. Increased diversification has tended to temper the "boom or bust" cycles of bumper crops and crop failures.

The wheat belt of Anglo-America lies west of the corn belt and occupies a region that is either too dry or too cold for corn. Given these climatic limitations, wheat offers the most profitable opportunity to the farmer. Wheat country, however, differs from corn country not only in climate but in other respects as well. Land use, for example, is less intensive in the wheat belt and yields per hectare are lower. It follows that a farm must be relatively larger in the wheat belt in order to serve as an efficient unit of production. Farms of 400 hectares (1000 acres) or more are not uncommon. The average farm in the wheat state of North Dakota covers 376 hectares (930 acres) as against 105 hectares (265 acres) in the corn state of Iowa. Also, because wheat farming is highly mechanized, it supports a small work force. Consequently, the density of population is low; North Dakota has an average population density of four persons per square kilometer (9 per square mile) compared with 20 per square kilometers (51 square miles) in Iowa. Fewer people means less demand for urban services, and wheat country has few large towns and virtually no large cities. The whole wheat belt from Kansas City to Denver and from Lubbock to Edmonton is an area of openess, a vast expanse of wheat fields. Occasional farmhouses are marked by windbreaks (that offer protection from the cold northwest winter winds) on their windward sides, but towns are few and far between. Such towns as do exist feature grain elevators and railroad tracks that stretch away towards the market (Fig. 10.12).

Wheat is highly transportable. It is consumed almost exclusively by humans, which means that wheat must move to those densely inhabited regions that constitute its market. The largest market areas consist of northeastern United States,

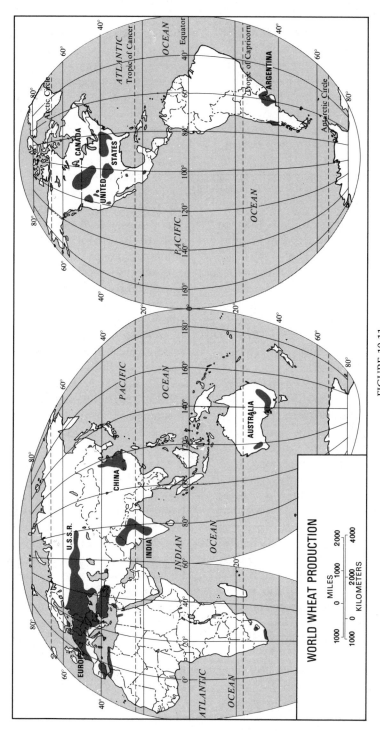

FIGURE 10.11
MAJOR WORLD AREAS OF WHEAT CULTIVATION.

States, such as the humid east, where it appears as part of the crop-mix of general farming. Other areas devoted to commercial wheat farming are the Pampas of Argentina and southeastern Australia. Because of the limited size of the domestic market, Australia and Argentina are major exporters of wheat, ranking third and fourth respectively after the United States and Canada.

COMMERCIAL GRAZING

Commercial grazing refers to the raising of livestock for meats, hides, or wool. Commercial grazing covers enormous areas, most of which are marginal for other purposes. It is dominant in the dry interior of Soviet Asia, throughout much of Australia, and in the drier regions of South Africa. It covers almost half of South America, reaching from the steppes of Patagonia through the grasslands of the llanos of Venezuela and Colombia (Fig. 10.1).

In North America ranching prevails in a broad belt stretching from northern Mexico through the western third of the United States into Canada. It is dominant in the Intermontane Basin between the Cascades and the Sierra Nevadas in the west and the Rockies on the east in areas too dry for crop farming. Ranching also covers the eastern flanks of the Rockies through the High Plains areas from western Texas to Montana. Some of the world's largest and most productive ranches are located here. Thousands of ranches have thousands of acres each. Cattle and sheep raising is widespread, with sheep tending to be concentrated in the southern areas, especially in west Texas. Throughout the ranchlands, large acreages of irrigated land are devoted to the production of fodder crops to supplement the feed supply of range animals during the winter months. Increasing numbers of cattle are fattened for market in area feed-lots, using feed produced locally by irrigation (Fig. 10.13).

Because of the low carrying capacity of the region, ranches are large and the ranching population is small and scattered, except where small clusters of population are found in association with irrigation or mining areas.

Commercial grazing tends to be associated with people living in advanced industrialized societies. Meat is an expensive item of diet and can be con-

Europe, India, the Soviet Union, Japan, and China. About 20 percent of the world crop of wheat enters international trade, a figure significantly exceeding that for either rice or corn, both of which tend to be consumed nearer to home.

All phases of wheat farming are highly mechanized including plowing, seeding, and harvesting. Because the harvesting machines involved are large, complicated, and expensive, and are used infrequently during the year, few farmers have their own harvesters. They contract instead with custom-combining crews that provide that service. The harvest starts early in summer in Texas when the winter wheat matures. The crop must be promptly harvested, threshed, and stored before the rain falls. Thousands of custom-combines move through the area annually; crews work in shifts around the clock, seven days a week, pausing only for excess humidity caused either by dew or rain. As the season advances, the crews work their way northward through Oklahoma and Kansas, then on to the spring wheat belt in the Dakotas and Montana, finishing in Canada in September before returning to their homes, usually in Texas, for the winter.

Wheat is grown in other regions of the United

FIGURE 10.13
CATTLE FEEDING YARDS AND AUTOMATED FEED MIXING MILL WITH SELF-LOADING FEED WAGONS, NEAR BAKERS-
FIELD, CALIFORNIA.

sumed in large quantities only by high-income groups such as are found in large numbers in the developed world; the largest market areas for beef are Anglo-America and northwestern Europe.

Generally speaking, the livestock of ranching areas consists of breeding stock and young animals. Calves are born on the ranches and kept there for a year or year and a half (Fig. 10.14). These yearlings are then rounded up and shipped to the feed-lot operators of the West or the Corn Belt, where they are fed-out to desirable characteristics of size, weight, and quality. They are then slaughtered and the meat products shipped to market. The largest market in the United States consists of the industrialized Northeast, where much western beef is ultimately consumed. The United States was formerly a net exporter of beef, but today consumption exceeds production, and the United States is a net importer. Major stocks come from Australia, Canada, and Latin America.

Beef cattle are also produced in considerable numbers in the southeastern United States. Here land is used much more intensively than in the West because of the higher carrying capacity of the improved pastures. A long growing season, combined with a generous supply of moisture, results in year-round grazing. Because carrying capacity is high, a cow and a calf can be grazed for a year on about one hectare. The mildness of the winters minimizes the problems of having to provide shelter and food supply for livestock during the cold months. Also, mature cattle are conveniently located with regard to market. All these factors have combined to stimulate the livestock industry in the Southeast and increase production.

Sheep graze in large numbers in Australia, the Soviet Union, New Zealand, Argentina, and Uruguay. They are raised either for meat or wool, or both. Australia and New Zealand lead the world in the export of wool and meat. Other major exporters are Argentina and South Africa. Major importers are Great Britain and Japan.

Mutton is a minor item of diet in the United States, and it is easy to lose sight of the number of sheep grazed. The largest concentration is located on the Edwards Plateau of west Texas, but sheep are grazed throughout the Rockies, frequently on lands that are marginal for cattle.

MARKET GARDENING

Market gardening is the raising of vegetables for commercial sale. In the United States, California is the leading state, producing almost one-third of the national supply. A considerable portion of California production is frozen or canned and then shipped to the northeastern part of the country, the major market area.

In the eastern states, the pattern of market gardening adapts itself to the cycle of the seasons. Thus, during the winter months southern Florida is a primary producer, particularly in the drained swampy areas around Lake Okeechobee, where vegetables are grown and sent north by truck. With the advance of spring, warmer temperatures permit plantings farther north, first in northern Florida, then in Georgia. Timing is important in getting the crop into the ground after the last killing frost, bringing it to maturity, harvesting it, and getting it on the road before competitive crops mature farther north. In late spring and summer, the center of activity shifts northward, particularly to New Jersey and the Delmarva Peninsula, where market gardening co-exists with dairying.

Market gardening represents intensive land use; returns per acre are higher than for any other form of agriculture. Therefore, market gardening can compete successfully for land in areas that are

FIGURE 10.14
A GRAZING SCENE ON THE OPEN RANGE IN IDAHO.

densely inhabited. Because every large city constitutes a market for fresh vegetables, almost every large city has a discontinuous fringe of "truck farms" that specialize in this type of production. Farming in New Jersey reflects this tendency. New Jersey is virtually surrounded by large urban markets. Competition for land is severe in the region, and considerable areas are devoted to commercial, industrial, and residential use. Market gardening survives, however, demonstrating that it is a viable form of land use.

LAND USE THEORY

Land use patterns reflect economic pressures. In a microstudy of a small area, one can assume that physical and cultural characteristics are homogeneous, allowing one to isolate theoretically the effect of a single variable as, for instance, distance from market. This approach was taken by Johann von Thünen in his classic study, *The Isolated State*, published in 1826. Taking the area surrounding a single market town, von Thünen concluded that economic activities would fall into a pattern of concentric circles, each centered on the market as focal point (Fig. 10.15). Land use would become less intensive as one moved away from the market. Thus, closest to town would be found a belt devoted to highly intensive dairying and vegetable farming. Next would lie a fringe of woodland, which supplied the daily firewood needed in the early 1800s. Beyond the woodland would lie a zone of crop farming of diminishing intensity, until, moving outward, one would reach the most remote belt of economic activity, occupied by pastures and grazing. Beyond the grazing area would be the wilderness "wasteland," lying outside the "economic reach" of the market and enjoying no interaction with it.

Attempts to verify von Thünen's theory suffer from the obvious difficulty of finding an area that is homogeneous in its cultural and physical characteristics. Even the building of a road distorts a landscape with regard to convenience of transport. Despite the difficulties, studies have been done that generally support von Thünen's hypothesis.

For a macrostudy encompassing a larger area, it is realistic to accept the fact that physical and cultural characteristics vary from place to place and

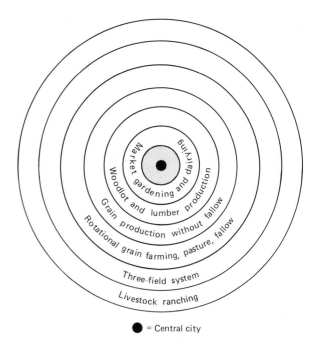

= Central city

FIGURE 10.15.
AGRICULTURAL PRODUCTION AFTER THE VON THÜNEN MODEL.

that these various characteristics exert economic pressures. Thus, the amount or distribution of rainfall will favor one crop against another. Other physical factors that might act as economic determinants are landforms (rugged or flat), the length of the growing season, temperatures, soil characteristics, altitude, and prevailing winds. Cultural phenomena that can affect land use might include population density, religious attitudes, land tenure systems, political environment, taste preferences, levels of technology, and cultural inertia.

Whatever the mix of economic pressures, a landowner will usually try to maximize returns from his land; in consequence, a coherent and recognizable pattern of land use will come into existence. Furthermore, this pattern is logical and lends itself to rational analysis. The task of geographers, therefore, is to try to understand the relationships that foster these patterns.

With this in mind, let us examine the pattern of farming activities in the United States. For the sake of simplicity four states have been selected that typify activities on an east-west axis of the country; New Jersey (market gardening), Iowa (corn belt), South Dakota (wheat belt), and Wyoming (ranching). An examination of these states

TABLE 10.1 CHARACTERISTICS OF LAND USE IN FOUR STATES[a]

	Type of agriculture	Population per km^2	Average Farm Size hectares	Value per Hectare	Annual Rainfall, cm
New Jersey	Market gardening	372	52	$11,250	114
Iowa	Mixed farming	20	105	4,476	83
South Dakota	Wheat	4	430	902	43
Wyoming	Ranching	1	1710	502	30

[a] Note that land use intensity falls off consistently from the densely settled East to the dryer, sparsely inhabited West.

Source. U.S. Bureau of the Census, 1978.

reveals a remarkable continuity of trends east to west (Table 10.1). Density of population falls off consistently, as does rainfall and the value of land per acre. Average size of farms, however, rises consistently, as larger and larger homesteads are required to organize an efficient unit of production. In short, intensity of land use falls as we move west from the Atlantic seaboard to the mountain states. Furthermore, farming activities in America fall into a pattern of zones, or belts, and these belts would seem to conform to the von Thünen model (as applied on a macroscale) with "megalopolis" acting as the market, or focal point (Fig. 10.16). We must not jump to conclusions,

however, for the von Thünen theory requires homogeneity throughout the study area; such conditions obviously do not hold across the broad reaches of the United States, in particular with regard to rainfall. It is more realistic to conclude, therefore, that farming activities in the United States fall into patterns that reflect economic pressures, that rainfall would seem to be one of these pressures, distance another, but other factors make themselves felt.

Today's farmers and ranchers supply us not only with food but also with fabrics, leather, cordage, chemicals, pharmaceuticals, and fuels; products that sustain people and their industries.

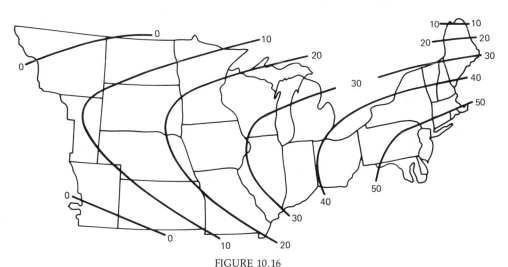

FIGURE 10.16

NET INCOME PER ACRE FROM FARMING. Source: P. O. Muller, "Trend Surfaces of American Agricultural Patterns: A Macro-Thünian Analysis," *Economic Geography,* July, 1973, p. 234.

LAND USE THEORY

Our industrialized-urbanized life is supported by the surplus produced by agriculture; it is also supported by the abundant nature of the earth's seas, mines, and forests. The products of these extractive industries, without which modern society could not exist, are examined in the following chapter.

KEY TERMS AND CONCEPTS

subsistence farming
commercial farming
subsistence gathering
subsistence herding
transhumance
shifting cultivation
swidden
intensive subsistence agriculture
sawah

Green Revolution
plantation agriculture
economic nationalism
mixed farming
dairying
commercial grain farming
commercial grazing
market gardening
Johann von Thünen

DISCUSSION QUESTIONS

1. In what ways have modern farming techniques shifted high technology societies to dependence on non-renewable resources?
2. In what areas are most commercial plantations found? Explain.
3. What environmental characteristics promote dairying in Pennsylvania?
4. Discuss differences between mixed farming practices in Europe and in the United States.
5. Identify major areas of intensive subsistence agriculture in the world. How do you account for the persistence of this activity in these regions?
6. In the United States, the intensity of land use tends to fall off as one moves west from the Atlantic seaboard to the mountain states. Why? Discuss.
7. The term "Corn Belt" suggests a form of monoculture within that region. Is this in keeping with the facts? Is the term useful or should it be changed?
8. It is said that slash-and-burn agriculture is simply a form of field rotation. Do you agree? Discuss.
9. Soybeans are the second leading money crop in the United States and are also produced in large quantities abroad. Explain and discuss.
10. What forms of agriculture appear to be the most ecologically sound? the least sound?

REFERENCES FOR FURTHER STUDY

Alexander, J. W., and L. J. Gibson, *Economic Geography*, Second Edition, Prentice-Hall, Englewood Cliffs, N.J., 1979.

Carter, V. G., and T. Dale, *Topsoil and Civilization*, University of Oklahoma Press, Norman, OK., 1974.

Gregor, H. F., *Geography of Agriculture*, Prentice-Hall, Englewood Cliffs, N.J., 1970.

Higbee, E., *American Agriculture*, Wiley, New York, 1958.

Muller, "Trend Surfaces of American Agricultural Patterns: A Macro-Thünian Analysis," *Economic Geography*, 49, Worcester, Mass., 1973.

Symons, L., *Agricultural Geography*, Praeger, New York, 1967.

THE EXTRACTIVE INDUSTRIES

PRIMARY PRODUCTION INCLUDES NOT ONLY AGRICULTURE but also the so-called extractive industries: fishing, forestry, and mining. These activities represent a gathering of the bounty of nature but involve no change in form. Of the extractive industries, fishing and forestry offer mankind promising opportunities because they are renewable and new uses are being developed for them. Fishmeal provides feed for livestock and is also used as a fertilizer. Fish products can be converted into a protein-rich flour for human use. New technologies in the forest-product industry permit the manufacture of products previously made from oil. Ethanol from wood is being mixed with gasoline while research is under way to refine wood chips into fuel oil. Wood by-products are used to make plastics and synthetic fibers, products until now made from oil. And wood ash is being used as a substitute for chemical-based fertilizers. With wise and consistent management, the world's fisheries and forests can remain abundant and available for generations to come.

FISHING

Commercial fishing takes place in both fresh and salt water, and is a major source of food for large numbers of people. Productive fishing, however, is unevenly distributed throughout the world, and some maritime environments offer greater opportunities than others. Consider Norway as an example. Norway's agricultural potential is limited. Aside from some alluvial deposits, the soils are thin, stony, and unproductive, and most of the terrain is hilly or mountainous. Temperatures are cool, and the growing season is short. Farming offers limited opportunities. But Norway has a deeply indented fjord coast that lends itself to the development of maritime skills. The rugged terrain makes land travel difficult and tends to "push" people to use the sea as a means of travel. The land is covered by forests, making timber available for the building of boats. The waters offshore teem with fish; the Norwegian Sea and adjacent areas constitute one of the richest fishing grounds in the world, and there is convenient access to markets via water-borne transport. Norway, with a population that barely exceeds four million, has traditionally been placed among world leaders in total catch of fish landed.

Ocean Fishing

Most commercial fishing occurs in the shallow waters that overlie the continental shelves. Marine life is dependent on minute life forms called *plankton* that drift freely in the water. Plankton exists either as plants (phytoplankton) or animals (zooplankton). It depends on the nutrient content of the water, and the offshore shallows are richer in nutrients than deeper waters. Rivers discharge

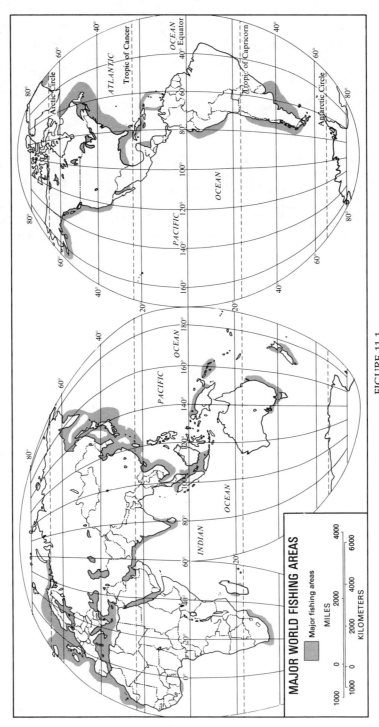

FIGURE 11.1
MAJOR WORLD FISHING AREAS.

minerals and organic materials to offshore waters. Upwelling of sea water dredges up nutrients from the ocean floor to the sun-drenched upper levels where phytoplankton absorb these simple chemicals and form them into protoplasm. Here begins the oceanic food chain. Phytoplankton are devoured by small life forms, which are in turn eaten by larger life forms. Fish congregate in waters rich in plankton, areas that coincide with the shallows of the continental shelves. Indeed, as far as fishing is concerned, most of the oceanic surface is a biological desert.

Commercial fishing is generally more successful in cool waters than in warm. Cool waters support a denser population of marine life because organic wastes break down quickly in warm water and are consequently removed from the food supply. Also, there is less mixing of tropical oceans and fewer minerals get to the surface to promote phytoplankton growth. Tropical environments are permissive in terms of biological experimentation, thus leading to a proliferation of species. Many different kinds of fish swim the tropical seas; nets cast in these waters will land catches containing many different species, not all of which are useful to humans. Sorting is required and efficiency impeded. Cool waters, on the other hand, are inhabited by fewer species of fish which tend to run in large schools, each school being made up of the same kind of fish. Handling is facilitated and production costs are lowered.

Commercial fishing employs a broad array of techniques that vary from simple to advanced. Methods differ according to the type of fish sought and the equipment available. Some fishermen still follow the traditional way, fishing from a dory with hook, line, and sinker. But large modern trawlers also ply the seas, where they work in groups. Sonar and helicopters are used to locate fish and large nets are hauled in by power winches. The catch is delivered to *factory ships* where it is prepared for market, sometimes being frozen or canned. The Soviet Union was the first nation to employ factory ships.

Fishing Areas

Five oceanic areas are extraordinarily productive for commercial fishing (Fig. 11.1). Four of the areas lie along a latitudinal zone in the northern hemisphere where table-grade fish are caught.

These areas include the western Pacific, the northeastern Pacific, the northwestern Atlantic, and the northeastern Atlantic. Another area off the west coast of South America specializes in catching fish for fishmeal.

The Northwest Pacific. Currently, the western Pacific leads the world in commercial fish production. The area extends from the Aleutians to Indonesia, embracing Japan, Taiwan, and the Philippines, including the various seas lying amongst these island groups. The continental shelf is broad, and conditions are ideal for marine life. The area is densely inhabited by people who are accustomed to the consumption of fish, which is important in an area whose land resources furnish inadequate supplies of protein. It is not surprising, therefore, that the three principal fishing nations (Japan, the Soviet Union, and China) all front upon this coast, and that two other nations of the region, South Korea and Thailand, are also included among world leaders (Table 11.1).

The most important fish caught in the area are salmon, which lead both in tonnage and value. Salmon are *anadromous*, meaning they spawn in fresh water, but the fingerlings work their way downstream to the sea, where they mature and spend most of their lives. Eventually they return to the same stream in which they were born, where they spawn and expire. The northwest Pacific provides good breeding ground for salmon, because of the many clear rivers that cascade down from the mountains of Hokkaido, Sakhalin, and mainland Siberia. Local varieties of salmon include chum, pink, red, silver, masu, and chinook. Further south the warmer seas support tuna, mackerel, and sardines. Shellfish are also collected.

Equipment used in the western Pacific varies considerably. The Chinese, Filipinos, and Indonesians tend to use small fishing craft and work close to home. Facilities to preserve the catch are inadequate and most of the fish are consumed immediately. In contrast, the Russians and the Japanese operate modern, powerful trawlers, or factory ships, that have the capacity for canning or freezing the catch. They are thus able to work far from base, and also to ship their product to distant markets. Large tonnages, however, are consumed by the home markets. Fish, in fact, provide more than one third of all protein intake in both Japan

TABLE 11.1 FISH CATCHES BY COUNTRY (millions of metric tons)

	1970	1977	Percent Change, 1970–1977
1. Japan	9.4	10.7	+ 14%
2. USSR	7.3	9.4	+ 29%
3. China	6.3	6.9	+ 10%
4. Norway	3.0	3.6	+ 20%
5. USA	2.8	3.1	+ 11%
6. Peru	12.5	2.5	− 80%
7. India	1.8	2.5	+ 39%
8. S. Korea	0.8	2.4	+200%
9. Denmark	1.2	1.8	+ 50%
10. Thailand	1.4	1.8	+ 26%
World total	70.0	73.5	+ 5%

Source. Yearbook of Fishery Statistics, 1977, Food and Agricultural Organization of the United Nations (New York), 1978.

and the Soviet Union (Table 11.1). In a recent development, both the Japanese and Russians have turned to collecting trash fish that are ground into fish meal and used as animal feed or fertilizer. About 25 percent of the world's fish catch is used for this purpose.

The Northeast Pacific. Productive fisheries extend from Alaska to the waters off Baja California. Salmon fishing is dominant in the north where conditions are ideal for their propagation and sustenance. Unfortunately, the life cycle of the salmon makes them vulnerable to overfishing, particularly during the spawning runs. As a result, landings have declined sharply. Today the United States catch is about one-quarter of what it was in the mid-1930s.

Halibut abound in Alaskan waters where they lie on the bottom and are taken on hooks and lines. King crabs, taken in the Gulf of Alaska, command a high price per pound and provide a valuable source of income to the local fishing industry. Farther south, off southern California and Mexico, fishing activities center on tuna and shrimp.

The Northwest Atlantic. The fishing grounds of the northwest Atlantic extend from the Davis Straits to the West Indies. A particularly active area lies astride the Grand Banks, a submarine plateau that stretches south and east from Newfoundland. The Banks lie at an average depth of about 90 meters (300 feet) and cover an area of 95,000 square kilometers (37,000 square miles). The warm Gulf Stream and the cold Labrador Current converge on the Banks. The Labrador Current provides a rich supply of nutrients, and the collision of the two currents creates turbulence that provides movement of the nutrient throughout the water over the Banks. Fishing fleets gather from near and far to harvest cod, haddock, halibut, hake, and herring. All methods and levels of technology are used. The Portuguese work from schooners and dories with hook, line, and sinker. Other Europeans fish these grounds, as do the Canadians, Americans, Russians, and Japanese. Some of these latter use large, sophisticated equipment and engage in "vacuum-cleaner" techniques. As a result, the Banks suffer from overfishing. The newly imposed 200-mile limit, however, places much of the Banks under the jurisdiction of Canada, and the imposition of controls may foster the recovery of fish populations.

Fishing extends along the east coast of the United States, with concentrations in the traditional regions of New England and Chesapeake Bay. In recent times, however, the focal point of American fishing has shifted south; Louisiana is the leading American state in weight of catch, and the Gulf of Mexico is the leading region (Table

11.2). The Gulf leads in landings of grouper, mullet, spotted trout, red snapper, and oysters. It provides three-fourths of the American catch of shrimp and two-thirds of the menhaden. Key West, Pascagoula, Morgan City, and Brownsville are a few of the many Gulf towns with fleets of fishing craft.

The Northeast Atlantic. The waters off western Europe comprise a traditional fishing ground that is second in productivity only to East Asian waters. The most active fishing is found in the waters around the British Isles, particularly in the North Sea where the Dogger Banks play a role similar to that of the Grand Banks in the Western Atlantic. There is also intensive activity in the Norwegian Sea, the Barents Sea, the Baltic, the Irish Sea, the English Channel, and the shallow waters around Iceland. The fish caught include herring, cod, plaice, sole, halibut, sardines, mackerel, and flounder. Virtually all European fishing fleets work in this area, and the stock of fish has been depleted by overfishing.

Adjacent areas of Europe are densely inhabited and constitute a large market for the fishing industry. Dietary habits are strongly oriented towards the sea, and per capita consumption of fish and shellfish is high throughout Great Britain, France, the Low Countries, West Germany, and Scandinavia. Large tonnages are exported; fish make up 90 percent of the value of exports of Iceland and also make an important contribution to the foreign earnings of Norway, Denmark, Portugal, and Spain.

The West Coast of South America. The Humboldt Current flows north along the west coast of South America, bringing cold water and nutrients from the Antarctic Basin to the waters off Chile and Peru. These waters support a large population of marine life. Tuna, bonita, and mackerel thrive as do anchovies, which run in huge schools. Because of the extraordinary richness of this resource, Peru has specialized in fishing, investing heavily in modern trawlers and processing plants. In particular they have exploited anchovies, landing enormous tonnages that are converted into fish meal. By the early 1960s Peru had emerged as the world leader in tonnage of catch. In 1972, however, disaster struck. A warm countercurrent known as "El Niño" appeared, flowing south along the coast of Peru and disrupting the marine environment. The anchovies disappeared, ranging elsewhere in search of cool water. Peruvian landings dropped from 12.5 million tons in 1970 to 2.5 million tons in 1977. Inasmuch as fisheries had generated more than one-third of Peru's foreign earnings, serious problems developed. Peru has been forced to reevaluate its dependence on what is apparently an unreliable resource.

Inland Fishing. Inland fishing accounts for about 10 percent of world tonnage. The activity

TABLE 11.2 U.S. FISH CATCH BY AREAS (millions of kilograms)

	1965	1978	Percent Change, 1965–1978
New England	319	300	− 6%
Middle Atlantic	162	91	−44%
Chesapeake Bay	269	272	+ 1%
South Atlantic	162	181	+12%
Gulf States	665	1040	+56%
Great Lakes and Inland Waters	64	57	−11%
Pacific Coast	521	791	+52%
Hawaii	9	7	−24%
United States	2171	2740	+26%

Source. *Fisheries of the United States, 1978*, National Oceanic and Atmospheric Administration (Washington, D.C.), April 1979.

is concentrated in Eurasia, where China, India, and the Soviet Union contribute more than three quarters of the world's total. In the Soviet Union many salmon are caught in fresh water. Sturgeon are also landed in the rivers and in the Caspian and Black Seas. Sturgeon is highly valued for its flesh as well as for its roe.

In southeast Asia, carp and eel are raised in penned enclosures, a practice called *aquaculture,* or fishfarming (Fig. 11.2). Fish meal is added to the water as a nutrient, and protein from fresh-water fish makes a significant contribution to local diets. Aquaculture has also been introduced into the United States, where fish farms produce an annual crop of about 15,000 tons, mostly in the lower Mississippi Valley. Currently aquaculture furnishes about four percent of the global fish harvest. Because it could be expanded, it represents an important potential source of food.

The Great Lakes once produced large tonnages of lake trout and whitefish, but overfishing, pollution, and the destructive effects of the parasitic sea lamprey have made such inroads that production has fallen off. The United States accounts for only two percent of world landings of freshwater fish.

Optimistic projections have forecast a tremendous growth in the productivity of fisheries and aquaculture, with the hope of adding greatly to world food supplies. Results, however, have been disappointing. In spite of intensified efforts, total annual landings have leveled off since 1970 at about 70 million tons. Some species of fish have suffered obvious depletion, and "vacuum-cleaner" techniques are now viewed as self-defeating. A large portion of potential fish production must be left uncaught so that the fish catch can be sustained. The international community needs to arrive at some acceptable understanding as to how this important resource is to be managed and conserved.

FORESTRY

Forests are one of mankind's greatest resources. They provide materials for shelter, fuel, foodstuffs, and manufactured goods. Many factors, however, influence the manner in which a particular stand of trees will be utilized. The only certainty is that the pattern of exploitation will conform to the constraints of the economic environment; relevant factors can include climate, soils, landforms, density of population, distance to market, and levels of technology.

Forest Products

Forest growth forms two major belts of greenery around the world. One such belt encircles the globe in the tropics, while the other extends across North America and Eurasia in the mid-latitudes of the Northern Hemisphere. Smaller areas are found in the mid-latitudes of the Southern Hemisphere and in mountainous areas (Fig. 11.3). In spite of widespread cutting, forests still cover about 30 percent of the earth's land surfaces.

Trees are designated as either *hardwoods* or *softwoods.* Hardwoods include many broadleaf trees such as oak, maple, and mahogany. Most hardwoods of the mid-latitudes are deciduous. Most hardwoods of the tropics are evergreen. Generally speaking, hardwoods are fine grained and slow growing. Many of them are prized by the furniture industry, but hardwoods also find use as structural timber and in the pulp industry.

Softwoods, or conifers, tend to be coarse grained and fast growing. The softwoods find their principal use as structural lumber and in the paper and pulp industry, for which they furnish the

FIGURE 11.2
AQUACULTURE REPRESENTS AN IMPORTANT POTENTIAL SOURCE OF FOOD. Fish farming ponds on Luzon Island, Philippines.

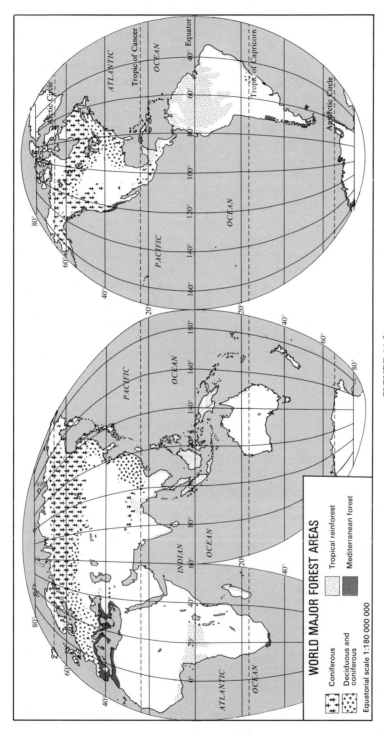

FIGURE 11.3
WORLD MAJOR FOREST AREAS.

WORLD MAJOR FOREST AREAS

Coniferous

Deciduous and coniferous

Tropical rainforest

Mediterranean forest

Equatorial scale 1:180 000 000

main raw material. *Tree farms*, where trees are planted, managed, and harvested like a crop, almost always specialize in softwoods, because of their rapid rate of growth.

Physical Characteristics

Trees grow in a variety of soils that have, however, certain characteristics in common. Trees are water-loving and grow best in humid areas. They tend to do well in soils that have good qualities of water retention, particularly in the substrata. Trees with deep taproots can reach to moisture in the subsoil, thus surviving brief droughts. A difference in the porosity of soils and the drainage characteristics of an area can have a major effect on tree growth.

Since trees grow in humid areas, most forest soils are leached, acid, and poorly suited for crop farming. This is particularly true in the humid tropics and also in the subpolar regions of the taiga. However, the hardwood deciduous forests of northeastern United States and Western Europe developed on highly desirable alfisols. As a consequence, both these areas have been densely settled by farming societies, and the forests have been cleared to make way for more intensive land use. Only vestiges remain of the great oak forests that once covered the United Kingdom and northern France. This is also true of the mixed stands that once dominated the northeastern United States.

Forestry lends itself particularly to hilly and mountainous regions. Flat lands are covered by commercial forests only where *opportunity costs* are not too high (that is, if the land cannot be put to more profitable use). Forestry, however, can generate income from sloping lands that are marginal for other purposes. For this reason, foresty frequently occurs in regions of rugged terrain.

In the mid-latitudes, most wood is absorbed by industry, while in the tropics most wood is consumed as fuel. On a worldwide basis, 55 percent of roundwood is used by industry; 45 percent as fuel. In the industrial world, half of all timber products is used as lumber, about one-third is

TABLE 11.3 ROUNDWOOD PRODUCTION BY CONTINENTS AND
LEADING COUNTRIES, 1976 (millions of cubic meters)

	Coniferous	Broadleaf	Total
Continent			
Asia (except USSR)	142	573	715
North America	380	122	502
Africa	12	315	327
Europe (except USSR)	200	103	303
South America	30	207	237
Oceania	12	18	30
Country			
USSR	322	62	384
USA	259	83	342
China	90	105	195
Brazil	24	140	195
Indonesia	—	129	129
India	5	122	127
Canada	105	11	116
Nigeria	—	69	69
Sweden	46	7	53
Japan	20	17	37
World	1098	1400	2498

Source. *United Nations Statistical Yearbook*, 1977, New York, 1978.

turned into pulp, and most of the remainder is made into plywood or veneer. Only four percent of the wood used in the developed world is burned as firewood.

Wood contains cellulose fiber which is the principal ingredient in pulp and paper. The pulp and paper industry, therefore, is a major consumer of wood, absorbing about one-quarter of all softwood cuttings, and lesser amounts of hardwood (Table 11.3).

There are two principal methods of rendering wood into pulp. In the mechanical method, wood is ground up under a flow of water; in the chemical method, the wood is first reduced to chips, and then treated with chemicals that dissolve the lignin and other non-cellulose materials. In both methods, the pulp is then run through rollers and dried. The cellulose fibers adhere together, forming paper. Newsprint and other cheap grades of paper are generally made by the mechanical process. Finer papers are generally made from chemical woodpulp and sometimes reinforced by cotton fibers, linen fibers, or esparto grass. Strong wrapping papers and cardboard are made from pure woodpulp by either the mechanical or chemical process.

Accessibility and International Trade

In order for timberland to be usable it must be accessible, meaning that the economic distance separating forest and market must not be excessive. Logging operations produce a bulky commodity of low value (Fig. 11.4). Transport expenses, therefore, make up a considerable percentage of total production costs. Consequently, every effort is made to minimize the costs of transport. Waterways can be helpful in this sense, as the least expensive way to move logs is to float them downstream, or move them by flume, barge, or ship. In the absence of convenient waterways, more expensive transport such as roads or railways must be used.

Sawmill operations account for a loss of bark, slab, and sawdust, averaging together about 40 percent of gross weight. Also, squared-up lumber is relatively more manageable and compact than raw logs. Therefore, sawmills tend to be located near the forest, frequently at waterside (Fig. 11.5).

The major markets for commercial wood products are the densely inhabited, industrialized re-

FIGURE 11.4
MODERN LUMBERING MACHINES INCREASE THE EFFICIENCY OF LOGGING OPERATIONS.

gions of Anglo-America, Europe, and Japan. Forests enjoying convenient linkages with those areas are heavily used for that purpose, while those more remote are non-competitive. It must be remembered that the overriding consideration in these linkages is that of cost rather than distance. Wood products can be moved thousands of kilometers to market if efficient, waterborne transport is available.

FIGURE 11.5
TEARDROP LOG BOOMS IN SAULT STE. MARIE, ONTARIO, CANADA.

Most forests are publically owned, particularly those which are inaccessible. Even in areas without controlled economies, government-owned forests comprise a large percent of the total (Europe, 45 percent; South America, 55 percent; and the United States, 42 percent). In the United States, 85 percent of publically owned lands are held by the Federal government; the remaining lands are held by state, county, or municipal governments.

International trade in forest products is dominated by two major flows. The first of these is an exchange of softwood products between the industrialized nations of the Northern Hemisphere. Canada exports lumber, pulp, and paper to the United States and Europe, and a similar flow runs from Scandinavia and the Soviet Union to Western Europe.

The second flow of forest products originates in the tropics and moves north to the industrialized areas. Included here are valuable woods such as mahogany, rosewood, ebony, and teak that are used in cabinet making and the furniture industry.

The United States is the world's largest net importer of wood products. Nearly two-thirds of the nation's newsprint and one-third of its softwood lumber is imported from Canada. This makes up about 70 percent of the total United States wood imports.

Tropical Forestry

The forests of the tropics produce a magnificent growth of trees, but commercial lumbering is inhibited by problems that increase the cost of removal. The proliferation of species that is common to all tropical forests creates difficulties because many species are worthless. The lumberman combs the forest, therefore, removing the valuable trees that are convenient to waterways, and floating them downstream to a sawmill. Unfortunately, tropical hardwoods tend to be dense; some require flotation gear least they sink. Also, tropical hardwoods tend to grow slowly; an area once cleared needs substantial time to recover, forcing the sawmills to become mobile. Distance to market tends to be considerable, raising costs of transport. Because of these difficulties, only the most valuable woods can absorb the high costs associated with production. Other trees are left untouched.

Tropical forests are used, however, as firewood, which accounts for more than 90 percent of round-wood consumption throughout the region. Much of this activity is at a subsistence level; the wood is used primarily for cooking, but at times is refined into charcoal. Firewood removals in the tropics are of such volume that Brazil, Indonesia, India, and Nigeria are listed among the top ten harvesters of roundwood in the world. Methods of harvest are haphazard, and broad areas are experiencing severe deforestation.

The tropical regions, however, have a great potential as a future source of wood and pulp, and projects are now underway to realize these opportunities. In the Brazilian Amazon, for instance, an American shipping magnate, Daniel Ludwig, has invested almost a billion dollars in a development along the Jari River. He has acquired land holdings larger in area than the state of Connecticut and has built a wood-chipping mill, a pulp plant, a chemical factory, and a power generating facility. In order to provide grist for these mills, he has created a vast timber farm planted with *Gmelina arborea*, a hardwood tree native to southern Asia that grows to a diameter of 40 cm (15 in.) in five years, nearly twice as fast as the American southern pine. Ludwig plans to produce 750 metric tons of pulp daily by the early 1980s. If this project is successful, it is probable that similar facilities will be created elsewhere, and the humid tropics will begin to contribute substantially to world supplies of wood and pulp.

Mid-Latitude Forestry

The Soviet Union leads the world in roundwood production. Vast stands of boreal forest cover 40 percent of its natural territory. Most of the trees are conifers, which account for 83 percent of the cut. Wood is commonly used as a construction material within the Soviet Union, and timber products also comprise an important item of international trade, with most exports going to Europe.

The United States ranks second as a producer of forest products. Softwoods are dominant, making up 74 percent of fellings. There are three distinct areas of production in the United States: the West, the Southeast, and the Northeast (Fig. 11.6).

The commercial forests of the West are the largest in the country, and the region leads in forest production. The timber consists of softwoods, in-

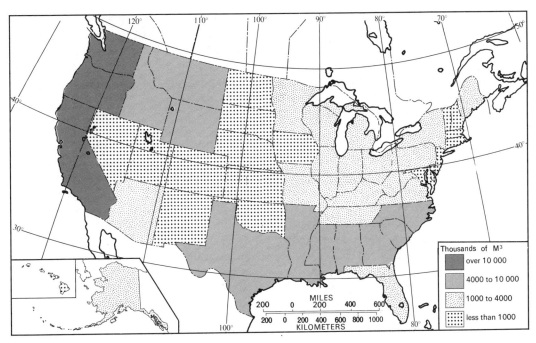

FIGURE 11.6
DISTRIBUTION OF SAW TIMBER PRODUCTION IN THE UNITED STATES.

cluding Douglas fir, spruce, pine, and redwoods. These forests are concentrated in the Cascade and the Sierra Nevada mountains, with lesser stands in the Rockies. Oregon, Washington, and California are the leading states in standing volume of commercial timber.

With regard to accessibility, western products are at a disadvantage in the markets of the East because of the distances involved, but they compete effectively in the Midwest. Obviously the western forests enjoy a distinct advantage in California and the expanding markets of the Southwest. One-half of American saw-timber fellings now come from the West.

The Southeast comprises a second major area of forest production in America, contributing more than one-third of national production. The regional leaders are Louisiana, Alabama, Georgia, Mississippi, and Arkansas. Climatic conditions include a long growing season, moderate temperatures, and generous amounts of moisture. Tree growth is rapid and access to market in the industrial Northeast is good. Farmland in the South has been bought in large amounts by paper companies and tree farms have been established. The area sustains a large number of pulp mills and paper factories (Fig. 11.7). Hardwoods also grow

in the region. Together with softwoods, they are used as structural lumber and veneer to supply the furniture industry centered in North Carolina. Also, hardwoods are used in making plywood and veneer, and by the pulp industry. This area produces one-third of the nation's softwood plywood, 60 percent of its pulpwood, and 56 percent

FIGURE 11.7
THE PULP AND PAPER INDUSTRY IS A MAJOR CONSUMER OF WOOD. This pulp mill is in Plymouth, North Carolina.

of its newsprint. It has been estimated that by the year 2000 the Southeast will produce the majority of the nation's wood supply.

The Northeastern forest has been decimated by removals, but it contributes, nevertheless, 14 percent of national cuttings. Most of the hardwoods have fallen victim to land-clearing operations, but woodlots on farms and in hilly areas are still productive. In addition, coniferous forests of northern New England and the Great Lakes states, although they suffered from mismanagement in the past, produce structural timber and pulpwood. Summers are short, and growth rates are slow, but the land attracts little or no competitive use. Indeed, many farms of the area have been abandoned and have reverted to tree cover. Access to market is good, and sawmills, pulp mills, and paper factories are scattered across the region. Maine is the regional leader, followed by Michigan.

Vast reaches of Canada are covered by coniferous forests, and that country is a major producer of wood products. Growth rates, unfortunately, are slow, and large regions are not used because they have poor stands of timber and are inaccessible. As a result, exploitation is confined to areas convenient to neighboring markets. Nevertheless, Canada is among world leaders in exports of timber, woodpulp, and paper. The United States receives large tonnages of newsprint from Canada.

Finland and Sweden also have large forests of conifers. They are close to Western Europe and supply that market with timber, pulp, and paper.

Conservation

Forests are a renewable resource. They must be properly managed, however, because they are vulnerable, and can be damaged by fire, disease, insects, or unwise exploitation. In the developed countries, most commercial forests are now managed so as to achieve sustained yield, but careless logging practices still cause some loss of topsoil and degradation of the environment. The problem of deforestation is more acute in the less developed countries, where demand for firewood causes tree cover to be removed from steep slopes, thus bringing about serious erosion. A number of countries in the tropics have established programs of reforestation, but progress is difficult in areas where the poor are tempted to cut trees before they grow to optimum size.

Forests provide not only a broad array of material products, but they aid in watershed protection and erosion control. They also provide habitat for wildlife and add to the esthetic quality of the landscape.

MINING

Mining is the collection of minerals or other substances, whether solid, liquid, or gaseous, either at or below the surface of the earth. Mining is an important component of the economic environment, but it presents difficulties to the geographer because it resists organizational efforts on a regional basis. Minerals, for example, differ in terms of origin; some occur in igneous formations, and some are found in sediments. Some minerals consist of organic remains. Minerals differ in value and in their ability to absorb transport costs. Moreover, mining techniques differ; operations in the industrialized world are highly mechanized, while much hand labor is still used in the developing world.

In spite of this, certain generalizations can be established relative to the mining industry:

1. Mining is not a large employer. It employs only 0.6 percent of the labor force in the United States and less than one percent worldwide. Moreover, the labor force in the mines is shrinking; United States mines employed only two-thirds as many miners in 1978 as in 1940. This same period, however, saw a marked increase in production as a result of mechanization.

2. As a rule, mining does not act as an agglomerating factor for industry. Frequently mines are located in empty, isolated regions. Only coal mines, by reason of the versatility of coal both as an industrial fuel and raw material, attract industries. It is no coincidence that the major industrial complexes of the world are, with few exceptions, close to coalfields.

3. Mining is conducted at the surface or underground. Other things being equal, surface mining, or *strip mining*, as it is called, is less expensive because it is applied to ore bodies that lie at or near the surface. The overburden is removed and the ore is withdrawn by enormous specialized equipment, including power shovels that can remove more than 100

cubic meters at a time (Fig. 11.8). Strip mining can be very destructive, leaving behind devastated areas covered by mounds of slag. This problem can be solved, however, as has been demonstrated in Jamaica, where companies mining bauxite are required to restore the landscape to its original condition. Underground mining is relatively more expensive. It also imposes environmental problems in terms of accumulation of slag, pollution of streams, and land subsidence.

4. Many minerals are abundant and widely distributed. Because of cost relationships, however, it is generally necessary that a mineral deposit be concentrated in order for mining to be profitable. A deposit that can be worked profitably is called an *ore*. Obviously the profitability of a particular ore body will depend not only on its chemical makeup but also on market conditions (Table 11.4). As an example, consider the American gold mining

FIGURE 11.8
STRIP MINING IS THE LEAST EXPENSIVE FORM OF MINING BECAUSE IT IS APPLIED AT OR NEAR THE SURFACE USING HIGHLY SPECIALIZED EQUIPMENT.

TABLE 11.4 MINERAL PRODUCTION IN THE UNITED STATES, 1978

	Quantity (thousands of metric tons)	Value (millions of dollars)
Fuels		
Coal	593,000	$13,800
Petroleum	432,000	26,200
Natural gas (teracalories)	5,400	12,500
Metals		
Iron	80,500	2,200
Copper	1,360	1,970
Molybdenum	60	600
Lead	531	398
Zinc	310	212
Nonmetals		
Cement	75,600	3,200
Stone	997,000	2,600
Sand and gravel	852,000	2,100
Phosphate Rock	49,000	900
Lime	20,200	743
Clay	50,700	694
Sulphur	5,700	500
Silicon	480	440

Sources. *Mineral Commodity Summaries 1979*, Bureau of Mines, U.S. Department of the Interior (Washington, D.C.) 1979; and *Monthly Bulletin of Statistics*, United Nations (New York), September 1979.

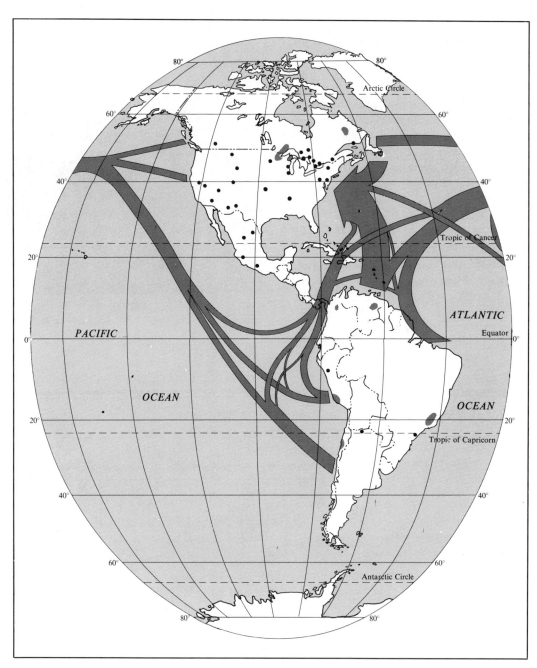

FIGURE 11.9
PRINCIPAL IRON ORE DEPOSITS.

industry. In 1934 the federal government set the price of gold at $35 per ounce. With this relatively high price acting as a stimulus, American gold mines swung into full production and gold began to accumulate in the U.S. Treasury. However, time passed and the United States experienced inflation, along with rising wages and power costs. The price of gold remained at $35 per ounce, however. Gradually the gold mines in America began to lose their margin of profit as costs began to exceed income. Many mines closed down

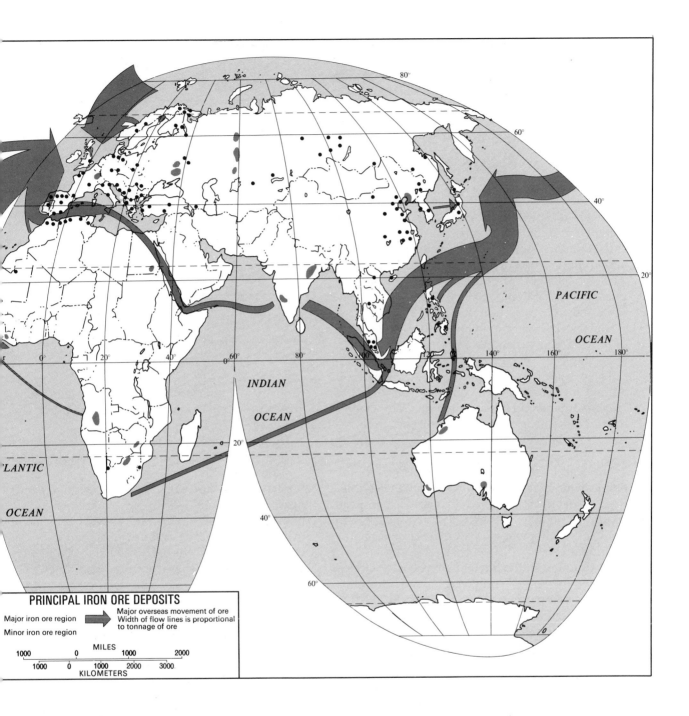

PRINCIPAL IRON ORE DEPOSITS

Major iron ore region

Minor iron ore region

Major overseas movement of ore
Width of flow lines is proportional
to tonnage of ore

MILES
1000   0   1000   2000

1000   0   1000   2000   3000
KILOMETERS

and United States production fell. By the 1960s, the industry was in serious decline. However, in 1974, the price of gold was unfrozen, and it floated to high free market levels. In response to this higher price, dozens of gold mines throughout the Black Hills and

the Rockies that had been boarded up for years returned to a flourishing and profitable operation.

5. Mining involves the collection of fuels, metals, and non-metals. Of these, fuels are by far the most valuable, comprising about

three-fourths of the value of total removals. Nonmetals, including stone, cement, and sand, make up 15 percent. Metals make up the remaining 10 percent. The remainder of this section deals with the extraction of selected metallic and non-metallic minerals. Fossil fuels are discussed in the following chapter.

Iron Ore

Iron is ubiquitous, making up six percent by weight of the continental crust. In order to be considered an ore, however, iron content must reach a minimum concentration of about 20 percent. Also, an absence of contaminants is desirable; phosphorous and sulfur are particularly troublesome, raising costs of production.

Iron ore is highly mobile and large quantities of it move in international trade. The efficiency of superships has reduced transport costs per ton-mile so that iron ore can now move halfway around the world as, for example, from Brazil to Japan. Large tonnages also flow from South America to both North America and Europe, from Africa to Europe, and also from Australia to Japan (Fig. 11.9).

The world's leading iron ore producers are the Soviet Union, Australia, the United States, and Brazil. Iron is also mined in several European countries, notably in France, Sweden, Spain, and the United Kingdom, so that the regional total is considerable.

Iron mining in the Soviet Union has expanded dramatically since World War II, and now accounts for 24 percent of world totals. Production is concentrated in the southern Ukraine, in the Ural Mountains, in central European Russia, and in Kazakhstan. The ores are high grade, averaging 60 percent iron. Reserves are substantial, particularly in Siberia.

Australia and Brazil have emerged in recent years as major producers of iron ore (Table 11.5). Domestic consumption is limited, so most of the ore is exported, with large tonnages going to Japan.

In 1978 production of iron ore in the United States totaled 81 million tons, of which 68 million tons originated in the Lake Superior district. The Mesabi Range of Minnesota is the most produc-

TABLE 11.5 IRON ORE PRODUCTION—1978
(millions of metric tons)

USSR	248
Australia	90
USA	81
Brazil	66
Canada	42
India	36
China	35
France	33
South Africa	24
Liberia	23
Sweden	22
Venezuela	19

Source. Monthly Bulletin of Statistics, United Nations (New York), September 1979.

tive single source, contributing 59 million tons. Other productive areas are found in Michigan, California, Wyoming, and Alabama.

The Mesabi Range was developed in the latter part of the nineteenth century and has produced hundreds of millions of tons of high grade ore. In the late 1940s, however, the high grade deposits neared exhaustion, leaving behind billions of tons of *taconite,* a low grade ore of about 20 percent metal. Since taconite resisted exploitation under the technology of the period, the steel industry began to look elsewhere for iron. New mines were opened in Canada and Venezuela, both of which now contribute to American supplies. Interest continued, however, in the taconite of the Mesabi, which could not be used because of its extreme hardness and low iron content. After much experimentation and research, a process was developed in which the rock was crushed and the metallic content increased to about 60 percent. These concentrates were then heated and formed into pellets, so as to provide ease of handling. These processes, called *beneficiation* and *pelletization,* have rendered the Mesabi taconites competitive; today they supply about three-fourths of the production of iron ore in the United States.

In western Europe, France is the largest producer of iron ore. The mines are located in northeastern France, mostly in the province of Lorraine. The ore averages only 30 percent iron and is contaminated with phosphorous, but finds extensive use because of the convenience of its location.

An important source of high-grade ore in Europe is found at Kiruna in northern Sweden. The ore runs as high as 65 percent iron. Most of the ore is sent by railway to the ports of Narvik in Norway and Lulea in Sweden and shipped to the industrial areas of western Europe. West Germany is the leading customer, followed by the United Kingdom.

Copper

Copper was the first metal used by man and is still in strong demand because it is a good conductor of electricity. It is also a component of brass, which is widely used for hardware and fittings. The United States is the leading producer of copper, accounting in 1978 for 20 percent of world production. The mines are concentrated in the Southwest: Arizona, Utah, and New Mexico being the leading states. The ores are thin, the average yield of copper averaging about 0.5 percent. Nevertheless, the total value of copper mined in the United States almost equals the total value of iron ore (Table 11.5).

The Soviet Union is second in copper production, with mining activities concentrated in the Ural Mountains. A region of copper mining stretches through the Andes Mountains in South America; Chile and Peru are respectively third and sixth as producers of copper. Zambia, which is fourth, has the distinction of having the richest ores, averaging about seven percent pure copper.

Stone

It is easy to overlook the importance of stone, but it has many uses and is mined in prodigious quantities. It serves, for example, as ballast in highway and railway construction, as a building material, and as a flux in the steel industry. In 1978 the total value of stone produced in the United States exceeded that of any single metallic ore. Stone is ubiquitous and also of low value per unit weight. This means that stone is not mobile; it must be used close to where it is quarried or competing alternatives will intervene. Pennsyl-

vania is the leading producer in the United States, but there are quarries in every state of the union.

The production of mineral fuels is also a component of the extractive industries. As in other mining activities, these energy sources are nonrenewable. Today's high-technology societies no longer find energy resources as available, or as inexpensive, as they once did. Rising demand and subsequent shortages have prompted a fever of exploration for new deposits of mineral fuels, especially petroleum, as well as alternative sources of energy. These topics are examined in the following section.

KEY TERMS AND CONCEPTS

plankton
factory ship
anadromous
aquaculture
sea lamprey
hardwoods
softwoods
conifers

deciduous
tree farms
strip mining
ore
taconite
beneficiation
pelletization

DISCUSSION QUESTIONS

1. Commercial fisheries tend to be concentrated in five regions of the world. Where are they? Explain the outstanding productivity of these regions.

2. What are the "Big Three" fishing countries of the world? Why do these countries depend so heavily on the resources of the sea?

3. Identify two major forested zones in the world. Distinguish between the two in terms of manner of usage by humans.

4. What factors affect how a forest is used or not used in a commercial sense.
5. What is a tree farm? Where are they found in the United States? Explain.
6. Discuss the limitations of the oceans as a source of food for humans. What dangers lie before us in this sense?
7. It is said that mining presents difficulties to a geographer because it is not regional in character. What does this mean? Discuss.
8. Generally speaking, iron ore is shipped to coal, rather than coal being shipped towards iron mines. Why? Why does coal act as an agglomerating factor for industry?
9. What is taconite? What role does it play in the steel industry in the United States?
10. What are the benefits and defects of strip mining? What can be done to protect the environment against the hazards that accompany strip mining?

REFERENCES FOR FURTHER STUDY

Alexander, John W., and Lay James Gibson, *Economic Geogrphy*, Second Edition, Prentice-Hall, Englewood Cliffs, 1979.

Dasman, F. R., *Environmental Conservation*, 4th ed., John Wiley and Sons, New York, 1976.

Griffen, P. F., A. Singh, W. R. White, and R. L. Chatham, *Culture, Resource, and Economic Activity*, 2nd ed., Allyn and Bacon, Boston, 1976.

Peach, W. N., and James A. Constantin, *Zimmerman's World Resources and Industries*, 3rd ed., Harper and Row, New York, 1972.

Russell, W. M. S., *Man, Nature, and History*, Natural History Press, Garden City, N.Y., 1969.

Wagner, R. H., *Environment and Man*, Norton, New York, 1978.

12
ENERGY

WHEN WAR BROKE OUT IN THE MIDDLE EAST IN THE FALL of 1973, the Arab oil producers placed an embargo on the export of oil to the United States to protest a military airlift and some $2 billion in economic assistance to Israel. This action was followed by economic disruption in most of the industrialized nations. The valves were reopened in March, 1974; oil began to flow again, but the price had quadrupled. The era of cheap energy was over.

THE EXPONENTIAL CENTURY

The fuels used as energy sources have changed with time. Early useful energy was obtained from waterwheels, windmills, and the burning of wood. By about 1750 the coal deposits of England began to be exploited to fuel the Industrial Revolution. Coal was used to generate steam in an improved engine patented by the Scottish inventor James Watt in 1769. Coal and the steam engine were to play major roles in the Industrial Revolution. Industrialization spread rapidly throughout the world, dividing it into the haves and have nots. The haves were those who had control of inanimate sources of energy. This industrial change has been driven by energy, derived at first from coal, and later from petroleum and natural gas. In 1900, coal was the dominant source of energy for heating and power in the United States. At that time oil, which had begun to be produced in commer-

cial quantities in 1859, was still being used primarily as an illuminant in the form of kerosene. Natural gas was virtually unknown except in the areas immediately surrounding the producing fields. Water power then accounted, and still accounts, for only a small share of the energy market.

Energy, like charity, begins at home. It begins with food intake, which in the countries where the population is well fed averages approximately 3000 calories per capita daily, or roughly 1 million calories annually. These calories are kilogram calories, 1000 times as large as the basic calories of the metric system. Except for their origins, they are otherwise identical with the calories of physics.

The 1 million kilogram calories that we consume annually provides us with the heat needed to keep our bodies warm and the power to move about and do our thing. But this is not a large amount of energy. It is only enough, for example, to drive an automobile about 500 miles, to do the cooking for a single family for 1 year, or to heat a small, New England home for something less than 1 midwinter week.

Source. Nathaniel B. Guyol, *Energy in the Perspective of Geography,* Englewood Cliffs, N.J.: Prentice-Hall, Inc., 1971, p. 1. Reprinted by permission of the publisher.

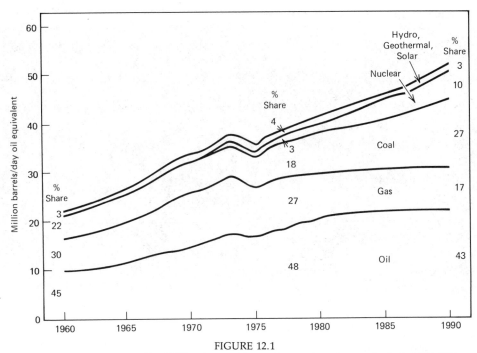

FIGURE 12.1

UNITED STATES ENERGY SUPPLY. Source: Exxon Corporation, *Energy Outlook, 1978–1990,*
May, 1978.

The decline in coal consumption since 1900 has been matched by an increase in the use of oil and gas. In 1947, coal accounted for about 48 percent of American energy requirements, oil and gas for another 48 percent, and hydropower for the remainder. By 1976, coal's share had declined to 18 percent while the combined share of oil and gas had increased to nearly 75 percent of the total energy consumed (Fig. 12.1). Nuclear power, expected to provide 10 percent of total energy needs by the year 1990, became of significance in the early 1970s, but the largest consumer energy in-

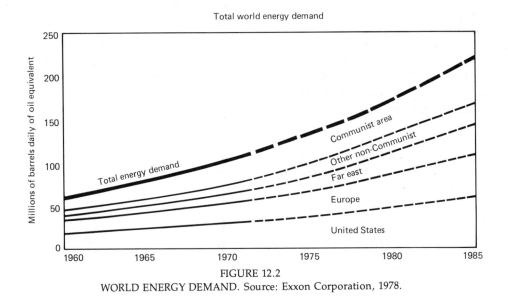

FIGURE 12.2

WORLD ENERGY DEMAND. Source: Exxon Corporation, 1978.

ENERGY

292

crease was in the use of oil and natural gas. The shift to a greater usage of oil and gas reflects the following factors: oil and gas are easier to handle, transport, and store than coal; they burn more cleanly; there has been a steady rise in the use of automobiles and in home ownership; there has been greater demand for oil and natural gas for industrial purposes and in agriculture; and there is an increased use of petrochemical-based products such as plastics.

Energy consumption in the post-World War II era has been characterized by a continuous rise in demand (Fig. 12.2). Indeed, energy consumption is doubling every 20 years, but not necessarily in those countries with high population growth rates. In fact, per capita energy consumption rates in the developed countries rose faster than those in the less developed countries. We have suddenly become too many and the world's resources too limited. We have too many cars and too much highway; we need too much food and too many houses. Not surprisingly, a large part of the energy crisis can be traced to population growth.

The United States is the world's largest energy consumer and the largest oil importer. With only six percent of the world's population, Americans consume one-third of the world's energy production at a rate equivalent to a barrel of oil per week per person (Fig. 12.3). Americans use twice as much energy per person as Germany or Sweden, nations with nearly comparable living standards. The United States imports one-third of all oil used in the world every day. At the beginning of 1978, we were using 18 percent more oil than in 1973 and our imports had risen from a third to a half of total consumption. Most imports are destined for the transport sector of our economy, where the job of moving people consumes 60 percent of transportation's total energy. Indeed, a seventh of all the oil used in the world every day is used on American highways. One energy authority has observed, "If at the time of the Arab oil embargo in 1973–74, American cars had been the same weight as other cars in the world and had they been driven at fifty-five mph or less, the embargo would have had no effect on the U.S. economy, for we would not have been importing oil."*

Patterns of Consumption

The transportation sector has become an increasingly important user of oil and gas. In 1947, transportation in the United States accounted for nearly 13 percent of energy consumption; by 1976, its share had doubled (Fig. 12.4). The transportation sector is almost entirely dependent on oil and will account for nearly half of total oil consumed through 1990. Along with transportation's rising share of energy consumption, there has been a drop in the percentage share of the industrial sector (manufacturing and mining). This partly reflects the change in the composition of national output; that is, away from manufacturing and toward less energy-intensive services. At the same time, rapid improvements in energy-use efficiency have taken place in the industrial sector. The transportation sector, however, remains a prime target for long-term as well as short-term strategies. In the United States the transportation sector consumes a quarter of total energy requirements. Obvious inefficiencies within the sector invite improvements such as reducing the weight

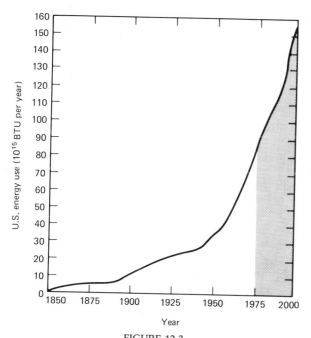

FIGURE 12.3
ENERGY DEMANDS IN THE UNITED STATES. Source: H. S. Stoker, S. L. Seager, R. L. Capener, *Energy: From Source to Use* (Glenview, Ill., Scott, Foresman and Co., 1975), p. 11.

* Earl Cook, *Energy: The Ultimate Resource?* Resource Papers for College Geography No. 77-4, (Washington, D.C.: The Association of American Geographers), p. 20.

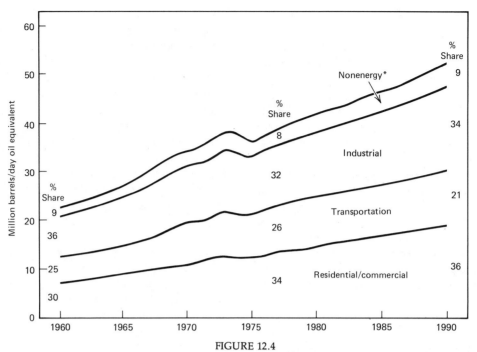

FIGURE 12.4

UNITED STATES ENERGY DEMAND BY CONSUMING SECTOR. *Consumers in the nonenergy sector use oil, gas, and coal as feedstock or raw material, rather than as fuel, in the manufacture of such products as petrochemicals, asphalt, lubricants, and steel. Source: Exxon Corporation, *Energy Outlook, 1978–1990*, May 1978.

of automobiles, lowering speed limits, and making greater use of mass transit facilities.

Comfort energy, the energy used to heat and cool living space, accounts for almost two-thirds of total residential energy consumption. The greatest growth in demand has been in air conditioning, but consumption by all appliances, except cooking, is increasing (Fig. 12.5). Natural gas and electricity are the most important sources of energy used in this sector. The rapid increase in electric space heating is worth special note. In 1968 only 3 million homes, 7 percent of the total, used electric heat. The Federal Power Commission's projection is for an increase to 12.5 million homes in 1980 and to 24 million in 1990. This is just one of the changes bringing about the rapid demand for electricity. Direct use of electric power in space heating has been doubling every 10 to 15 years, and this doubling rate is expected to continue throughout the remainder of the century.

Commercial usage includes buildings, small businesses, schools, colleges, hotels, restaurants, and the like. The largest energy use is for space

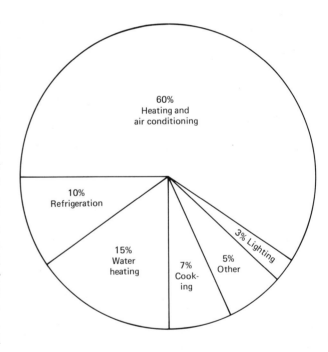

FIGURE 12.5
RESIDENTIAL ENERGY USE.

heating, cooling, and lighting. This latter is the fastest growing sector in energy use.

SOURCES OF ENERGY

Five categories of energy are available on the earth. One is the chemical potential energy in *fossil fuels* (coal, petroleum, and natural gas), which are the remains of plants and animals formed millions of years ago. Energy is also available from *nuclear fuels*, chiefly uranium. *Geothermal energy* uses the hot interior of the earth as an energy source, while *solar energy* may be used directly to capture radiation on mirrors or solar panels, or indirectly through the use of windmills or hydroelectric plants. Finally, *tidal energy* utilizes energy generated by the rotation of the Earth-Moon system.

Most of the natural resources we exploit for energy are *nonrenewable*: once used, they cannot be replenished. Thus, the energy potential in coal and oil is erased permanently. In contrast, some resources are *renewable*: wind, water, and wood are examples. Water, for example, can be pumped from an underground supply so long as withdrawal does not exceed infiltration into the underground reservoir. And the sun, the ultimate source of all energy on earth, is almost limitless in potential.

Solar and tidal energy can provide a continuous flow of power and may thus be considered renewable sources of energy. Chemical and nuclear energy are drawn from a limited stockpile of materials and are considered nonrenewable. Geothermal energy falls in between. Although the ultimate source of this energy (the hot interior of the earth) is so large that it is for all practical purposes infinite, the actual geothermal reservoirs are not infinite and can be exhausted by drawing the heat energy out of them faster than it comes from the interior of the earth.

In the short run, fossil fuels are the most important. In the United States, for example, they make up nearly 95 percent of all energy produced. The remaining energy is predominantly hydroelectric and nuclear. The traditional fossil fuel sources are being rapidly depleted, and a population that depends on nonrenewable energy resources faces a serious problem. Depletion may be delayed by finding new deposits or new sources of nonrenewable energy, but it cannot be averted. The solution to the problem is to develop renewable energy sources.

Fossil Fuels

Coal. Coal is a sedimentary rock consisting chiefly of decomposed plant matter that once grew in ancient swamps. Buried plant deposits, subjected to heat and pressure, gradually lost their hydrogen and oxygen, leaving carbon to become concentrated in the residue. The higher the percentage of carbon in the remaining deposit, the higher the value of the coal. The highest grade of coal, *anthracite*, contains 90 to 95 percent carbon. *Bituminous*, or soft, coal has about 80 percent carbon; *lignite*, the lowest grade of coal, has about 70 percent carbon. Coal is the most abundant of the fossil fuels. High grade bituminous coals of coking quality are those most desired by industry. Coal occurs widely but unevenly around the world, with the Soviet Union, the United States, and China being particularly rich in this resource (Fig. 12.6). Most of the world's coal resources are located in the Northern Hemisphere. Experts believe that most of these resources have been discovered and that at present rates of consumption coal reserves will last for several centuries.

Even before environmental factors were considered important, coal lost heavily to oil and natural gas, partly because these are cleaner burning fuels, but also because, weight for weight, they contain a good deal more energy than coal. The loss was particularly heavy in the transport sector, since oil can be used in internal combustion engines. Nevertheless, coal production in the United States is expected to more than double from 678 million tons in 1977 to about 1.5 billion tons in 1990.

Oil. Oil is formed by a process similar to that producing coal. Oil is thought to have been formed from sea animals as well as plant remains that sank to the bottom of the great seas that covered the earth in earlier geologic periods. Oil was formed by pressure of subsequent layers of sediment and the effects of heat. Oil is only a few tens of millions years old as against hundreds of millions of years for coal.

World coal resources

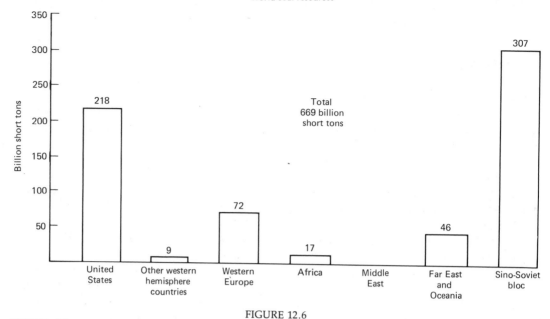

FIGURE 12.6

WORLD COAL RESOURCES. Source: U.S. Department of Interior, *Energy Perspectives 2*, Washington, D.C., June 1976, p. 17.

Oil was first produced in Romania in 1857. In 1859, the first oil well in the United States was drilled at Titusville, Pennsylvania. This epoch-making event spelled the doom of candles and whale-oil lamps and ushered in the transportation revolution.

Oil reserves are generally calculated on the basis that two barrels are left in the ground for every barrel recovered. Efforts are being made to improve recovery, mainly through the use of detergents or heat (steam) to step up the flow. The success of these efforts has still to be demonstrated. The largest known oil reserves are in the Middle East, which also leads the world in oil production (Table 12.1). In 1980 world oil reserves were estimated at 642 billion barrels. The Soviet Union ranks first in national production, followed by Saudi Arabia and the United States (Table 12.2). Only in Communist nations are consumption and production of oil in equilibrium; elsewhere consumption exceeds supply or supply exceeds consumption. From the pattern of production and consumption one may deduce that there is a great

TABLE 12.1 ESTIMATED OIL RESERVES (1,000 bbl)

Middle East	361,947,300
USSR, China, Soviet bloc	90,000,000
Western Hemisphere	89,772,500
Africa	57,072,100
West Europe	23,776,400
Asia-Pacific	19,355,200

Source. The Oil and Gas Journal (December 1979), pp. 70–71.

TABLE 12.2 CURRENT OIL PRODUCTION[a]

Soviet Union	11.7	Libya	2.2
Saudi Arabia	9.3	Indonesia	1.6
United States	8.7	United Kingdom	1.6
Iraq	3.4	Abu Dhabi	1.5
Iran	2.9	Mexico	1.5
Nigeria	2.4	Canada	1.5
Venezuela	2.3	Algeria	1.6
Kuwait	2.2		

[a] Millions of barrels per day.

Source. The Oil and Gas Journal (December 1979), pp. 70–71.

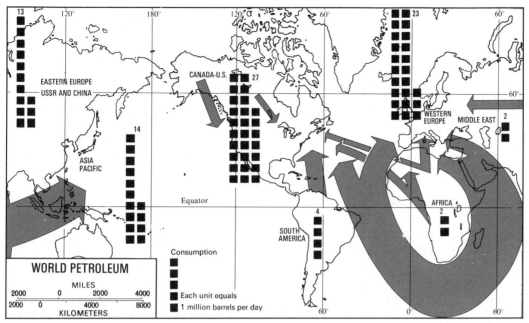

FIGURE 12.7
WORLD PETROLEUM.

deal of petroleum moving about in the world (Fig. 12.7).

Oil prices climbed steeply since 1973, in spite of which demand has increased steadily. A White House report says that for the world even to maintain its current rate of consumption and keep its reserves intact, it "would have to discover another Kuwait or Iran roughly every three years, or another Texas or Alaska roughly every six months."*

Fortunately there are some bright spots. Mexico, for instance, is expected to produce 5 to 6 million barrels of oil per day in the 1980s. The combined output of British and Norwegian fields in the North Sea may reach 4 million barrels a day at about the same time; and Egypt's production may reach 2 million barrels. British and Norwegian prospects are most promising in the eyes of continental Europe, which has no comparable resources. The United Kingdom hopes to fill its own oil and gas needs by the 1980s. Norway achieved self-sufficiency in 1976, and now exports small quantities.

For a time, North Sea output may stabilize European requirements for Middle East oil, so that only the United States and Japan will draw in-

creasingly on that area. However, British experts believe their North Sea fields will be running dry by the 1990s; similarly, some Norwegians believe their fields will peak in the mid-1980s and fall off toward the end of the century.

In Mexico vast new oil fields have been discovered in the states of Vera Cruz and Tabasco. It is anticipated that Mexico will become a major exporter of petroleum by the early 1980s. However, with the exception of Mexican oil, no new additional sources of oil are likely to be available before 1990, because of lead time required to bring new fields into production.

Natural Gas. Natural gas is formed by the same processes that form oil. Natural gas occurs with oil and often helps provide the pressure that drives the oil from the well. Such gas is called "associated" gas. Because it is more mobile than oil, gas can move through cracked rock or sandstone and collect in separate isolated pockets beneath the surface. This gas is referred to as "nonassociated" gas. Because of its mobility and its discovery with and without oil, it is extremely difficult to estimate the size of undiscovered gas resources.

Natural gas now provides about one-fifth of the world's energy, with demand growing by about

* The National Energy Plan, White House press release, April 19, 1977, p. viii.

TABLE 12.3 NATURAL GAS[a]

USSR	900,000
Middle East	740,330
United States	194,000
Algeria	132,000
Canada	85,500
Netherlands	59,500

[a] Estimated Reserves, 10^9 cubic feet.

Source. Oil and Gas Journal (December 1979), pp. 70–71.

7 percent annually; currently estimated reserves would last only about 50 years at the present rate of production (Table 12.3). Natural gas is unevenly distributed. Although 70 countries produce natural gas, four countries (the United States, the Soviet Union, Canada, and the Netherlands) account for about four-fifths of world marketed production.

Alternative Sources of Petroleum

Oil shale, another source of petroleum, consists of solid organic matter enclosed in fine-grained sedimentary rock. Liquid and gaseous hydrocarbons can be separated from such deposits by distillation, but the process is expensive. Commercial production of this oil, though practiced in some countries since the last century, has never been on a large scale. Shale oil was not competitive while inexpensive petroleum was readily available. Now that the price of crude oil has risen so sharply, recovery of oil from shale is beginning to be economically feasible.

The principal deposits of oil shale are in the United States, Brazil, the Soviet Union, China and central Africa. The world's largest and richest known deposits are in the United States; across portions of Colorado, Utah, and Wyoming. They are collectively known as the Green River oil shales (Fig. 12.8). Experts suggest that 50 percent of the oil from these deposits will be recovered; production capacity is expected to reach 1.5 to 2 million barrels per day by the end of the century. Shale recovery has become a subject of environmental controversy because of the need for extensive earth removal and for the use of large quantities of water, a scarce commodity in the western states.

Tar sands are another potential source of oil. In tar sands the oil, rather than impregnating shale rock, has been formed in sand and binds this material together. The tar is exceedingly viscous and thick and cannot be pumped. It must be processed in a manner similar to shale; that is, mined and then heated. The largest deposits of tar sands appear to be those in Alberta, Canada, where the *Athabasca tar sands* covers an area of 50,000 square kilometers (19,000 square miles). An estimated 300 billion barrels of oil might eventually be recovered from this deposit (Fig. 12.9). Other deposits are found in Albania, Romania, the Soviet Union, Venezuela, Colombia, and the United States.

Fossil fuels are nonrenewable. The two fuels on which we now rely most heavily, oil and gas, are also the least abundant. Our traditional sources of these fuels are thus limited and will be essentially depleted early in the twenty-first century. Coal, being more abundant than oil and gas, will meet long-range demands, and so may oil shales and tar sands. But the prediction that the world will run out of fossil fuels within 100 or 200 years will not be modified if we continue to use fossil fuels at the present increasing rates.

Other Sources of Energy

Water Power. Water power is a form of solar power. Water power is a renewable resource since

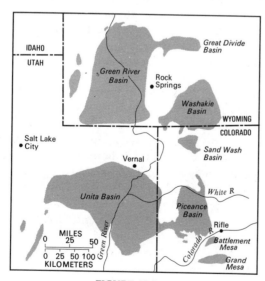

FIGURE 12.8
GREEN RIVER OIL SHALES.

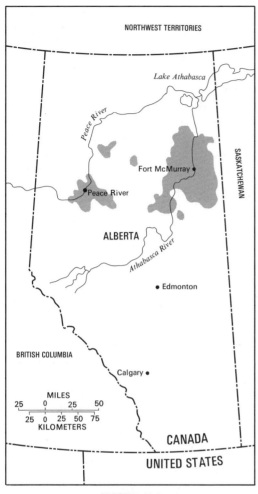

FIGURE 12.9
ATHABASCA TAR SANDS.

power potential has been developed. Africa and South America, areas lacking large reserves of coal, have the greatest water power potential. About 25 percent of the United States' water power capacity has been developed. In the United States, much of the remaining potential is protected by legislation such as the Wild and Scenic Rivers Act.

Nuclear Energy. The basic chemical unit is the atom. At its center is a small, dense core called the nucleus—1/10,000 the size of an atom. Nuclei are made up of the elementary particles, protons and neutrons. The energy binding these particles together in the nucleus is the basis of nuclear power. *Fission* and *fusion* are two important nuclear reactions in which energy is created. In a fission reaction a heavy nucleus, such as uranium, is split into two lighter ones. In this process mass is converted into energy. This is the energy obtained from the hot interior of a nuclear fission reactor. The fusion reaction involves the fusing together of very light nuclei. In a typical example, four hydrogen nuclei, by a complicated series of reactions, form a helium nucleus in a process similar to that that produces the energy of the sun. In fusion, some mass is converted into energy. Fis-

it is based upon the hydrologic cycle. The sun's heat evaporates water from the sea and the winds distribute the water vapor over the landmasses, where it falls as precipitation and flows into streams and rivers. That continuous cycle suggests that energy obtained from flowing water is continuous. However, water power has the disadvantage that dams necessary for the generation of electricity have limited lifetimes, because streams carry a suspended load of sediment that silts-up reservoirs (Fig. 12.10).

Water power has been used for thousands of years. Water wheels were used by the Romans and were, with wind, a source of power long before the steam engine. Only in the twentieth century has water power been widely used to generate electricity. About 6 percent of the world's water

FIGURE 12.10
WATER POWER IS AN EXPRESSION OF SOLAR POWER. It has the advantage of being a renewable resource since it is based upon the hydrologic cycle. Hoover Dam and Lake Mead are along the Nevada-Arizona border.

SOURCES OF ENERGY
299

sion and fusion are the basis of the atomic bomb and hydrogen bomb respectively. Technology has been developed to control fission and permit the slow release of energy for conversion into electric power. This is the process used in scores of nuclear power plants around the world. We cannot yet control the fusion process for peaceful purposes. If the technological problems can be solved, and if controlled hydrogen fusion can be achieved, then fuel for the process will present no problem because the oceans are an essentially limitless source of hydrogen.

To many people throughout the world, nuclear power is seen as the only real opportunity to reduce the dependence of their economies on foreign oil (Fig. 12.11). The world's coal reserves are 14 times greater than the remaining oil and gas reserves, and nuclear fuel resources will probably exceed coal resources if we adopt "breeder" reactors: reactors that breed, in the form of plutonium, more fuel than they consume.

Some serious questions about nuclear energy remain to be answered. One sensitive area in the nuclear fuel cycle involves the enrichment of uranium. Nuclear reactors for generating power usually burn low-enriched uranium as fuel; however, the same process also can produce highly enriched uranium, suitable for making nuclear bombs. Another sensitive area in the fuel cycle involves reprocessing, during which the spent fuel from a reactor is broken down into waste materials and reusable materials. Among the later is plutonium, a highly toxic man-made substance that also can be used for making nuclear bombs. The final disposal of radioactive wastes has been a particularly vexing problem in nuclear fission. This, together with the realization that uranium is a depletable resource, has added to hopes that nuclear fission may one day be supplanted by fusion, with the advantages of fewer environmental problems, a virtually limitless supply of fuel, and greater reactor safety. The accident at Three Mile Island in southeastern Pennsylvania in 1979 strongly reinforces the arguments of those who oppose the expansion of nuclear power as an energy source in the United States (Fig. 12.12).

Geothermal Energy. In a sense, geothermal energy is nuclear energy. The interior of the earth is made up of hot material which comprises a reservoir of heat energy. This heat energy is stored kinetic energy. We are not certain how the energy

FIGURE 12.11
SOME SERIOUS QUESTIONS ABOUT NUCLEAR ENERGY REMAIN TO BE ANSWERED. Three Mile Island, site of the worst commercial nuclear accident in United States history, is located on the Susquehanna River eleven miles southeast of Pennsylvania's state capital, Harrisburg.

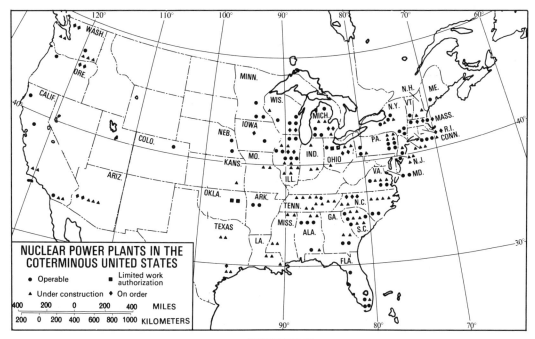

FIGURE 12.12
NUCLEAR POWER PLANTS IN THE UNITED STATES.

got there but we are fairly certain that it is the radioactivity of some of the material in the earth which keeps the rock hot. There is probably enough uranium, potassium, radium, and other naturally radioactive materials in the earth's interior to produce the relatively small amount of heat needed to make up for that lost through the surface.

Commercial development of geothermal energy is possible in areas of recent or current volcanic activity, where the magma is close to the surface and can serve as a source of heat. Energy from the interior leaks out in various ways. Steam and heated water may leak to the surface in the form of hot springs and geysers. Underground reservoirs that feed these springs may be tapped by drill holes to bring the steam or hot water to the surface to generate electricity.

Potential geothermal resources exist around the world (Fig. 12.13). The first steam well was drilled in 1904 at Larderello, Italy, and has been in use ever since. Domestic and industrial heating by geothermal heat was initiated in Iceland in 1925, and more than 100,000 people now live in houses heated in this manner. New Zealand began major exploration of hot spring and geyser areas in 1950, and successful results there proved that commer-

cial steam can be developed from areas containing very hot water rather than steam at depth (Fig. 12.14). The best known geothermal field in the United States is The Geysers in California, where steam is piped to steam turbines and used to generate electricity. Large resources of geothermal energy also exist in Japan, the Soviet Union, several countries in Central America and East Africa, and many countries around the Mediterranean Sea. Scores of countries are actively involved in geothermal exploration and development.

It is too early to judge whether natural steam has the potential to satisfy an important part of the world's requirements for electric power, but in favorable locations it is already an attractive source of cheap power. Geothermal energy, like nuclear energy, is an energy of the future.

Solar Energy. Solar energy has been in use for some time. We can now buy solar hot-water heaters as well as homes that are heated, at least in part, by solar radiation. Serious technological barriers need not prevent the greater use of solar heating and cooling; rather, what is needed is the standardization and mass production that will make solar systems viable economic alternatives to our present energy systems.

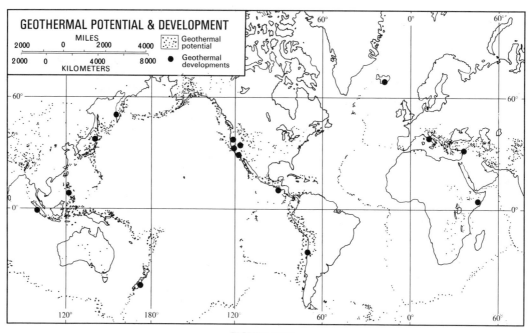

FIGURE 12.13
GEOTHERMAL POTENTIAL AND DEVELOPMENT.

The direct conversion of solar energy into electricity, however, is a more difficult challenge. Two techniques can be used. One is a solar-thermal system that uses mirrors to focus sunlight onto receivers that heat a fluid which in turn is used to drive a turbine. The second technique involves the use of photosensitive silicon cells to convert sunlight directly into electricity. At this time both methods remain inefficient and expensive when compared to available alternatives.

The amount of solar energy that reaches the earth's surface represents the greatest energy source available to man. In a very sunny climate, the average daily solar radiation falling on one acre of ground is equivalent in energy content to about 20 barrels of oil (the average for the continental United States is about 11 barrels per acre). Solar energy's diffuse and intermittent nature, however, forces enthusiasts to await the development of inexpensive collection and storage facilities (Fig. 12.15).

Tidal Energy. Tidal water wheels, used for grinding grain, have been used for centuries. Today's interest in tidal power lies in its potential as a source of electricity. Sea-water flowing in and out of a restricted bay can turn generators just as river water does. The generation of electricity is only practical, however, where unusually high tides occur.

The difference between high tide and low tide along most of the world's coasts is too small—about 2 meters (6 feet)—to exploit tidal energy. However, in the Rance River Estuary in northwest France tides rise and fall as much as 13 meters (44

FIGURE 12.14
THE EARTH'S HEAT IS CONVERTED INTO ELECTRIC POWER AT THIS NEW ZEALAND GEOTHERMAL FACILITY.

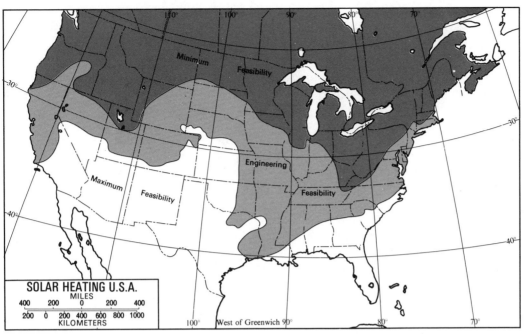

FIGURE 12.15
SOLAR HEATING IN THE UNITED STATES.

PRESENT AND FUTURE ENERGY SOURCES

Energy Source	*Developmental Status and Prospects for Future Use*
Fossil fuels	
Petroleum	Now widely used. Supplies limited—possibly exhausted in 30–40 years.
Natural gas	Now widely used. Supplies limited—possibly exhausted in 10–20 years.
Coal	Now widely used. Supplies somewhat limited—possibly exhausted in 300–500 years.
Hydroelectric	Now in use. Number of sites for future development is limited.
Solar	Now in limited use. Practicality somewhat dependent on geography, weather patterns, etc.
Nuclear	
Conventional fission reactors	Now in limited use. Low-cost fuel supply possibly exhausted in 30–40 years.
Fast-breeder reactors	Now in late stages of development. Greatly extends potential fuel supply of fission reactors.
Fusion reactors	Feasibility still to be proven.
Tides	Now in very limited use. Number of suitable sites for future development is limited.
Geothermal	Now in very limited use. Number of suitable sites for future development is limited.
Wind	Now in very limited use. Number of suitable sites for future development is limited.
Ocean thermal gradients	Not now used. Feasible, but dependent on geography.

Source. H. Stephen Stokes, Spencer L. Seager, Robert L. Capener, *Energy: From Source to Use,* Glenview, Ill.: Scott, Foresman and Co., 1975, p. 47. Reprinted by permission of the publisher.

feet). The first tidal dam and generator was built there in 1965. The Soviet Union later built a tidal-power plant near Murmansk on the White Sea. Although no tidal conversion facility has been built in the United States, the Passamaquoddy Bay area on the far northern coast of Maine is a possible site for a tidal plant. While tidal-power may be significant locally, it could never be very important on a worldwide scale.

Some other possibilities include the energy of wind to turn windmills that can supply small energy needs for homes or farms; energy from the *thermal gradient* of the ocean, where the difference in temperature between cold bottom waters and warm surface waters could, theoretically, be converted into power; and the energy from pounding waves along seashores, and the giant ocean currents.

THE FUTURE OF ENERGY

What are the solutions to our energy problems? Numerous suggestions have been made: remove price controls on oil and gas, break up the oil companies, build breeder reactors, develop solar energy, build windmills, and make use of gasohol. Three viewpoints are frequently heard. One takes the business-as-usual approach and places its faith in the private sector and marketplace. Proponents of this view favor price deregulation, weakening of environmental restrictions, and increased exploration and exploitation of existing resources. An opposite viewpoint prefers to encourage and subsidize conservation and efficiency; the goal here is less consumption, with punitive taxes on large cars, for example. The third theme suggests we rely upon our ingenuity and develop exotic sources of energy, with the sun most often mentioned as the panacea for our energy ills. There are, of course, positive aspects to each point of view. And each is being considered by the U.S. Department of Energy, a Federal agency that has developed plans for energy research, development, and demonstration. The plans have a priority scale (because some courses of action are closer to realization than others) and a time scale consisting of near-term (to 1985), mid-term (1985 to 2000), and long-term (post 2000) programs.

For the near-term, priorities are established as follows: 1) expand production of coal, our largest energy resource; 2) increase the percentage of oil and gas recovered from existing fields; and 3) build conventional nuclear reactors. As to demand, steps should be taken to increase efficiencies and conserve existing resources. Mid-term priority is given to the production of liquid fuels from coal and to the extraction of oil from the Green River deposits of oil shale. Geothermal and solar resources are given second priority during the mid-term. Because new energy programs take 8 to 10 years to develop, those suggested for the mid-term should have begun in the late 1970s. The long term programs will take us into the next century when "inexhaustable" sources will replace our diminishing fossil fuels. The breeder reactor, fusion process, and solar energy will be given highest priority to ensure us of an energy-abundant future.

Previous generations of Americans have faced major challenges—settling the frontier, industrialization, war, depression. This generation is discovering that it faces a challenge that is equally great—the energy crisis. Meeting this challenge will require sacrifice, hard work, skill and imagination on the part of the American people. It will require a new national ethic that values energy efficiency and condemns energy waste. And it will require a degree of cooperation that the United States has attained only in meeting the great challenges of the past. As the President stressed in his address on April 18, 1977, "This difficult effort will be 'the moral equivalent of war'—except that we will be uniting our efforts to build and not to destroy." The prospect of America organizing to meet the energy crisis is not grim. It is exciting.

Source. The National Energy Plan, Washington, D.C., 1977.

The products of our fields, forests, mines, and the sea may be viewed as the bounty of nature. When mankind harvests this bounty, it enters into primary production, the first stage in which goods are created. Secondary production, or manufacturing, is that which involves a change in the form of the products thus harvested. Wheat made into bread, iron into steel, and steel into an automobile are all processes representative of manufacturing, our next topic of discussion.

fossil fuels

nuclear fuels

geothermal energy

solar energy

tidal energy

nonrenewable
resources

renewable resources

anthracite

bituminous

lignite

oil shale

Athabasca Tar Sands

nuclear fission

nuclear fusion

DISCUSSION QUESTIONS

---

1. Which energy sources are considered renewable and which are nonrenewable?
2. How do you explain the decline in coal consumption since 1900 and the increased demand for oil and natural gas?
3. What role does the United States play in total world energy consumption?
4. What is meant by the term "comfort energy"? What are the most important sources of this type of energy?
5. Which fossil fuels are the least abundant? What alternative sources of petroleum are available and what environmental concerns are associated with their development?
6. Discuss the various energy sources available that are alternatives to fossil fuels.
7. What problems are associated with nuclear power development?
8. What is the current status of solar energy development?
9. What areas of the world have the largest petroleum reserves? What country is the largest producer of petroleum?
10. Discuss the developmental status and prospects for future use of the various sources of energy.

REFERENCES FOR FURTHER STUDY

Carr, D., *Energy and the Earth Machine*, W. W. Norton and Co., New York, 1976.

Cook, E., *Man, Energy, Society*, W. H. Freeman and Co., San Francisco, 1976.

Guyol, N., *Energy in the Perspective of Geography*, Prentice-Hall, Englewood Cliffs, N.J., 1971.

Halacy, D., *Earth, Water, Wind, and Sun: Our Energy Alternatives*, Harper & Row, New York, 1977.

Stokes, H., S. Seager, and R. Capener, *Energy: From Source to Use*, Scott, Foresman and Co., Glenview, Ill., 1975.

Turk, A., *Environmental Science*, W. B. Saunders Co., Philadelphia, 1978.

MANUFACTURING

MANUFACTURING REFERS TO THOSE PROCESSES BY which goods are changed in form. Clay made into a vase, cotton woven into cloth, and wheat made into bread are all examples of manufacturing. Such processes contribute to human society by providing the basic necessities of life as well as by fostering comfort, convenience, and well-being. Indeed, manufacturing is so important that countries are commonly classified according to the level to which they are industrialized or the extent to which their manufacturing sector is modern in terms of capacity and technology. This is the sense in which we speak of the developed countries and the less developed countries. Although the criteria are arbitrary and imprecise, there is general agreement that a list of developed countries would include the United States, Canada, the Soviet Union, western Europe, Japan, Australia, and New Zealand. Other areas, particularly in Latin America and peripheral Europe, including Brazil, Argentina, Mexico, Ireland, and Spain, are intermediate in terms of development; their technology has evolved beyond the traditional, less developed phase, but they have not yet emerged as modern industrial states.

The less developed countries tend to be agrarian. Although most people in such countries work on the land, returns are meager and little surplus is produced. Accordingly, opportunities for capital accumulation are limited and little is available for investment in housing, clothing, education, or medical services. Moreover, lack of capital inhibits improvement of productive facilities. Although some manufacturing exists, it is generally organized in units that are small, inefficient, and poorly equipped. Much work is done by hand in these so-called cottage industries, but *productivity* (output per man hour) is often low because of obsolete and inadequate equipment. When modern technology is introduced into a traditional society of this kind, the result is a complete transformation of the way of life.

THE DEVELOPMENT OF MANUFACTURING

The term Industrial Revolution has been used to refer to the application of inanimate power to the processes of production. Historically, the first step was the development of a workable steam engine. This occurred in Great Britain in 1769. Steam engines were first employed to pump water from mines, but soon after were used to power textile machinery, thus giving Britain a commanding lead in the textile industry. Steam engines were also applied in locomotives and ships, thereby increasing the distance from which raw materials could be collected and the distance finished goods could be sent to market.

Another innovation was the development of a

307

process enabling the manufacture of inexpensive steel in large quantities. This was known as the Bessemer process. Man had long known how to make fine quality steel: Damascus sword blades were famous in the Middle Ages, and Arab armourers brought their skills to Spain where they became strongly established in Toledo and Seville. But steel was expensive because it could be produced only in small quantities. But after the Bessemer process was introduced in England in the early 1860s, British steel production climbed from 100,000 tons to 2,000,000 tons per year within a few years. Steel thus became available to make ships, railroads, bridges, and a host of manufactured items; steel is the common structural material of the modern world.

Further improvements in manufacturing include mass production and the introduction of interchangeable parts. The introduction of factory production lines (where tasks are divided into separate operations, most of which are easily mastered), represented a shifting of skills from worker to machine. Modern automobiles, for example, are made mostly by unskilled labor; the requisite skills are built into the factory equipment and a high-quality product is produced at low unit cost.

Improvements in the manufacturing process resulted in an increase in productivity, the benefits of which have spread throughout most sectors of society in the developed world. Thus, in the developed countries, living standards tend to be high. Unfortunately, industry has its negative side, and large-scale industrial development has had a negative effect upon the environment. Heavy industry can be dirty and noisy; it can pollute the air and water; it can be a bad neighbor. In recent decades resistance to these unattractive aspects of industry indicate a change in public attitudes, signaling a tendency to question the value of an industrialized society. While recognizing the validity of such inquiry, we should not lose sight of the benefits that industry has brought to the human family; the good should be retained, and the bad reduced.

Modern industry is uneven in its distribution. Four major industrial areas can be identified in the world (Fig. 13.1). These are east central Anglo-America, western Europe, the Soviet Union, and the Far East. Only minor agglomerations exist elsewhere. However, the Industrial Revolution is an ongoing process, and modern technology is slowly spreading throughout much of the Third World where it reflects the aspirations of the people of less developed countries. Stimulated by rising expectations, the poorer countries have developed a natural tendency to copy the successful techniques of the developed countries. Local governments tend to support the drive toward modernization. However, lack of capital is almost always a problem. Here the multinational corporations play an important role. Although based in the industrialized world, these companies are international in scope, and their size gives them tremendous power. For example, the biggest multinationals, such as General Motors or Royal Dutch Shell, have larger annual incomes than any country in South America except Brazil. In order to minimize costs, the multinationals tend to locate their *labor-intensive* operations in low-wage areas, usually found in the less developed nations.* Most television sets sold in America are now manufactured in Taiwan, Hong Kong, South Korea, and Mexico. Leica cameras are manufactured not in West Germany but in Singapore. A substantial number of Volkswagens are manufactured in Mexico. The tendency of the multinationals to utilize low-wage labor of the Third World contributes substantially to the spread of technology.

LOCATION FACTORS

A *location factor* is any aspect of an environment that affects the location of an industrial activity. Important location factors include market, raw materials, labor supply, energy, climate, political environment, amenities, and locational inertia. It is useful to distinguish between *heavy industry* and *light industry*. Heavy industry, as in a steel mill or petroleum refinery, generally involves the processing and assembly of bulky raw materials. Large, heavy equipment is used, and the sites are usually spacious, frequently covering hundreds of hectares. Also, heavy industry tends to locate at waterside or along a railway line so as to minimize freight charges. Light industry refers to processes such as the assembly of television sets or the man-

---

* A labor-intensive operation is one in which labor expenses comprise a major portion of total costs, as opposed to a capital-intensive operation in which capital equipment would be the major item of cost.

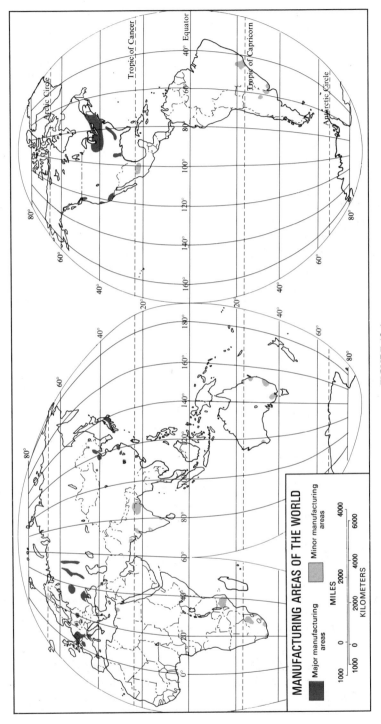

FIGURE 13.1
MANUFACTURING AREAS OF THE WORLD.

ufacture of wrist watches. Light industry is frequently organized in small units, and costs for freight and raw materials are a minor item of expense. Heavy industry and light industry each respond to different sets of location factors.

Market

Manufacturers generally prefer to be as close to their markets as possible. This consideration is important where products are perishable or bulky, and freight charges are a significant part of cost. A bottling plant for fresh milk, for example, or a sand and gravel works, is required to locate near the market so as to minimize transport costs. Proximity to market has other advantages. The manufacturer can maintain a close relationship with his customers and respond quickly to shifts in demand. This factor is important to subcontractors and is evidence of the so-called agglomerating tendency of industry. A market area usually includes a supply of labor that can also be useful to an employer.

Raw Materials

The location of raw materials is of major consequence in heavy industrial operations where freight charges tend to be an important item of expense. In order to maximize efficiency, the plant should be located so as to minimize the cost of assembling materials and of shipping the finished products to market. Several different cases are possible:

1. If one or more ubiquitous materials are used, the plant will locate at the market.*
2. If one raw material is used and is fixed (non-ubiquitous), and there is no change in weight between source and market, the plant can locate anywhere between source and market.
3. If one fixed raw material is used and there is a gain in weight between the source and the market, the plant will locate at the market.
4. If one fixed raw material is used and there is a loss of weight between the source and the market, the plant will locate at the source.

---

* A ubiquitous material is one found everywhere. Air, or sand and gravel, are normally considered ubiquitous.

5. If two or more fixed raw materials are used and there is a loss of weight for at least one, the site of minimum transport costs will lie at some intermediate point, the location of which will vary according to circumstance.

Raw materials are an important factor in locating such facilities as concentrators, smelters, refineries, sawmills, sugar mills, canneries, freezing plants, and slaughter houses (Fig. 13.2).

Labor

Labor supply is a location factor of consequence but is obviously more important in some processes than in others. A distinction is made between skilled and unskilled labor because these two categories respond differently to the job market. Unskilled labor is virtually ubiquitous; it can be recruited wherever there is a population. Moreover, because of the weakness of their bargaining position, unskilled production workers are mobile; they go wherever job opportunities exist. Skilled production workers, however, are neither ubiquitous nor mobile. They tend to agglomerate in a spatial sense and exhibit considerable spatial inertia. Therefore, an industry needing highly-skilled workers may be required to locate where a pool of labor exists. Such constraints account for the manufacture of firearms in New England, the tailoring industry in New York, and the making of fine watches in Switzerland. A low-skill industry, however, is freer to locate where it chooses, confident that workers will follow or that local people can be quickly trained to acceptable standards of performance. These considerations provide at least a partial explanation of why the cotton industry, once so strongly established in New England, was able to move south to the Piedmont.

Many manufacturers are eager to locate in areas where low wages prevail. Consider the following example. A corporation wishes to produce television sets for sale in the United States. The manufacturing process involves two operations; first, the making of components, which is easily mechanized and is therefore capital-intensive; and second, the assembly of these components, which resists mechanization and is therefore labor-intensive. In order to minimize costs the corporation builds two plants, one in California and one just south of the border in Mexico. The components

FIGURE 13.2

MEGALOPOLIS REPRESENTS A SIZABLE MARKET FOR THE PETROLEUM PRODUCTS REFINED AT THIS FACILITY NEAR PHILADELPHIA.

are manufactured in California where labor is expensive but capital is available. The components are then shipped across the border where they are assembled in the low-wage environment of Mexico. The assembled television sets are re-imported into the United States, custom duties being paid only on value added, which is based on the low wages paid in Mexico. These sets can then sell at a competitive advantage in the American market. In consequence of this situation, factories specializing in just such labor-intensive operations have come into existence along the northern border zone of Mexico.

Similar arrangements make it possible for multinational corporations to benefit from the low-wage labor of the Orient. Factories in Hong Kong, Taiwan, South Korea, and the Philippines assemble optical goods, television sets, clothing, toys, computers, electronic equipment, and cameras that command a large portion of world markets.

Energy

An energy-intensive industry is one in which energy is a major item of cost. The refining of aluminum provides an excellent example. Two steps are involved in the refining process. First the ore, *bauxite,* is crushed and soaked in caustic soda, thus concentrating the *alumina,* which is pure aluminum oxide. Second, the alumina is subjected to electrolysis, a process that extracts the pure aluminum, which is then drawn off in a molten form. Twenty-two kilowatt-hours of electricity are required to produce one kilogram (2.2 pounds) of aluminum.

Because of their heavy demands for electricity, aluminum mills tend to locate near sources of cheap power, particularly hydroelectric facilities. Thus, they are located in Oregon, Washington, and British Columbia, and also in Siberia, Switzerland, and Norway. The largest aluminum mill in the world is that owned by Aluminium of Canada at Arvida, on the Saguenay River in Quebec. Recently, however, there has been a shift towards the use of electricity generated by burning fossil fuels, and aluminum mills have been built near the oilfields in Texas and Louisiana, and also near the coalfields of northern Appalachia. Dramatic increases in the price of petroleum may, however, reverse this trend.

Climate

Climate affects industrial location in a number of ways. A long, harsh winter can impede water transportation by freezing waterways. Also, low temperatures raise heating costs not only for fuel but also by requiring the building of more substantial buildings and a greater investment in heating equipment. Other things being equal, industry prefers a moderate climate.

Water, which is an aspect of climate, is also an industrial commodity of vital importance. It is used in tremendous quantities for cooling and washing, as a solvent, and as a raw material. More than three hundred metric tons of water, for instance, are used in making one metric ton of steel. Water also serves (via waterways) as a preferred mode of transport, particularly for heavy industry. As a further consideration, large populations can exist only where water is available in generous amounts. (And people, of course, constitute both market and labor for industry.) More industry is congregated in humid regions than in dry ones.

Political Environment

The political atmosphere works to encourage or discourage industrialization. As a basic necessity, the businessman needs political stability so that he can engage in long-term planning. Sudden or repeated shifts in the "rules-of-the-game" make difficulties for the businessman, and create what is known as an unfavorable investment climate. One advantage provided by the political environment in the United States has been two centuries of continuous and stable government, with an orderly transfer of power from one administration to the next. The business community has been able to feel a reasonable degree of confidence in the future of American society. This feeling encourages savings and investment and has fostered the emergence of America as a leading industrial

power. Such confidence does not prevail in Bolivia, for example, which has had many revolutions in recent years.

Political attitudes also affect location decisions in other ways. A tax policy, for instance, at either the national or local level, can foster or inhibit industrialization. An intangible known as community attitudes can express itself through the medium of zoning ordinances or it can have an effect on the availability of land. Local statutes on the control of pollutants may be permissive, or they may be so stringent as to be prohibitive. This latter point is of particular importance for the location of heavy industrial sites; many communities do not want such sites because of concern about degradation of the environment. Petroleum refineries, in particular, are viewed as undesirable and, local economic benefits notwithstanding, petroleum companies are often told to look elsewhere. Whatever the matrix of variables, the political environment, at both the national and local level, is important and cannot be ignored in making investment decisions for industrial facilities.

Amenities

Amenities refer to the quality of life available in a community. People in modern industrialized societies are more demanding in this sense. This factor is of particular importance to high-technology industries that need skilled scientists and engineers. Personnel of this calibre not only command high salaries but also tend to insist on pleasant living conditions. They want good schools, clean air, and sunshine, and they prefer access to a variety of recreational facilities. The right mix of amenities can provide a considerable advantage in recruiting high-quality personnel.

Today people in many industrialized areas are asking searching questions about their manner of life, about the satisfaction it holds for them, and about the effect it has on their children. These attitudes have triggered an exodus from many metropolitan areas in the United States. These areas are experiencing a net loss of people, plants, and industry, often to southern areas extending from the East Coast to California. The latter are less crowded and the winters are mild. Many people are swayed by such considerations, and employers are conscious of the advantages of less congested areas in the South.

Locational Inertia

Once a factory is established there is a tendency not to move it unless there is a strong motive to do so. Several reasons underlie this *locational inertia*—habit, fear of the unknown, and the severing of familiar relationships. It is also generally less costly to expand an existing plant than to relocate it. These factors combine to resist relocation. Many industrial facilities, therefore, are found in places selected for reasons no longer relevant. Such locations can be explained only in terms of locational inertia.

SELECTED INDUSTRIES

There are many different manufacturing industries. Rather than attempt to study all of them, three representative industries will be examined to determine the course of their development, their significance, and the factors that led to their current distribution. The industries selected are steel, textiles, and computers. Steel was selected because it is a fundamental industry in the modern world. Its products affect every aspect of an industrialized society and its production, or lack thereof, is a distinguishing difference between developed countries and less developed ones. The textile industry was selected as representing a traditional long-established industry that produces a commodity useful to all mankind. Computers were selected as representing new developments at the frontiers of human technology.

The Steel Industry

Steel is the leading structural metal of the modern age, although it is being challenged by plastics and light metals in some industries. In 1978 world production of steel was 660 million tons, as against 12 million tons of aluminum, its closest competitor. Steel is still a growth industry; world capacity expanded by 75 percent between 1963 and 1978.

Several factors explain the extraordinary dominance of steel. First, steel is made from iron, and iron is relatively abundant. Second, steel is cheaper to produce than any competing metal. It is strong and easily worked. Finally, steel is versatile; it can be combined with various alloy metals to create a steel, for example, that is extremely hard (as for cutting tools); a steel that is flexible and resistant to metal fatigue (as for springs); a steel high in tensile strength (as for cables); a corrosion-resistant steel (for piping or tubing or kitchen equipment); or a steel that is tough and resistant to shock (as for armor plate). No other metal can rival steel in its wide range of uses. So important is steel that its production is used as a measure of business activity and as an index of national power.

The Manufacturing Process. The major raw materials used in the manufacture of steel are iron ore, scrap, coke, and limestone. Iron ore is an oxide of iron. Coke is made by heating coal in the absence of air. This process, known as dry distillation, drives off volatile substances, leaving behind coke, which is virtually pure carbon.

The first step in the manufacture of steel is the smelting of iron ore into pig iron, which takes place in a blast furnace. Iron ore, coke, and limestone are dropped into the furnace from above, the coke is ignited, and hot air is forced through the mass from below. As the temperature rises, the impurities in the iron ore are absorbed by the limestone, which serves as a flux. Molten pig iron is drawn off from the bottom of the furnace (Fig. 13.3).

Pig iron is high in carbon content, which makes it hard but brittle. Pig iron can be used as cast iron, but since it lacks tensile strength, its usefulness as a structural material is limited. In order to arrive at a product that is tougher, stronger, and

FIGURE 13.3
THE PRODUCTION OF PIG IRON IS THE FIRST STEP IN THE MANUFACTURE OF STEEL.

more reliable, a further refining process is required. This consists first of burning the impurities out of the pig iron, a process that results in the creation of wrought iron, which is nearly 100 percent pure iron. Wrought iron is used for making decorative scroll work, banisters, and furniture, but is too soft for many uses. To make steel, alloys must be added. These alloys include carbon, manganese, nickel, cobalt, or tungsten. The resulting product is steel.

Steels vary in accordance with the alloys that are added and the manner in which the steel is heated, cooled, and worked. Steel is usually ordered for a specific purpose for which it must meet stringent requirements. Precise control is, therefore, of vital importance in the modern steel industry.

Size of Plant. Two contradictory trends are in evidence with regard to the size of new steel mills.

There is a tendency to build units of ever larger capacity, and also to build small mini-mills designed for limited local markets. The large mills, with capacities in excess of 2 million tons per year, achieve cost efficiencies through economies of scale. They use a *continuous process* operation based on the output of enormous blast furnaces, some of which can produce more than 10,000 metric tons of steel per day. These mills are usually located at tidewater so as to avail themselves of waterborne transport (Fig. 13.4). The economic reach of these plants is worldwide. Japanese mills, for example, use iron ore from Brazil and Australia and coal from the United States; a new mill at Dunkirk, France, uses coal from the United States and ore from Africa and Sweden.

Mini-mills, with capacities as low as 50,000 metric tons per year, melt local supplies of scrap metals in electric furnaces. They are competitive in local restricted markets.

FIGURE 13.4
STEEL MILLS USUALLY LOCATE ALONG NAVIGABLE WATERWAYS TO AVAIL THEMSELVES OF INEXPENSIVE WATERBORNE TRANSPORT. Bethlehem Steel Corporation's Burns Harbor, Indiana, plant.

Location of the Steel Industry. The iron and steel industry responds to several location factors, of which market is probably the most important. Nearness to market facilitates interaction with customers, and the surrounding area serves also as a source of scrap and labor. In addition, a large market will probably have well-established transport links with other areas, and these are helpful in gathering raw materials and shipping finished goods.

ECONOMIES OF SCALE

The principal of *economies of scale* states that industry is most efficiently organized in large units. Large-scale production permits the use of highly specialized equipment, which in turn brings about lower unit costs. Consider, for example, the manufacture of automobiles. In a large factory, such as the River Rouge plant of the Ford Motor Company, fenders for an auto are shaped by a large powerful hydraulic press. The operation is quick and simple and thousands of fenders are made each day at low unit cost. Automobiles can also be handcrafted in a small shop, but the fenders would then have to be shaped by hand, which is time-consuming and expensive. Assuming similar differences of technology throughout the manufacturing process, it follows that the River Rouge plant can turn out thousands of autos per day at low unit cost, whereas the small shop can turn out only a few cars per year at very high unit cost.

The earliest steel mills exhibited a strong tendency to locate near coalfields because early blast furnaces consumed more coal than iron ore. The iron ore, therefore, moved to the coal. Modern technology, however, has reduced the requirements for coal. Moreover, improvements in methods of transportation have lowered freight charges to the point that steel mills have been "liberated" from the coalfields. A discernable trend in recent decades has been the tendency of steel mills to locate at tidewater where accessibility to ocean-borne transport is available. This trend is discernable in the presence of large new mills along the Atlantic seaboard of the United States, along the shores of the Mediterranean Sea and the North Sea, and along the Inland Sea of Japan (Table 13.1).

Textiles

The textile industry produces commodities used by all and finds ready markets in both the developed and less developed countries. Cloth may be made from either natural or synthetic fibers or from mixtures of both. It is a pioneer industry and forms the largest source of industrial employment in many less developed nations.

Because it is versatile and inexpensive, cotton is the world's most widely used fiber. Wool is another important fiber; it provides warmth without weight, can absorb moisture without feeling wet, and is used to make higher-priced garments. Synthetics can be made from cellulose or other organic materials such as coal, petroleum, or vegetable oils. Synthetics can even be made from spun glass. Cotton and wool are agricultural products and tend to be raised in areas exterior to manufacturing regions. This is particularly true of wool, as the grazing of sheep is an extensive form of land use. Synthetics, however, are industrial products originating in chemical plants. This fundamental

TABLE 13.1 LEADING STEEL PRODUCERS, 1978
(millions of metric tons)

USSR	151
USA	124
Japan	102
West Germany	41
China	34
Italy	24
France	23
Britain	22
Poland	19
Czechoslovakia	16
Canada	16
Belgium	13
Brazil	12
Spain	11
India	10
World	660 (estimated)

Source. United Nations, *Monthly Bulletin of Statistics* (New York, May 1979), pp. 78–79.

difference in source of raw material affects the spatial relationships within the industry and the trade linkages among sources, factory, and markets. These linkages tend to be longest for the woolen industry, intermediate for cotton, and shortest for synthetics.

Cotton. The cotton plant requires a long growing season with abundant moisture and high prevailing temperatures. A dry period is desirable during harvest. The United States is the world's leading producer, followed by China, the Soviet Union, India, and Egypt. Cotton farming in America is concentrated in the southern part of the country; the leading states are Texas, California, Mississippi, and Arkansas. About one-third of the American cotton crop is exported, with major commodity flows going to Japan and Europe.

Cotton manufactures in America were first established in New England, but the industry started to move south in the latter decades of the nineteenth century. Mills were established in the Piedmont area, and North Carolina, South Carolina, Georgia, and Alabama emerged as national leaders. The forces behind this migration illustrate a classic case study in the field of economic geography.

What factors led to the early leadership of New England in cotton manufactures? Innovative enterprise played a significant role. In the latter half of the eighteenth century, the British had a monopoly on the design and construction of advanced textile machinery, and they forbade the export of this technology. Samuel Slater, however, who worked in British mills, learned how the machines were made. When he migrated to America in the 1780s, he was able to reproduce them from memory. He acquired financial backing, and in 1790 founded his own mill. The British monopoly was broken, and New England established early technological leadership in the textile industry in the United States.

Other factors also encouraged industrialization in the region. New Englanders had made money in shipping; therefore, venture capital was available. Competing alternatives were not attractive; farming, for instance, was unpromising due to rough topography, thin soils, and long cold winters. Industry, however, provided a productive outlet for the creative energies of the people, and a manufacturing sector came into existence. Textiles, both cotton and woolens, formed an important part of this sector.

After the Civil War, however, the situation began to change. The center of gravity of the market moved as increasing numbers of Americans occupied lands to the west. The United States lost its coastal orientation and began to assume continental dimensions. Railroads were built, creating a land transport network and thus supplementing and replacing the maritime linkages that had until then been dominant. The cotton mill owners in New England began to realize that the location of their mills offered limited access to market and, because cotton came from the South, limited access to raw materials. It was perceived, however, that the Piedmont was convenient to both.

The Piedmont enjoyed other advantages. Power was readily available, either from the rivers that coursed through the region or from the coalfields of Appalachia. An abundant supply of water was available for washing and processing. The climate was humid, thus preventing the undue breaking of threads caused by electrostatic build-up. Winters were mild, thus affecting savings in heating costs. Inexpensive land was available for factory expansion and taxes were low. Moreover, the labor situation was advantageous; wage levels were lower than in the North and the southern workers resisted unionization (Fig. 13.5). The exodus to the South began in the 1890s and proceeded steadily to the point that the Piedmont surpassed New England in number of spindles in 1924. Today 93 percent of the nation's active cotton spindles are found in the South.

Cotton manufactures in America hit a peak output in the 1920s but have been in decline since. This trend has accelerated since World War II, and is due to a considerable degree to the importation of goods from low-wage areas of the Far East. Since 1950, for example, cotton spindle hours in American mills have dropped by more than 50 percent, and during the same period employment has fallen by half.

Computers

Unlike steel and textiles, which are traditional, long-established industries, computers have become prominent only in recent decades. Computer manufacturing is a high technology industry

FIGURE 13.5
A HIGHLY-AUTOMATED SOUTHERN TEXTILE MILL. Note the absence of a large labor force.

that stands at the forefront of science. Beginning with the relatively simple concepts of mechanical adding machines and calculators, computers have increased their capacity enormously through the use of electronic circuitry. The first such adaptation used the vacuum tube as a switching device; these so-called *first-generation* electronic computers came into existence in the 1940s. However, they were phased out in the 1950s after the development of the transistor, which was smaller, faster, and more reliable. In the 1960s transistors were replaced by micro-circuitry; silicon chips now in use are smaller than a collar button but have the functional equivalence of 16,000 transistors. It is projected that by the 1980s the next generation of computers will use magnetic-bubble memory units of similar size with a functional equivalency in the millions. Because of these improvements in design, it is possible to build computers that are exceedingly adaptable. Such computers are used to maintain records in banking and finance, tax rolls, inventory and traffic control, ballistics guidance, the automation of production lines, and in space technology. Men walked upon the moon only because they were aided by computers.

Computer manufacturing qualifies as light industry, keyed primarily to technological leadership. Research and development absorb a high percentage of the total receipts of the industry. To remain competitive a company must attract and retain highly skilled scientists and engineers. We find, therefore, agglomerations of production facilities centered around major research areas—one near Boston and the Massachusetts Institute of Technology, one around Los Angeles and the California Institute of Technology, another in the "Research Triangle" area of North Carolina. Americans dominate the industry; indeed, one company, International Business Machines, regularly accounts for more than half of the value of shipments throughout the free world. Electronic computers and equipment constitute a major item of American exports and make a positive contribution in terms of balance of trade. To give some measure of the size of the industry, IBM had gross sales in 1977 of $18 billion, and ranked seventh among American corporations.

MANUFACTURING REGIONS OF THE WORLD

Manufacturing is concentrated in the mid-latitudes of the Northern Hemisphere, and particularly in four regions: Anglo-America, Europe, the Soviet Union, and the Far East. This section describes the location and characteristics of these regions.

Anglo-America

Manufacturing in the United States is concentrated in a zone bounded by a line connecting Richmond, Boston, Milwaukee, and St. Louis (Fig. 13.6). Within this area, which contains about 70 percent of American industrial capacity, various subregions can be identified. Of these, metropolitan New York City, which includes adjoining areas of New Jersey, Long Island, and Westchester County, forms the largest single node. Manufacturing in and around New York City is broadly based and includes both heavy and light industry. There are petroleum refineries, chemical plants, smelters, automobile assembly plants, and meat packing plants. Factories turn out machinery, metal products, electrical and communications equipment, aircraft manufactures, and textiles. New York City is the largest center of apparel manufactures in America and also leads in print-

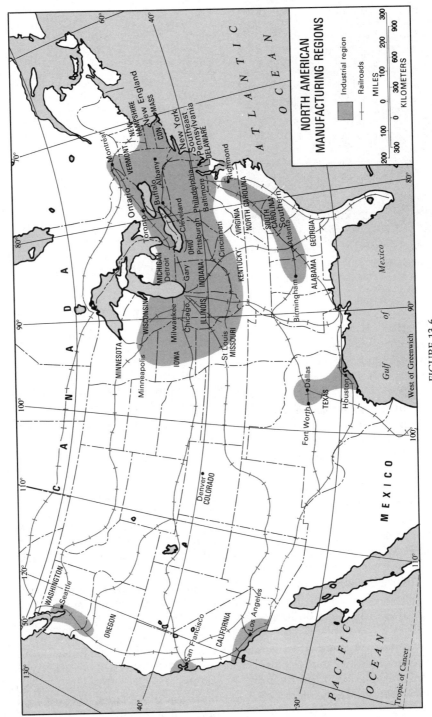

FIGURE 13.6
NORTH AMERICAN MANUFACTURING REGIONS.

NORTH AMERICAN MANUFACTURING REGIONS

Industrial region

Railroads

MILES

KILOMETERS

ing and publishing. The city contains more than 50,000 factories; the value added by manufactures within this zone exceeds that of any other urban center in the world.

The factor that accounts for New York City's prominence as an industrial center is its role as a leading seaport. New York City functions as a gathering place for raw materials. It is America's largest city and is located in the center of the densely populated northeastern megalopolis, an area that constitutes a large market for factory output. The large population of the city provides a ready supply of labor, and because New York City serves as a hub of transport routes (both land and sea), there are convenient linkages with extra-regional markets, both overseas and domestic.

Specialized heavy industry in the United States is concentrated in the area to the south of the Great Lakes along an axis that runs approximately from Philadelphia to Chicago. This region constitutes the productive heartland of America. Good soils support a large farming population, and many cities are located here. The large population of the area constitutes both market and labor supply for industrial plants. Extra-regional linkages are well developed and finished goods are easily distributed to markets. The local resource base is good; high quality coal is abundant, and iron ore is conveniently accessible. There is petroleum, natural gas, timber, and nickel. The availability of coal and iron ore stimulated early steel production and the core area of that industry lies within the region. The steel industry in turn supplies components for a host of heavy industrial facilities that produce farm machinery, construction equipment, trucks and automobiles, railroad equipment, and electrical goods. Chemical and meat packing plants and flour mills are characteristic of the area.

The southern Great Lakes States, richly endowed by nature and inhabited by highly skilled people, provide the mainspring of the American economy and explain to a considerable degree the enormous gross national product enjoyed by the nation.

New England, which once led the country in industrial output, has declined in relative importance. As America changed from a series of settlements along the Atlantic seaboard to its current continental configuration, New England, situated in the northeastern corner of the country, devel-

oped locational disadvantages. Nevertheless, specialized industries depending upon the skilled labor force of the area have managed to survive, and regional factories turn out woolen goods, firearms, hardware, machine tools, shoes, rubber goods, clocks and watches, and other types of light industrial products. Regional industry, however, is on the defensive, as current migration patterns show that industrial activities are leaving the area.

The South is the fastest growing industrial region in the United States. Of these, the Piedmont covers the largest area. Extending from Virginia to Alabama, this region, plus adjoining sections of the coastal plain, dominates national production of cotton goods, tobacco products, woodpulp, and paper. There are steel mills in Birmingham, Alabama, and aircraft assembly plants in Atlanta, Georgia. A thriving chemical industry produces synthetic fibers. There are food processing plants, furniture factories, textiles and apparel plants, automobile assembly plants, aerospace facilities, and computer manufactures. Moreover, the area is experiencing additional industrial growth as American manufacturers exhibit a tendency to leave crowded metropolitan areas for regions less congested and milder in climate.

Another industrial region within the South lies along the coastal plain in Texas and Louisiana. Based primarily on local resources of petroleum, natural gas, and sulphur, this area includes petroleum refineries and the production of chemicals. There are also aluminum refineries, aircraft manufactures, food processing, and aerospace facilities.

California is industrialized to the extent that it leads the nation in value of shipments. It is the leading center for aircraft manufactures and food processing (Fig. 13.7). There are assembly plants for automobiles, agricultural machinery, and construction equipment. Electrical machinery is produced, as well as instruments, chemicals, plastics, clothing, and pharmaceuticals. Motion picture production forms a profitable industry in Los Angeles. Although separated by distance from the great eastern markets of America, California is today the most populous state in the Union. Industry in California, therefore, has two options: either to produce for the local market or to produce goods of high value that can absorb the costs of a long trip to market.

FIGURE 13.7

THE AIRCRAFT INDUSTRY REQUIRES A SKILLED LABOR FORCE AND VAST ASSEMBLY FACILITIES. Boeing Aircraft plant near Seattle, Washington.

The industrial core area of Canada lies in the St. Lawrence lowland and the Ontario Peninsula, roughly along a line that stretches from Montreal to Windsor. This region accounts for about 90 percent of Canadian manufacturing. Although Canada has a small population (24 million), the resource base is strong and levels of technology are high; Canada, therefore, is a modern industrial power of considerable consequence.

Manufacturing activities in Canada are concentrated in basic metals, machinery fabrication, foodstuffs, and forest products. Canada has a vigorous and productive mining sector that provides raw materials to its industries and for export. Canada is eleventh in the world in the production of steel (Hamilton is the leading center), third in the production of aluminum (behind America and Japan), and is also among the leaders in copper, zinc, nickel, and uranium production.

Canadian factories turn out a wide variety of metal goods. Windsor, Ontario, just across the river from Detroit, Michigan, produces trucks and automobiles. Canada produces tractors, railroad equipment, agricultural machinery, and electrical equipment. Much of this activity is closely tied to American manufacuring, and there is a large flow of components back and forth across the border.

Canada is a major exporter of foodstuffs, all of which require some preparation for market. Consequently, there are flour mills, slaughter houses, fish canneries, and facilities for the processing of fruit, dairy products, and vegetables. These food processing plants tend to be widely dispersed, some being located in the prairie provinces, others on the Pacific coast.

Much of Canada is dominated by an enormous coniferous forest, which provides the basis for a thriving wood-products industry. Exploitation of

MANUFACTURING

320

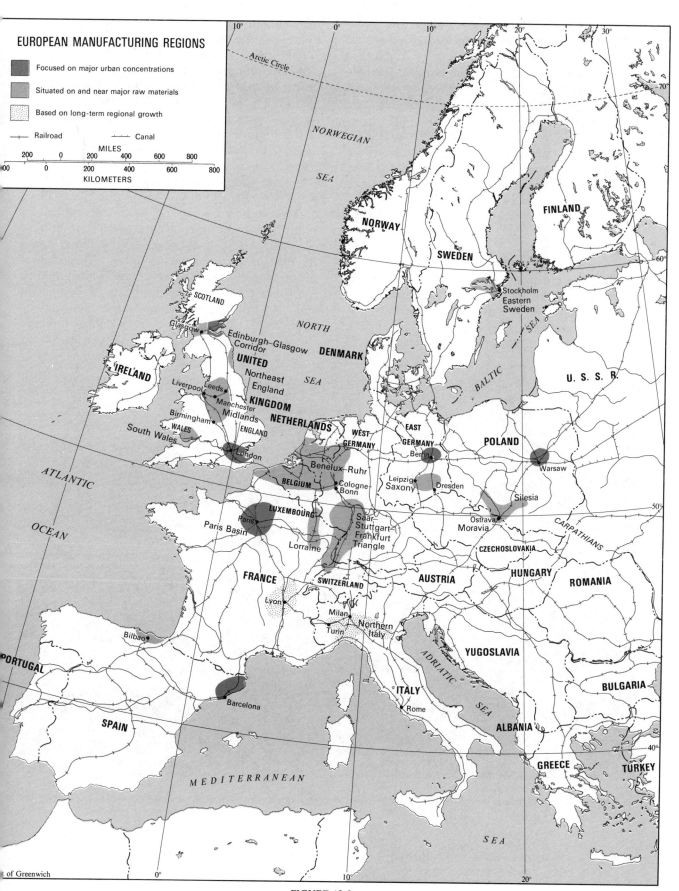

FIGURE 13.8
EUROPEAN MANUFACTURING REGIONS.

the accessible woodlands is extensive and saw-mills abound. Canada is a world leader in the production of pulp and paper with large quantities exported to the United States and Europe. The forest products industry is a large employer and makes a substantial contribution to Canada's balance of trade.

Canada's industrial core area shows a definite orientation towards population and transport facilities. In addition, the area's rail network and water facilities connect the core area to overseas markets via the St. Lawrence Seaway.

Europe

Western Europe, the world's industrial hearth, is compartmentalized into small national economies whose individual industrial statistics are not impressive. Viewed as a whole, however, Europe is an industrial power of major dimension, fully on a par with the industrial superpowers of the modern age (Fig. 13.8). Europe no longer dominates the industrial world but remains a viable economic competitor.

West Germany is the largest single industrial power in western Europe. Although truncated by territorial losses following World War II, the country has reestablished its primacy in the area. Industrial production in West Germany today is more than double that of any of its neighbor states.

German industry is centered in the valley of the Ruhr River, a tributary of the Rhine. The valley, measuring about 60 by 120 kilometers (40 by 75 miles), is nearly covered by a continuous conurbation of industrial development that forms a productive manufacturing complex. The Ruhr's prominence as an industrial center is attributed to its mineral resources, accessibility, and population. The Ruhr Valley contains large, high-quality deposits of coal with seams that are thick and easily worked. The navigable Ruhr River flows into

FIGURE 13.9
THE RHINE RIVER. Europe's most important inland waterway carries vast tonnages of bulky industrial materials.

MANUFACTURING
322

the Rhine, the principal traffic artery of western Europe (Fig. 13.9). Much of western Europe is criss-crossed by a dense network of canals connecting the various rivers. The Ruhr Valley thus enjoys convenient access via waterborne transport either downstream to the ocean or through canals and rivers to interior regions. Railways, highways, and pipelines add flexibility to the system.

The Ruhr lies at the center of densely inhabited northwestern Europe. The highly-skilled population provides both labor and market for the mills of the area. Production in the Ruhr includes iron and steel, metals fabrication, chemicals, and textiles. Iron ore is lacking but is imported from Sweden, France, and Africa.

Other industrial regions in Germany include the Saar, with its coal deposits, and the coastal regions fronting on the North Sea, that have new steel mills, shipyards, flour mills, and textile plants. Automobile production, widely scattered about the country, is important; sales of Volkswagens regularly account for about one-third of the value of German exports.

France and Italy constitute two other industrial powers of consequence in continental Europe. French industry is found in two major agglomerations, one of which is broadly based and centers on Paris, and the other, specializing in heavy engineering and textiles, lies in the northeast in Alsace, Lorraine, and Picardy. France's large deposits of iron ore provide the basis for the French steel manufactures around Metz and Nancy. Recently, steel mills have been built at Dunkirk on the English Channel and also near Marseilles on the Mediterranean, demonstrating the tendency of the modern steel industry to locate at tidewater. Marseilles is beginning to emerge as a subordinate industrial complex based on its accessibility as a seaport and the availability of oil and iron ore from North Africa.

Italian industry is based in the north, mainly in a triangle formed by Milan, Genoa, and Turin. Milanese industry is broadly based, while Turin specializes in automobile manufactures. Genoa, Italy's principal seaport, is also the site of a new steel mill, as are Pisa and Taranto, located farther south along the Italian coast.

Although the United Kingdom is no longer the leading industrial power of the world, it remains an important manufacturing nation. During the Colonial era, raw materials were imported from abroad, processed in British mills, and exported as finished products. The system was predicated on technological leadership, which Britain then enjoyed. The techniques have spread, however, and with them have gone Britain's advantage. Increasing competition has affected price levels and has cut drastically into demand for British goods. Britain has been forced to retrench to make up for its decline in industrial output.

Five industrialized zones are generally recognized in Britain. These are London, which has a broad array of diversified industries; the Midlands, the tranditional heart of British heavy industry, specializing in steel and textiles; South Wales, with steel, chemicals, and the smelting of nonferrous metals; the Northeast, with steel, shipyards, and metals fabrication; and the Scottish Lowlands, centered on Glasgow, that produce steel, ships, and textiles.

In recent years Britain has enjoyed a North Sea oil boom which has strengthened her resource base and promises to pay dividends for years to come.

The Soviet Union

Major industrial agglomerations are found throughout the Soviet Union (Fig. 13.10). Industrialization lagged in Russia under the Czars. After the Bolshevik Revolution in 1917, the manufacturing sector was rapidly expanded in an attempt to turn an agrarian nation into an industrialized power. Beginning in the 1920s, a series of five-year plans was invoked. Progress was impressive and national productivity increased until interrupted by World War II. Industrial expansion has continued since the war and today the Soviet Union has a gross national product second only to that of the United States. The country leads the world in steel production and vies with the United States for leadership in aerospace accomplishments. The Soviets, who were the first to place a satellite in orbit around the earth, also have an impressive record in manned spaced flights. They have achieved a remarkable record in industrial accomplishments, and their success presents a challenge to the industrialized world.

The Soviet Union is so large that internal distances often impede economic integration. The

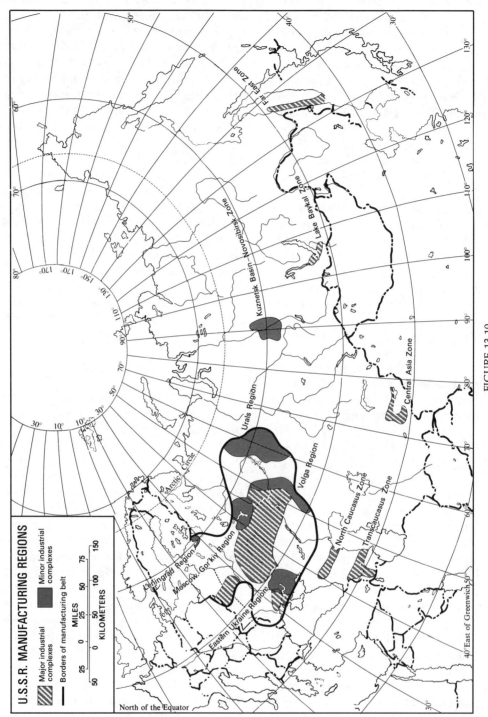

FIGURE 13.10
U.S.S.R. MANUFACTURING REGIONS

response to this problem has been to decentralize manufacturing by establishing separate industrial centers that are essentially self-sufficient, thus minimizing the effects of distance. There are still instances in the Soviet Union, however, where coal is hauled so far by railroad that much of the cargo is consumed by the locomotive. Inefficiencies of this sort inhibit production, an economic fact of life from which the Soviets, with their planned economy, are in no way exempt.

The largest industrialized region in the Soviet Union is centered on Moscow. This area is not particularly well endowed with raw materials, but it lies at the center of the most densely inhabited part of the country and acts as a hub with highways and railroads radiating in all directions. Also, a system of canals connect the rivers, so that Moscow has convenient waterborne linkages north to Leningrad and the Baltic Sea, and south to the Black and Caspian Seas. Moscow thus fulfills its industrial role in keeping with its position as the largest city in the country. Moscow has both light and heavy industry. It processes wood, wool, cotton, and flax, and manufactures electrical equipment, aircraft, machine tools, chemicals, hardware, and iron and steel goods. Gorky, lying 300 kilometers (200 miles) to the east, forms part of the same complex. Gorky is the Detroit of the Soviet Union and specializes in the manufacture of trucks and automobiles. It is worth noting that the Americans manufacture more automobiles than trucks, but the reverse is true of the Soviet Union, which produces more trucks and fewer autos, automobiles being viewed as luxury items.

The Donetz Basin in the southern Ukraine contains the traditional center of heavy industry in the nation. Coal and iron ore are available locally, and the area contains almost half the national capacity in steel production. The industrialized Donetz Basin produces farm machinery, railroad equipment, construction equipment, fertilizers, electrical goods, textiles, clothing, and foodstuffs.

To the northeast, the Volga region is emerging as an important industrial district. Local energy sources include petroleum and natural gas, while the Volga River has been harnessed by a series of dams to produce hydroelectric power. Local manufactures include metal products, chemicals, leather goods, and wood products.

The Ural Mountains contain an industrial complex of great importance to the Soviet economy. The local resource base is strong and includes iron, timber, bauxite, copper, chromium, and nickel. Local supplies of coal are inadequate and coal is shipped in from Siberia. The Urals complex specializes in heavy industry, with emphasis on the production of engineering equipment, mining machinery, heavy electrical goods, machine tools, and armaments.

The Soviets are making a concerted effort to develop their frontier lands to the east. The industrial agglomerations of Kuznetz and Karaganda have been created in response to this effort. The Kuznetz Basin has large deposits of high quality coal; it produces one-tenth of national steel production, plus finished metal goods including farming equipment, heavy machinery, and railroad equipment. Also produced are chemicals, food products, beverages, building materials, and textiles.

The Karaganda area also produces coal, some of which is shipped west to the Urals while the rest is used locally to support a burgeoning industrial sector including an iron and steel industry, metals fabrication, and consumer goods.

The Far East

Japan ranks as the third industrial power in the world; its gross national product is exceeded only by that of the United States and the Soviet Union (Fig. 13.11). Most of the Japanese industrial capacity has been established since World War II, and represents a triumph over an environment of limited resources. The country has no petroleum and only inadequate supplies of coal and iron ore. It has some timber and some non-ferrous metals, but most of the raw materials consumed by Japanese industry must be imported. Thus, the Japanese import oil from the Middle East and Indonesia, coal from the United States, iron ore from Australia, Brazil, and Chile, cotton from the Americas, rubber from southeast Asia, soybeans and corn from the United States, and phosphates, copper, zinc, nickel, and tin. All these commodities are processed in Japanese factories. Japan in turn exports manufactured goods, deriving profits from the value added.

Most Japanese industry is found in a discontin-

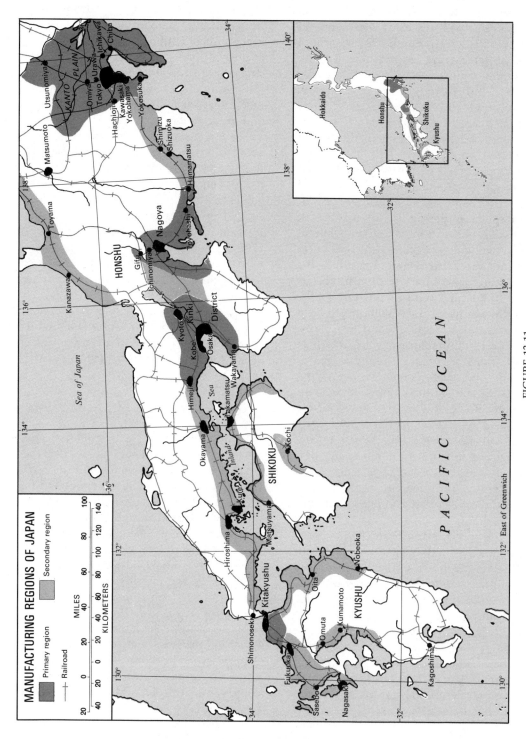

MANUFACTURING REGIONS OF JAPAN

Primary region
Secondary region
Railroad

MILES
KILOMETERS

FIGURE 13.11
MANUFACTURING REGIONS OF JAPAN.

uous belt that runs along the southern coast of Honshu from Tokyo through the area of the Inland Sea to Kitakyushu on the Island of Kyushu. Tokyo-Yokohama comprises the leading industrial center in the country. The manufacturing facilities are broadly based and include both heavy and light components. Chiba, which lies in the suburbs east of Tokyo, is a major center of iron and steel production. Kofu to the west produces optical goods. The Osaka-Kobe conurbation specializes in heavy industry. Here are found steel mills, shipyards, the manufacture of engineering equipment and electrical apparatus. Textiles are also produced.

Nagoya forms another major industrial concentration. Local manufactures include steel, fabricated metal goods, trucks, and automobiles. Pottery and ceramics and electrical goods are also produced in this area.

Farther west lie many smaller industrial centers located along the northern coast of the Inland Sea. Kitakyushu, a conurbation of several cities in northernmost Kyushu, represents the western terminus of the Japanese industrial core area. Coal mining in the area led to the establishment of steel mills that support shipyards and other forms of heavy manufacturing (Fig. 13.12). Textile mills and factories are also located here.

Such modern industry as exists in China is concentrated in the northeastern and eastern sections of the country (Fig. 13.13). The Dongbei Pingyuan (Manchurian Plain) has large deposits of coal and iron ore and the largest complex of heavy industry in the country. The region has a modern transport network, and Shenyang (Shen-Yang) and Anshan (An-Shan) have emerged as major centers of steel production and general manufacturing. Other cities in east China, such as Shanghai (Shang-hai), Tianjin (Tientsin), and Qingdao (Tsingtao) have modern industrial facilities, particularly textiles and light manufactures.

China needs to increase its industrial capacity, but this requires a massive program of investment. Because of the isolation of the regime from foreign sources of capital, there has been little investment from abroad. This isolation, however, is showing signs of ending, and capital inflows from Japan, the United States, and Europe promise to aid the Chinese in their effort to modernize.

In addition to the heavy industrial facilities of China, there are many light industrial plants scattered in various areas of the Far East. Manufacturers have been attracted to the area by low wage rates, and factories have been established in South Korea, Taiwan, the Philippines, Hong Kong, and Singapore. These factories turn out optical goods, electronic equipment, clothing, tobacco products, toys, and plastic goods. Many of these plants are controlled by multinational corporations based in Europe or the United States.

Just as manufacturing, or secondary production, is dependent upon raw material of primary production, so too manufacturing is dependent upon the service industries of the tertiary (third) sector of the economy. Such services include the buying, selling, and shipping of goods from areas of production to areas of consumption. The following chapter on trade and transportation examines the economic motive behind trade, the conditions that foster and impede its development, the principal trade routes, and the transport facilities available for shipping the principal commodities that enter international trade.

FIGURE 13.12
MODERN AND HIGHLY EFFICIENT INDUSTRIES HAVE BEEN ESTABLISHED IN JAPAN SINCE WORLD WAR II. The nation ranks as the third industrial power in the world.

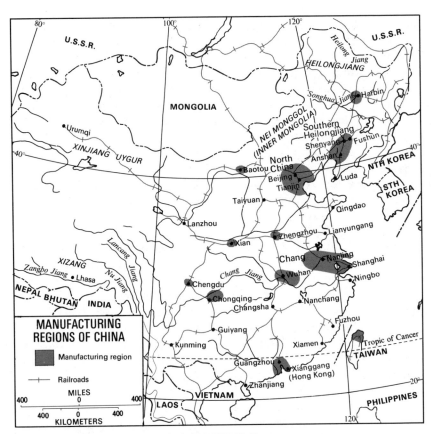

FIGURE 13.13
MANUFACTURING REGIONS OF CHINA.

The map labels (reading across the map) include:

U.S.S.R. · MONGOLIA · Urumqi · XINJIANG UYGUR · NEI MONGGOL (INNER MONGOLIA) · Heilong Jiang · HEILONGJIANG · Songhua Jiang · Harbin · Southern Heilongjiang · Shenyang · Fushun · North China · Baotou · Anshan · NTH KOREA · Beijing · Tianjin · Luda · STH KOREA · Taiyuan · Qingdao · Lanzhou · Zhengzhou · Lianyungang · Xian · Chang · Nanjing · Shanghai · XIZANG · Lancang Jiang · Nu Jiang · Zangbo Jiang · Lhasa · Chang Jiang · Wuhan · Ningbo · NEPAL · BHUTAN · INDIA · Chengdu · Chongqing · Nanchang · Changsha · Guiyang · Fuzhou · Kunming · Xiamen · Tropic of Cancer · TAIWAN · Guangzhou · Xianggang (Hong Kong) · Zhanjiang · VIETNAM · LAOS · PHILIPPINES

MANUFACTURING
REGIONS OF CHINA

■ Manufacturing region
┼ Railroads

MILES
400   0   400
KILOMETERS
400   0   400

KEY TERMS AND CONCEPTS

productivity
Industrial Revolution
cottage industries
traditional society
mass production
interchangeable parts
production line
multinational
  corporations
energy-intensive
  industry
labor-intensive
  industry
capital-intensive
  industry

gross national product
  (GNP)
location factor
heavy industry
light industry
agglomerating
  tendency
bauxite
alumina
investment climate
economies of scale
continuous process
pioneer industry
Ruhr

DISCUSSION QUESTIONS

1. What location factors explain the migration of cotton manufactures from New England to the Piedmont?

2. In what way are the multinational corporations fostering the spread of industrial technology into the Third World?

3. Discuss the factors of industrial location as they relate to the steel industry.

4. "The political environment in the Unites States has encouraged industrialization." Do you agree or disagree? Discuss.

5. In what ways do developed countries differ from less developed ones?

6. It is said that "the industrial revolution is an ongoing process that continues to spread out-

ward from its original source." Do you agree or disagree? Discuss.

7. What problems are peculiar to the Soviet Union as it strives to industrialize? Are these problems being solved, and if so, how?

8. It is said that "the steel industry is experiencing a drift to tidewater." Is this true, and if so, where? Explain.

9. New York comprises the largest single industrial node in the United States. Why? What factors come to bear? Can you identify parallel examples in other countries?

10. It is said that "the Industrial Revolution brought about not only the factory system but also, and of equal importance, improvements in the efficiency of transport." Do you agree? Discuss.

REFERENCES FOR FURTHER STUDY

Alexandersson, G., *Geography of Manufacturing,* Prentice-Hall, Englewood Ciffs, N.J., 1967.

deBlij. H., *Geography: Regions and Concepts,* 2nd ed., John Wiley, New York, 1978.

Hamilton, F. I., ed., *Spatial Perspectives in Industrial Organization and Decision-Making,* John Wiley, New York, 1974.

Morrill, R. L., and J. M. Dormitzer, *The Spatial Order: An Introduction to Modern Geography,* Duxbury, North Scituate, Mass., 1979.

Smith, D., *Industrial Location,* John Wiley, New York, 1971.

14
TRADE AND TRANSPORTATION

TRADE PROCEEDS IN ACCORDANCE WITH THE COMPARA-tive advantage enjoyed by different regions in the production of different goods. Consider the following example. Assume that people need cherries and oranges. Cherries grow well in Pennsylvania, while oranges thrive in Florida. Oranges could, of course, be grown in Pennsylvania but only in greenhouses. Cherries could be grown in Florida, but the climate is too hot and returns are poor. Therefore, rather than growing both crops in both places, it is more sensible for Pennsylvanians and Floridians to specialize in raising cherries and oranges respectively, and then trade with one another.

The foregoing presents the simple essentials of the *theory of comparative advantage,* which is the basic impulse behind trade. Trade facilitates regional specialization in accordance with local advantages and helps to eliminate the inequity of resource distribution. It thereby fosters efficiency and increases production and consumption. Trade will not take place, however, unless there is *complementarity* and *transferability.* Complementarity requires a surplus in one place and demand in another. Transferability means that a trade item can "reasonably" be moved from a supply area to a demand area.

At the international level trade has political overtones. International trade is controlled by nationalistic governments, each of which maintains its own set of attitudes and relationships with re-gard to the international community. Cuba does not trade with the United States, for example, because of a coolness between the two governments. For the same reasons, the United States trades vigorously with South Korea, but not at all with North Korea. Politics and economics march hand-in-hand in international affairs.

HISTORICAL PERSPECTIVES

Trade is common to all peoples and is prehistorical in origin. Trade routes connected the ancient cities of Sumeria. The Bible records the export of wheat from Egypt. The Phoenecians are probably the best known traders of the ancient world and were famous for their enterprise as early as the second millenium B.C. They were a great seafaring people; from their homeland in what is now Lebanon, they established trade links and colonies throughout the Mediterranean world. Carthage and Marseille were established by the Phoenicians, whose ships also reached Britain and the Azores. They were leaders in marine technology and the first to use the North Star as a navigational aid. Exports from Phoenicia included linens dyed with the famous royal purple, embroideries, metalwork and glass, wine, salt, and dried fish. Imports included papyrus, ivory, amber, silk, slaves, spices, gold, silver, tin, and precious stones. The Phoenician civilization was, however, thrust aside by the rise of the classical civilizations of Greece and Rome. Rome had distinct geographic advantages as a power center—a central position within the Mediterranean and a strong, productive land base. Exploiting these advantages, Rome became both a land power and a sea power and established dominance over much of the known world. It built a network of roads throughout the Empire, thus facilitating circulation and encouraging exchange. Trade flourished during the *Pax Romana*.

With the decline of Roman civilization, Europe entered the period of history known as the Middle Ages. As Roman authority broke down and the legions were withdrawn, people in the provinces became dependent on their own resources. There was a return to localism, in both political and economic terms, and the Empire disintegrated into a mosaic of feudal states. Travel in Europe became difficult, and trading leadership passed into the hands of the Islamic peoples of the Middle East and North Africa.

The pattern of modern trade began to emerge with the Renaissance and the Age of Exploration. The Iberians led the way in establishing new sea routes across the Atlantic to the New World and around Africa to the East Indies. European power and culture spread across the world, and Europe became the focal point for the trading system that evolved. Spanish galleons brought gold, silver, and sugar from the Americas. The Portuguese and the Dutch carried silks, spices, pepper, and perfumes from Asia. A vigorous traffic in slaves carried by Europeans from Africa to the Americas, was organized. Europe became wealthy and powerful. The ships of the period were small, however, never exceeding a few hundred tons displacement. Ocean travel was hazardous, and only luxury goods of high unit value could absorb the freight rates then prevailing.

The Industrial Revolution of the eighteenth and nineteenth centuries brought profound changes in transport as well as in manufacturing. Specifically, the new technology permitted the construction of merchant vessels with steel hulls, powered by steam engines, and propelled by screw propellers. Steam ships were large, strong, safe, and required relatively small crews. They could proceed directly from port to port, ignoring the vagaries of the winds. The steamship represented a vast improvement in transport efficiency, and its appearance signalled a new phase in maritime trade. For the first time, it became possible to move bulk commodities of low unit value for long distances and do so economically.

Sailing vessels were replaced by the realities of economics, and were superseded by plodding, coalburning freighters, many of them British. These steamships carried manufactured goods from the factories of Britain to the markets of the world, returning with raw materials such as lumber, cotton, wheat, wool, and hides. Britain held a dominant position as a maritime power at the time; in 1901, for example, 34 percent of the gross merchant tonnage in the world sailed under the British flag. Ship design has advanced in the twentieth century, resulting in the building of superships and further increasing efficiency.

As a parallel development, the past century has seen dramatic improvements in land transport. The developed regions in particular have been criss-crossed by railroads and highways, and similar facilities are now under construction in some areas of the less developed world. Pipelines have been laid to carry bulk commodities across vast distances. The net effect of these changes is to reduce the *friction of distance* (distance as a cost factor) on land as well as by sea, thus fostering the organization of a worldwide system of production and exchange (Fig. 14.1).

THE SCOPE OF INTERNATIONAL TRADE

The basic pattern of world trade can be categorized as follows: A wide variety of exchanges occur between the developed countries, including both manufactured goods and bulk commodities; a flow of bulk commodities moves from the less developed world to the developed countries; and a counterflow of manufactured goods moves from the developed countries to the less developed nations. Also, the free-market developed countries dominate international trade in terms of dollar value, and regularly serve as origin for 70 percent of the total value of world exports (Table 14.1). The less developed countries of the Third World originate 21 percent of international trade. The centrally planned countries are relatively inactive. With more than one-third of the population of the earth, this enormous area accounts for less than 10 percent of the value of world exports.

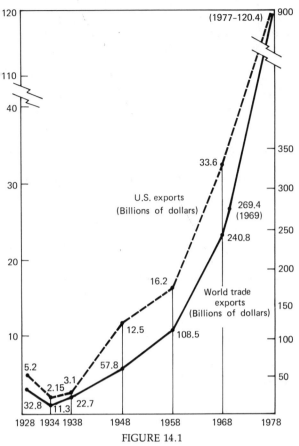

FIGURE 14.1

TRENDS IN U.S. AND INTERNATIONAL TRADE, 1928–1978.
Source: Adapted from *The Trade Debate*, U.S. Department of
State Publication No. 8942, Washington, D.C., 1978.

On the basis of tonnage, petroleum is the dominant commodity carried on the high seas. Fully half the tonnage moving across the oceans is made up of petroleum or petroleum products. This flow originates in the Third World, mainly in the Persian Gulf, and tends to move north to the free-market core areas. Western Europe is the largest single market for petroleum shipments.

A useful index of foreign trade is obtained by expressing national exports as a percentage of gross national product (GNP). This index is called the *export coefficient* and is derived by dividing exports by GNP and multiplying the result by 100. The export coefficient gives insights into the relative importance of the foreign trade sector in the national economy. Generally speaking, there is an inverse ratio between size of territory and the export coefficient because the chances are that a large country will have a broad resource base, thereby reducing import dependency. Therefore, a large country may approach *autarky*, or economic self-sufficiency, although this is a goal no country in fact achieves. Even the Soviet Union, with more than 22 million square kilometers (8.6 million square miles) of territory and stretching from the subtropics to the Arctic, finds it necessary to import. The small Benelux region, however, is criss-crossed by international boundaries, and, accordingly, Belgium, Netherlands, and Luxembourg have high export coefficients.

TABLE 14.1 MOST ACTIVE TRADING NATIONS, 1978 (billions of dollars)

	Exports	Imports	Balance of Trade
1. United States	$141	$183	$ − 42
2. West Germany	142	121	+ 21
3. Japan	97	79	+ 18
4. France	77	82	− 5
5. United Kingdom	72	79	− 7
6. Italy	56	56	− 0
7. Netherlands	51	54	− 3
8. USSR	52	51	+ 1
9. Belgium-Luxembourg	44	46	− 2
10. Canada	46	43	+ 3
World Total	1178	1203	

Note the extent to which the free-market industrialized nations dominate world trade.
Source. United Nations, *Monthly Bulletin of Statistics* (New York, June 1979), pp. 107–115.

THE SCOPE OF INTERNATIONAL TRADE

Trading activity also tends to coincide with high income levels. A high income level suggests a productive economy capable of producing an exportable surplus, thus generating the foreign credits necessary to buy imports. The developed countries, with their high income levels, tend to be active traders. Most less developed countries, however, have impoverished economies and little exportable surplus; hence they have only a limited ability to import. Per capita trade figures are low throughout most of the Third World, with the exception of some of the oil-exporting countries.

The United States

The United States is the world's largest trading nation; American exports and imports regularly make up about 13 percent of world totals. Because

TABLE 14.2 U.S. EXPORTS OF PRINCIPAL COMMODITIES TO MAJOR WORLD AREAS, 1976 (millions of dollars)

	Total	Canada	European Community	Japan	Less Developed Countries	Centrally Planned Economies
Corn	5,223		1,839	752	713	1,370
Chemical elements and compounds	4,408	656	1,497	374	1,470	32
Automotive parts and accessories	4,213	3,094	157	40	683	0
Wheat[a]	3,880	0	367	504	2,003	414
Soybeans	3,315	87	1,548	522	2,589	420
Automobiles	3,224	2,444	83	76	579	—
Civilian aircraft	3,211	73	509	170	1,778	4
Coal, coke, and briquettes	2,988	760	715	1,032	238	11
Electronic computers, parts, and accessories	2,588	408	1,179	240	310	32
Tractors and parts	2,222	687	218	8	808	147
Power machinery and switchgear	2,138	350	317	71	1,174	14
Telecommunications apparatus	1,997	272	352	75	1,118	3
Textiles other than clothing	1,970	250	576	58	532	23
Aircraft parts and accessories	1,933	93	512	104	1,000	1
Professional, scientific and controlling instruments	1,931	368	663	157	485	19
Internal combustion engines other than for aircraft	1,928	934	244	30	522	5
Mechanical handling equipment	1,859	390	316	24	883	64
Iron and steel-mill products	1,833	571	188	19	854	57
Total	114,997	24,109	25,406	10,144	40,372	3,638

[a] For marketing year 1976–1977.

Source. U.S. Department of State, The Trade Debate, Publication No. 8942 (Washington, D.C., May 1978).

of the enormous size of the American economy, however, these international exchanges involve only about seven percent of national income.

The export sector of the United States is highly diversified; it includes both raw materials and manufactured goods (Table 14.2). More than one-fifth of American agricultural production is exported. Farmers in the United States are highly competitive, and American corn, soybeans, wheat, cotton, and tobacco find ready markets abroad. Manufactured items exported by Americans include aircraft, computers, farm equipment, construction equipment, power-generating equipment, communications equipment, and materials-handling equipment. The United States also exports large quantities of military hardware.

In 1978, the United States imported $183 billion worth of goods, with petroleum expenditures accounting for one-third of the total. Saudi Arabia, Nigeria, Libya, and Algeria are the principal suppliers. The United States also imports oil from Venezuela, Iran, and Canada.

The United States imports large quantities of alumina, copper, nickel, tin, zinc, natural rubber, beef, wool, coffee, sugar, and cocoa. It is also dependent upon imports of such strategic minerals as cobalt, manganese, chromium, and titanium. A wide variety of manufactures are imported, including automobiles, cameras, optical goods, television sets, radios, stereos, computers, shoes, and clothing. Much of this flow of manufactures comes from low-wage areas of the Far East, including Japan, Taiwan, Hong Kong, South Korea, Singapore, and the Philippines.

Canada is America's single most active trading partner and usually accounts for about one-fifth of total U.S. trade. The United States, in turn, takes more than two-thirds of Canada's exports, and more than two-thirds of Canadian imports come from the United States. The Canadian and American economies are complementary. Canada serves as a source of raw materials for the United States, supplying newsprint, wood pulp, petroleum, lumber, iron ore, nickel, and aluminum; the United States supplies Canada with a broad array of goods including aircraft, auto parts, cotton, computers, tobacco, machinery, and railroad equipment (Table 14.3).

Within the Western Hemisphere, the United States trades actively with Mexico, Venezuela, and Brazil. It regularly absorbs two-thirds of Mexico's

exports and provides Mexico with two-thirds of its imports.

A major vector of trade flows across the Atlantic between Europe and Anglo-America. This flow makes up about one-quarter of the dollar value of United States trade. American exports to Europe include bulk commodities such as coal, chemicals, corn, soybeans, wheat, and tobacco, and manufactured goods including aircraft, computers, textiles, and engineering instruments. In return, Europeans export automobiles, woolen goods, wines and spirits, specialized foodstuffs, clothing, sports equipment, toys, watches, and clocks to the United States.

Trade patterns between the United States and the less developed countries have undergone significant change in the past several years. A decade ago, United States imports from those countries were concentrated heavily in raw materials and agricultural products. Since then, manufactures have been imported in greater quantities, so that now nearly a quarter of U.S. imports from those countries consist of such goods. The less developed countries are more important as a group to American trade than is often realized. Of the goods the United States exported in 1977, about 35 percent went to the less developed countries while imports from those areas represented about 45 percent of total purchases.

TABLE 14.3 MAJOR U.S. TRADING PARTNERS, 1977
(billion of dollars)

U.S. Exports to:		U.S. Imports from:	
Canada	$25.7	Canada	$29.4
Japan	10.5	Japan	18.6
West Germany	6.0	West Germany	7.2
United Kingdom	5.4	Saudi Arabia	6.4
Netherlands	4.8	Nigeria	6.1
Mexico	4.8	United Kingdom	5.1
Saudi Arabia	3.6	Mexico	4.7
France	3.5	Libya	3.8
Italy	2.8	Taiwan	3.7
Iran	2.7	Indonesia	3.5
South Korea	2.4	Algeria	3.1
U.S. Total	$120.2		$146.8

Source. Office of International Economic Research, U.S. Department of Commerce (Washington, D.C., 1978).

Western Europe

The free-market countries of western Europe constitute the most active trading area in the world. This region regularly accounts for about 40 percent of the total exports and imports of the world. Several reasons underlie this extraordinary level of activity. Western Europe is comprised of many small national economies, and exchanges between these small countries qualify as international trade. Intra-regional exchanges of this sort account for nearly half the European activity. Also, Europe is heavily dependent on foreign sources for many commodities. Petroleum is imported from Africa and the Middle East; grain from the United States, Canada, Australia, and Argentina; coffee, sugar, cocoa, rubber, and copra from the tropics; cotton from Egypt, the United States, and Mexico; and iron ore from Brazil. A host of commodities enters Europe from every quarter of the globe. These goods provide not only sustenance for the peoples of Europe but also grist for European factories, which produce the manufactured products that serve as items of export. These exports, in turn, provide the purchasing power that enables Europeans to import.

The high volume of European trading activity depends ultimately on an advanced level of technology in two fields: transport and manufacturing. Efficient transport facilitates circulation, making it possible to collect materials from abroad, ship them within Europe, and distribute goods to market, all at low cost. Modern factories enable Europeans to process materials efficiently, thereby producing goods that are competitive in foreign markets. Sale of these goods provides the purchasing power that makes it possible for Europeans to pay for their imports.

Japan

Since World War II Japan has enjoyed extraordinary economic growth. It has the third largest GNP in the world after the United States and the Soviet Union, and is in third position as a trading nation after the United States and West Germany. This record represents a considerable accomplishment, inasmuch as Japan has a limited resource base. The country has a population of more than 100 million crowded into a mountainous area the size of California. Farmland is limited, as are supplies of coal and petroleum. Nevertheless, the Japanese people have turned their island-nation into a powerful, aggressive, and competitive world trader that has achieved formidable market penetration in large areas of the world.

The Japanese system is to import raw materials, process them in their highly efficient factories, and export high-quality manufactured goods. They import coal from the United States and iron ore from Brazil and export iron and steel goods that undersell the products of United States' mills. They import cotton and export textiles and clothing at highly competitive prices. Japanese products, including automobiles, cameras, outboard motors, motorcycles, and electronic equipment, are inescapably in evidence in the markets of the United States, Latin America, Africa, Europe, and Asia.

The Centrally-Planned Economies

The Soviet Union is the most active trader of the centrally-planned states, usually accounting for about one-third of total trade volume. The bulk of Soviet exchanges are with the satellite states of Eastern Europe, but they also trade with free-market areas. The Russians export timber, wood pulp, paper, oil, natural gas, cotton, fish, and metallic ores; they import tropical foodstuffs, grains, and high-technology equipment.

China, the most populous centrally-planned state, is relatively inactive as a trader. The Chinese economy suffers from a lack of capital goods, and their ability to import is limited by low productivity. Moreover, the long political isolation of the Beijing (Peking) regime has limited their access to international credit sources, inhibiting their ability to trade. Recent developments, however, suggest that China is emerging from its isolation and may evolve into a more active trader during the 1980s.

Barriers to Trade

Attitudes towards trade differ from time to time and provide a continuing topic for debate in political circles. The basic issues separate those favoring *free trade* (low or no tariffs) from those favoring *protectionism* (high tariffs). Both views have found popular support at one time or another. During the 1930s, for example, when the world was in the grip of a deep depression, individual countries tried to "export" their unemployment with protectionist policies. Since World War II, however, free trade has been in vogue; current views hold that barriers to trade accomplish nothing except to increase real costs for everyone.

A common device used to limit or control trade is a *tariff:* a tax imposed on goods entering or leaving a country. Tariffs can be classified as either *revenue* or *protective.* The main purpose of a revenue tariff is to generate revenue for the government, while the main purpose of a protective tariff is to protect domestic industry by excluding competing products from abroad. Generally speaking, revenue tariffs are important only in less developed countries where they may bring in a substantial portion of total national receipts.

Non-tariff barriers to trade can include *quotas* and *licenses.* A quota specifies the amount of a commodity that can be imported, sometimes with respect to a particular country of origin. Licensing arrangements can impose standards in terms of quality, inspection procedures, or engineering specifications that can be used to reduce trade flows. Indeed, from a businessman's perspective, non-tariff barriers are frequently more difficult to cope with than tariffs. A tariff is simply a cost of doing business and can be passed along to a consumer, but a businessman trying to sell in a foreign market may find himself faced with a host of requirements and regulations sufficient to discourage the most optimistic entrepreneur.

TRADING BLOCS

In accordance with the principle of economies of scale, modern industry is most efficiently organized in large units. A large market, however, is needed to absorb the output of a large factory; anything that interferes with access to a market of appropriate size is counterproductive. National boundaries, which frequently restrict trade flows, can be a case in point.

As an example, consider the United States, the most productive society in the world today. One advantage America enjoyed during its formative years was the enormous size of its domestic market. It stretched from the Atlantic to the Pacific and from Mexico to Canada and was occupied by a large number of skilled, productive people. All this was gathered under one government, so there were no artificial barriers to trade within the entire region. American industrialists could think big. They could build enormous, efficient factories, confident that the American market could absorb their industrial production.

It is instructive to contrast the foregoing with the European experience. Europe consists of a mosaic of small states, the largest of which, France, is smaller than Texas. European manufactures suffered as the region was compartmentalized into small economies, each isolated from the other by tariffs and regulations. It is generally agreed that restrictive trade policies during the 1920s and 1930s were a contributing cause of the Great Depression that began in 1929 and continued for 10 years. However, by the mid-1940s the disadvantages of the traditional European economic structure had become obvious, and steps were taken to correct the situation. The first step was the amalgamation of Belgium, Netherlands, and Luxembourg into the Benelux Customs Union in 1948. This agreement eliminated customs barriers on commodities moving among the three signatories, and specified that the three would have common tariffs on goods brought in from elsewhere.

The Benelux arrangement was an immediate success. It was a sensible arrangement. Its principles were soon applied by the Schuman Plan to the European Coal and Steel Community. This agreement, established in 1952, permitted the unification of the steel industries of the Benelux countries plus France, Italy, and West Germany. The provisions of the agreement were that all tariffs on coal, iron ore, and scrap iron were erased among the six, and that the six would also act as a single unit when trading in the world steel market.

The success of the Schuman Plan led to a broadening of its concepts in the framework of the European Economic Community (EEC). The EEC, also known as the European Common Market, was established in 1958 in accordance with the Treaty of Rome. Its provisions called for the original members (Benelux, France, Italy, West Germany) to agree to a gradual removal of barriers to trade among members, to common external tariffs, common policies for economic development within the area, and free movement of labor and capital within the area.

The creation of the EEC brought into existence a market embracing many climatic regions, well endowed with resources, and inhabited by a large population of skilled, productive people. This large, free market permitted regional specialization, thus fostering efficiency and encouraging production. The reemergence of Europe as a strong center of economic activity can be attributed at least in part to postwar attitudes that made for ease of circulation.

Since 1973, the EEC (now known as the European Community) has acquired three new members, Great Britain, Ireland, and Denmark, giving it a total population of 265 million (Fig. 14.2). Other European countries are scheduled to join the European Community. This trading organization has also established common governmental structures in the Council of Ministers and the European Parliament. These agencies are restricted

in power but indicate a desire for a more binding unification. Further evidence of this is the authorization of a common currency to be issued in the 1980s.

Other regions of the world have also moved in the direction of economic integration. The Council for Mutual Economic Assistance (Comecon), for instance, unites to some extent the economies of the Soviet Union and the satellite states of eastern Europe. The Latin American Free Trade Association (LAFTA) does the same for that area; it includes all countries in South America (except the three Guyanas) and Mexico. Faced by problems of inadequate domestic markets, Latin Americans moved to integrate in 1960. Essentially, the charter of LAFTA agreed to the abolition of artificial trade barriers between member countries. Results, however, have been disappointing to date.

It is instructive to inquire into the marked differences between the European experience and the Latin American experience. In Europe, the members of the European Community are highly developed and form a coherent economic unit. They are in close proximity to one another and, in the absence of artificial restraints, would naturally trade with one another. In Europe, removing the artificial restraints permitted regional specialization in keeping with the principle of economies of scale. The member states of LAFTA, however, sprawl across an enormous area. They are all less developed countries that traditionally have exported raw materials, receiving manufactured goods in return. It follows that Latin Americans must, at this stage of their development, trade with areas that have a surplus of manufacturing capacity—Anglo-America, Europe, or Japan. Removing the artificial barriers to trade between Ecuador and Brazil, for instance, does little to stimulate trade between the two, as exchanging bananas for coffee would accomplish little in satisfying the needs of either. Thus, the traditional pattern of trade involving LAFTA members persists; this trade is still with the industrialized core-areas of the Northern Hemisphere. This situation will continue until industrial capacity in Latin America is increased to provide some regional surplus of manufacturing capacity. The European Community works because the economies of the member states are complementary; LAFTA is less successful because the economies involved are competitive.

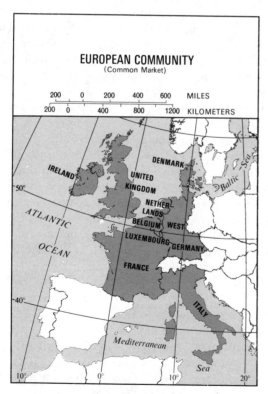

FIGURE 14.2
EUROPEAN COMMUNITY (COMMON MARKET).

Balances in International Trade

In the long run, the foreign trade sector of a country must be in balance. This means that the total flow of credits in and out of a country must be roughly equal in value. Two terms are used in this sense: *balance of trade* and *balance of payments*. Balance of trade means simply exports minus imports. A so-called favorable balance of trade is one in which exports exceed imports. Balance of payments, on the other hand, is a broader concept that includes all transfers of credit into or out of a country. It includes, therefore, balance of trade but also invisible export items such as tourist expenditures, payments for services (shipping, banking, insurance), profits or interest from investments abroad, military expenditures, foreign aid, new overseas investments, and pensions paid to nationals living abroad.

A country, like an individual, must earn (export) in order to buy (import). What is perhaps less obvious is that balance must exist throughout the international community so that trade can proceed in an orderly manner without any unhealthy accumulation of debts or credits in any one country. A serious imbalance in either direction causes stress and inhibits the flow of trade (as in a poker game where, if, one player wins all the money, the game stops). As an example, consider the situation in the oil-producing region of the Middle East. The countries bordering the Persian Gulf have been selling prodigious amounts of oil and buying very little in return, giving them a favorable balance of trade that reached $100 billion by the 1980s. An ocean of dollars is flooding into these countries; it accumulates in banks to the point that the stability of the monetary system is threatened. At the same time the oil-consuming nations are hard pressed to find the money necessary to fill their energy needs. A marked imbalance like this cannot continue indefinitely without affecting exchange rates between currencies, which will in turn affect trade flows between nations. Favorable exchange rates mean that the national currency stays strong and imported goods are available cheaply.

TRANSPORTATION

Modern industry and modern transport go hand in hand; one cannot exist without the other. Modern industry requires the assembly and distribution of large quantities of commodities that require, in turn, modern transport facilities. Well-developed transport systems, therefore, mirror the dense concentration of industry in the developed world. A dense network of railroads and highways is found in the industrial core-areas of the United States and western Europe. In areas of the world where industrial technology is limited, the network density is sparse. Thus the less developed regions of the world have substantially fewer miles of highways and railroads. China is deficient in transport arteries while Argentina and southern Brazil, with intermediate levels of industrial development, have moderately developed transport sectors.

Modern transport is efficient, meaning that costs are low on a ton-mile basis. Of all modes of transport, waterborne transport is the most economical. Pipelines are less economical than waterborne transport, railroads and trucks come next in order. Air transport is the most expensive form of transportation except for animate means. (ie, the use of camels, donkeys, and the like).

Other factors beside cost may be of importance in determining the mode of transportation selected. Speed, the nature of the cargo, the distance to be traveled, and the availability of alternatives can influence the decision. If speed is an urgent necessity, air transport will be preferred unless the distance is short, when highway transport will be used. With regard to cargo, high-value goods will be handled differently from low-value goods because of differences in their ability to absorb transport costs. Thus, wristwatches would ordinarily be sent from Switzerland to the United States by air; timber would not.

Transport activity also responds to non-economic stimuli. Military transport is relatively free from economic considerations. Also, political pressures can distort traffic flows in an uneconomic manner.

Water-Borne Transport

Shipping is slow but no other mode can approach it in terms of cost efficiency. Shipping is particularly useful for moving bulky materials; waterways are used to haul enormous quantities of petroleum, coal, metallic ores, wood products, and foodstuffs. The oceans provide a broad and

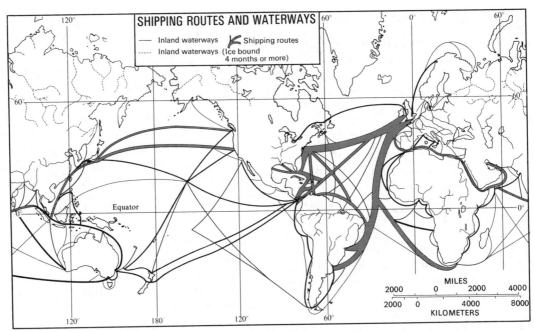

FIGURE 14.3
SHIPPING ROUTES AND WATERWAYS.

FIGURE 14.4
CONTAINERIZED SHIPPING TERMINAL AT NORFOLK, VIRGINIA.

TRADE AND TRANSPORTATION

FIGURE 14.5
ROTTERDAM, THE WORLD'S LEADING SEAPORT, HANDLES MORE TONNAGE THAN ANY OTHER PORT.

convenient highway while rivers, lakes, and canals carry enormous tonnages as well (Fig. 14.3).

During the twentieth century ocean shipping has experienced dramatic improvements in design. These improvements include increases in ship size, as well as the introduction of mechanized features that aid in shiphandling, loading, and unloading. With regard to size, the T-2 tanker, the workhorse of the U.S. Navy during World War II, weighed 16,800 tons. By 1962 the largest tanker afloat weighed 130,000 tons, and by 1973 the figure had increased to 326,000 tons. In 1978 the world's largest ship, the Bellamya (of French registry), weighed 554,000 tons. It measured 415 meters (1,360 feet) long by 63 meters (207 feet) wide and had a draft of 28 meters (92 feet) when fully loaded. A vessel weighing 1 million tons may soon be afloat. The future size of these mammoth ships will probably be limited by harbor depth rather than by other considerations.

Improvements in ship design include automation of engine room operations and the use of radar to aid in docking. Containerized cargoes facilitate loading and unloading and permit a

FIGURE 14.6
TRUCK TRAILERS CARRIED PIGGY-BACK CONSTITUTE A LARGE COMPONENT OF MODERN RAILROAD TRAFFIC.

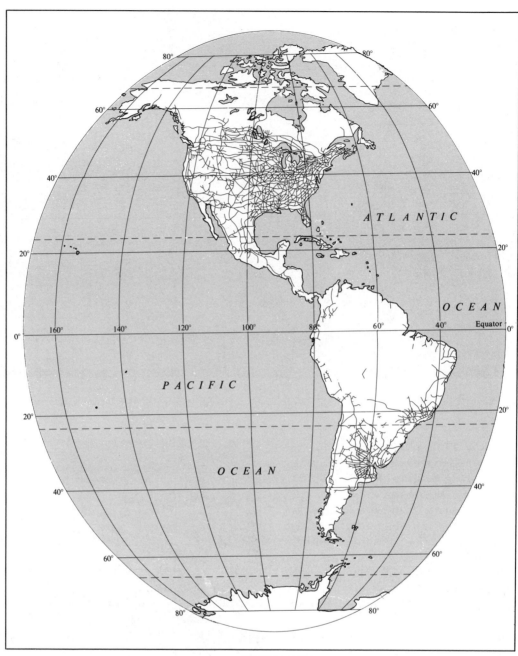

FIGURE 14.7
RAILROAD TRACKS.

reduction in the time a vessel must spend in port (Fig. 14.4). Large and small cargo carriers require roughly the same size crew; therefore, operation of the new superships has resulted in substantial economies; freight charges per ton-mile have been reduced to a negligible factor for common commodities. For example, a gallon of oil can be hauled from the Persian Gulf to the United States for less than one cent.

Inland waterways make a considerable contribution as arteries of transport. The Great Lakes, for example, together with the St. Lawrence and the Mississippi-Ohio Rivers, form an important transport network that connects the interior re-

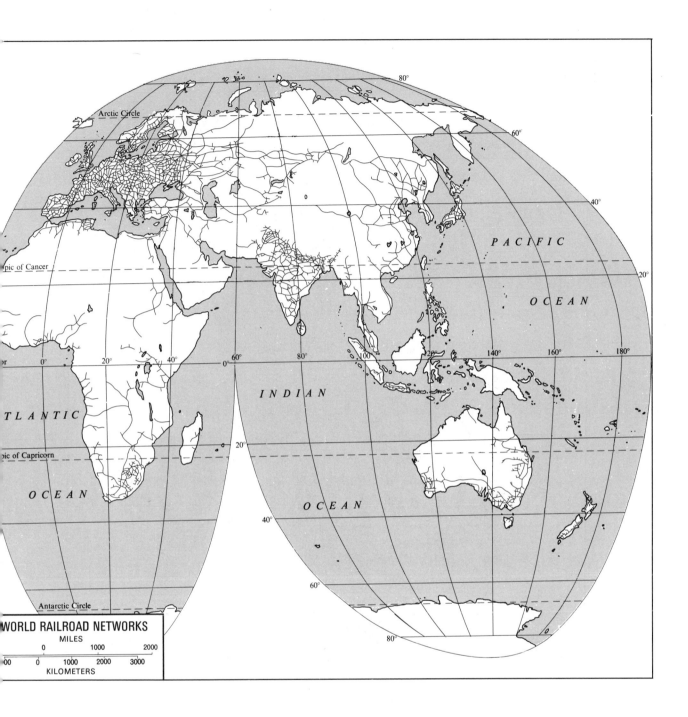

MILES

	0	1000	2000
00	0	1000 2000	3000

KILOMETERS

gions of Anglo-America with the rest of the world. Enormous tonnages are carried on these waterways, which make several inland cities important ports. Chicago is the eighth port in America in terms of cargo handled, and Duluth is tenth.

Inland waterways are also heavily used in Europe, where many rivers are linked by canals. Rotterdam is the world's leading seaport (Fig. 14.5). Located on the Lek distributary of the Rhine River, it profits from the transit trade generated along the Rhine from the Atlantic coast to central Europe. Increasing industrialization throughout Europe accounts for Rotterdam's tremendous expansion in recent times.

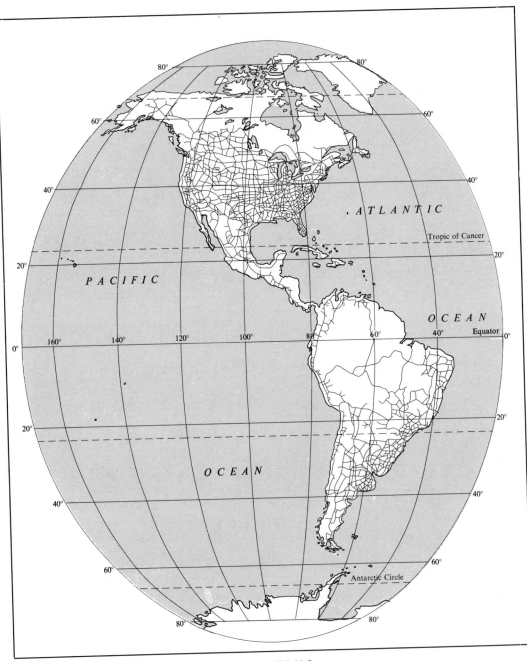

FIGURE 14.8
WORLD ROAD NETWORKS.

In the Third World, rivers are frequently used as a means of transport for lack of competing alternatives. The Amazon, for example, carries people and goods because land transport facilities in the area are poorly developed. The same is true of the Congo, the Chang Jiang (Yangtze), and the Irriwaddy.

Railroads

Rail transport lends itself to the movement of bulky goods in a relatively short time at moderate cost. Because terminal expenses are high and because railroads lack flexibility in terms of pick-up and delivery, cost efficiency increases with length of haul. Initial investment requirements are large

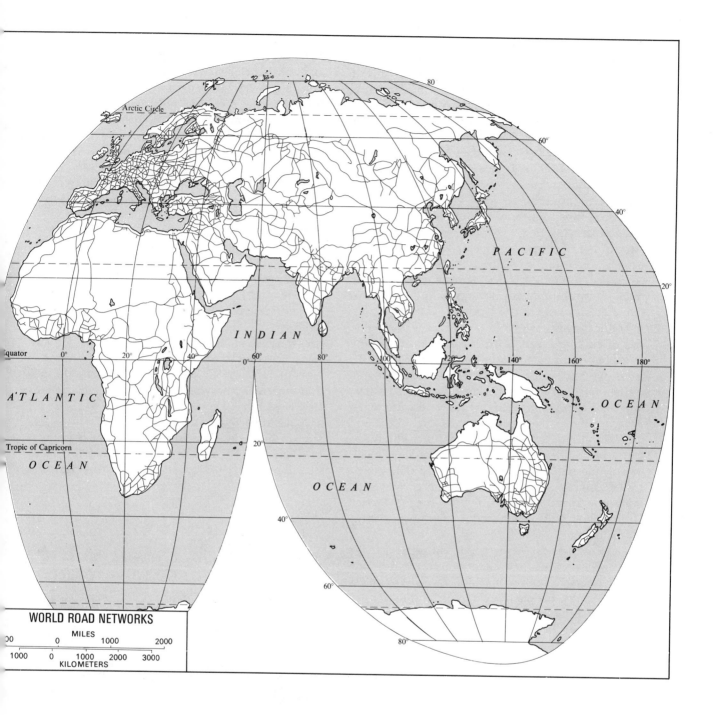

WORLD ROAD NETWORKS

MILES
00 · 0 · 1000 · 2000
1000 · 0 · 1000 · 2000 · 3000
KILOMETERS

and include the acquisition of right-of-way. Railroads typically carry coal, metallic ores, grains, timber, chemicals, automobiles, and heavy engineering equipment. Also, truck trailers carried piggy-back constitute a large component of modern rail traffic (Fig. 14.6).

A world map of railroad tracks shows two major areas of concentration, each of which contains about one-third of the world's total mileage (Fig. 14.7). One such concentration covers the eastern half of the United States and adjoining parts of Canada; areas respectively first and fourth in the world in miles of line. The railnet in eastern America is fine (consisting of many lines close together), but coarse (consisting of few lines far apart) in the western part of the country. Another major railnet, containing about one-third of the world total, covers Europe. This network stretches east from the Atlantic Ocean, becoming progressively more coarse towards Siberia. The Soviet Union, with about 125,000 kilometers (80,000 miles) of trackage, is second only to the United States with 300,000 kilometers (220,000 miles).

Only minor railnets exist elsewhere in the world. India, with 100,000 kilometers (60,000 miles), is most outstanding, followed by Argentina, Australia, South Africa, and Brazil. Vast stretches of the tropics are virtually trackless. Venezuela, for instance, which is half again as large as France, has only 1,000 kilometers (650 miles) of track. Only about half a dozen rail lines cross the equator anywhere in the world.

Traffic density, which is a measure of the intensity of rail line usage, is expressed in units of ton-miles per route-mile. Traffic densities vary sharply from one region to another. In the United States traffic densities are low because the railnet is fine and because pipelines, highways, and waterways provide alternative modes of transport. In the Soviet Union, however, traffic densities are high because the railnet is coarse and also because competing alternatives are in short supply. Therefore, traffic moving across the vast distances that separate the manufacturing regions in the Soviet Union must be concentrated on the few existing trunk lines.

As to rail passenger traffic, densities tend to be low in the United States except in the Northeast where people commute by train. Elsewhere in the United States, people travel by air or automobile. However, in Europe, Japan, India, and China, automobiles are fewer and rail passenger densities higher.

The rail networks of the world were for the most part established during the nineteenth century, the great era of railroad building. In the twentieth century, transport development has emphasized highway construction. Thus, much of the Third World seems to be moving directly from the oxcart to the highway with no intervening railroad phase.

Highways

Highway transport is flexible and convenient and is suited to moving light loads for short distances at high speed. Terminal costs are minimal, but costs for maintenance, fuel, and labor are high and are tied closely to length of haul. Trucks compete effectively with railroads over short hauls, but over long distances rail transport is more efficient. The practice of piggy-back freight, however, combines the advantages of both modes and enables truckers to be competitive over long hauls.

Trucks do not serve well in moving large quantities of heavy materials. Poor fuel efficiencies and the destructive effect of heavy loads on roadbed and equipment is the reason.

Good road networks coincide with high-technology societies (Fig. 14.8). Thus, good highway systems are found in the United States, Western Europe, and Japan, with intermediate facilities in India, Pakistan, South Africa, and Argentina. Even in high-technology societies, highway facilities vary in quantity and quality because network densities tend to correspond to population densities; the eastern United States, for example, has a fine highway network while the sparsely inhabited west has a coarse network.

A trend in recent decades has been the extension of highway networks into remote regions of the Third World. For example, during the 1970s Brazil pushed road construction in the Amazon

FIGURE 14.9
AIR TRANSPORT IS FAST BUT EXPENSIVE. It is best suited to passenger traffic but can also be used to move high value goods.

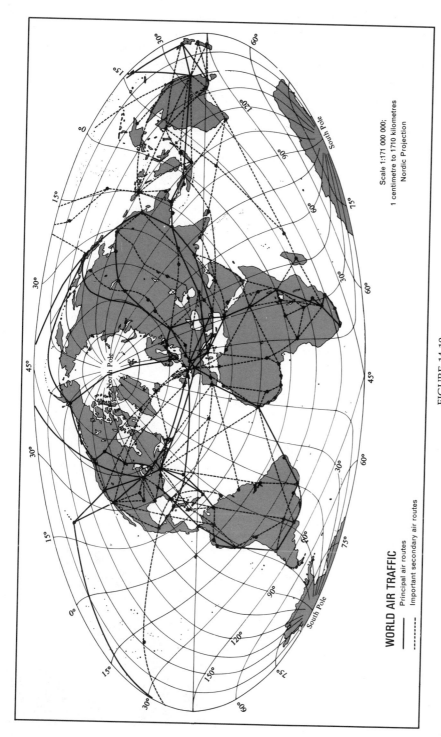

WORLD AIR TRAFFIC

Principal air routes
Important secondary air routes

Scale 1:171 000 000;
1 centimetre to 1710 kilometres
Nordic Projection

FIGURE 14.10
WORLD AIR TRAFFIC.

basin so that now one can drive from Brazil to either Peru or Venezuela, something not possible until then. Roads are also penetrating the interiors of Asia and Africa, opening these regions to the rest of the world.

In high-technology societies, automobiles are the usual mode of personal transport. The increased mobility provided by cars has led to urban sprawl and a decentralization of the settlement pattern. In the United States, for example, towns are laid out in keeping with the assumption that everyone will enjoy the use of a car. This matter is discussed more fully in the following chapter on urban geography.

Air Transport

Air transport is fast but expensive. It is best suited to passenger traffic but can also be used to move high-value goods that are not excessively heavy (Fig. 14.9). Terminal costs are high, so efficiency increases with distance.

Commercial air routes reach literally across the globe (Fig. 14.10). They dominate transoceanic passenger service and have also captured a significant portion of intercity passenger traffic. As traffic has increased, problems have risen at major air centers such as New York, Chicago, and London. Noise pollution, handling of baggage, and tie-ups due to bad weather are among these problems. Thus, one can go from downtown Washington, D.C. to downtown New York City almost as quickly by rail as by air. It is only over hauls in excess of 600 kilometers (400 miles) that air travel provides any appreciable advantage in time.

Air service also reaches into remote regions of the Third World. Today, it serves outposts in the Amazon, villages in Greenland, and the interior of New Guinea. No other mode of travel is so convenient in this sense.

The large aircraft that have recently come into service show improved fuel efficiency, making them more suitable for the hauling of air freight. Air freight is a growing sector of air transport activity.

Pipelines

In favorable conditions, pipelines can be a most economic form of transport for liquids, gases, and even solids. Initial costs of construction are high but a pipeline, once in place, needs few operational personnel and maintenance costs are low (Fig. 14.11). Constancy of use is needed, however, in order to amortize the original investment.

Pipeline networks are concentrated in three regions: Anglo-America, Europe, and the Middle East (Fig. 14.12). In the United States networks connect the oil and gas producing regions of Texas, Louisiana, and Oklahoma with market areas in the northeast. Other pipelines carry oil and gas from western Texas and New Mexico to California. The new trans-Alaska line brings oil from the Arctic to the southern coast. In Canada a large line carries petroleum from Alberta into the densely inhabited St. Lawrence region. Mexico, now emerging as a major producer of petroleum and gas, has a network connecting the oil fields on the Gulf Coast with Mexico City. This network continues north to tie in with the American grid.

In Europe pipelines are used to deliver petroleum from seaports to interior market areas. In the Middle East they connect the oil fields of the Persian Gulf and Iraq with loading points on the Mediterranean.

FIGURE 14.11
ONCE A PIPELINE IS IN PLACE, FEW OPERATIONAL PERSONNEL ARE NEEDED AND MAINTENANCE COSTS ARE LOW.

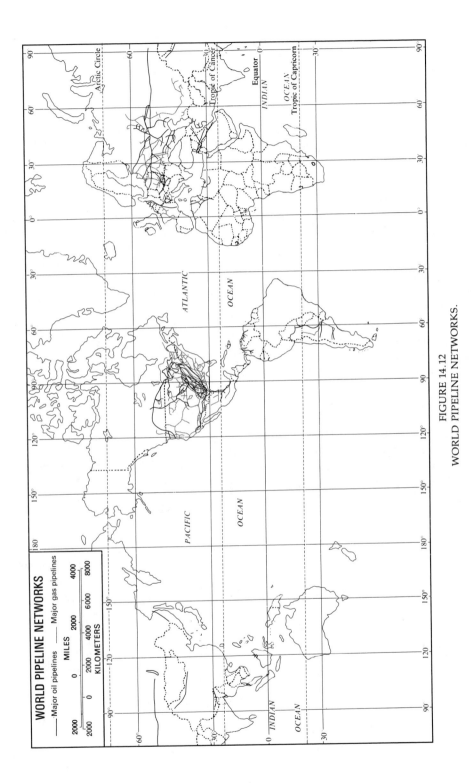

FIGURE 14.12
WORLD PIPELINE NETWORKS.

TRANSPORTATION
349

Pipelines are also used to move municipal water. Along with enormous aqueducts, pipelines carry water to New York City from the Catskill Mountains, and in Los Angeles bring it from the Sierra Nevada Mountains. Southern California is a water-deficit area and water is brought in more than 800 kilometers (500 miles) from the northern part of the state.

Solids can be carried by pipeline if they are pulverized and suspended in water to form a *slurry*. This technique is used to connect coal mines with power plants. The slurry can contain as much as 50 percent coal by weight. At the power plant the coal is dried and blown into a furnace. In a refinement of this technique the coal is suspended in oil and the mixture burned. Iron ore, copper ore, limestone, and wood pulp are also moved by pipeline.

Arteries of transportation are the modes by which raw materials are funneled into the industrialized-urbanized centers of the world. The finished products are then distributed throughout the cities' tributory areas. Within the cities people are engaged in buying and selling, and here trade and transportation reach their highest forms of development. The world's urban population is growing rapidly. Indeed, we live in such an increasingly urbanized world that urban society may soon represent society at large. The role that cities play in our everyday life, the way cities change, and their morphology and function are topics to be discussed in the following chapter on urban geography.

KEY TERMS AND CONCEPTS

comparative advantage	tariff
complementarity	protective tariff
transferability	revenue tariff
friction of distance	quota
export coefficient	balance of trade
autarky	balance of payments
free trade	slurry
protectionism	

DISCUSSION QUESTIONS

1. What is comparative advantage and in what way does it provide an impulse to trade?
2. Tonnages of export commodities moving in international trade have increased dramatically since 1900. Why?
3. Discuss the role of petroleum in international trade.
4. What factors have underlain the drive since World War II to create trading blocs?
5. Name five leading seaports of the world in cargo tonnage. What does this tell us about the gross pattern of international trade?
6. Characterize the gross pattern of international trade in terms of three major flows.
7. Discuss the cost relationships between various modes of transport. Which mode is cheapest and which is most expensive?
8. The eastern part of the United States is covered by a fine rail network while the western part is covered by a coarse network. Discuss.
9. Italy generates more exports and imports than does the Soviet Union. Why?

REFERENCES FOR FURTHER STUDY

Barnet, R. J., and R. E. Muller, *Global Reach*, Simon and Schuster, New York, 1974.

Bhagwati, J., *International Trade*, Penguin Books, Baltimore, 1969.

Foust, J. B., and A. R. deSouza, *The Economic Landscape: A Theoretical Introduction*, Merrill, Columbus, Ohio, 1978.

Stanley, William R., "Some Geographic Trends in World Shipping," *Geo Journal*, 2.2, Wiesbaden, 1978.

Taaffe, E. J., and H. L. Gauthier, Jr., *Geography of Transportation*, Prentice-Hall, Englewood Cliffs, N.J., 1973.

Thoman, R. S., and E. C. Conkling, *Geography of International Trade*, Prentice-Hall, Englewood Cliffs, N.J., 1967.

15
URBAN GEOGRAPHY

THE CITY IS AN INGENIOUS HUMAN INVENTION. ONLY when people began to live together in an urban setting did mankind begin to achieve any semblance of civilization. The word "civilization" comes from the Latin word for city, *Civitas*. Civilization is, therefore, "city living"; history itself begins with the city as a mature form. Cities have always been the cradle of human progress and human thought. It was in urban areas that mankind first began to write. It was here also that the great religions took their eventual forms. Here the world's artisans and writers have congregated, and painters and poets have often turned to the city to find a patron. The world's greatest artistic achievements are found in the museums, churches, and theaters of cities. The city is also the most effective means invented to facilitate ease of human communication and the exchange of goods and ideas. Accordingly, the prime occupations in cities are trade, manufacturing, and services. The city, in short, is the singular edifice that depicts mankind's greatest progress.

At the same time, the city represents some of our greatest failures. These failures are understandable when we consider that it took tens of thousands of years for people to evolve from migratory beings to a social species. Indeed, 99 percent of mankind's existence has been devoted to a migratory life dedicated to producing daily food supplies. With our rural and sparse heritage, it is not surprising that we often find city life bewil-

dering. Because of the relative newness of large-scale urban life, we have had little experience in tackling problems such as traffic congestion, blighted neighborhoods, air and water pollution, and urban sprawl.

What is a city? How did cities arise? What functions do they perform? These and other questions will be considered in the following sections.

HISTORY BEGINS AT SUMER

City life could emerge only when mankind turned from a migratory food gathering economy to an economy based on producing food. Neolithic technology involved a combination of agriculture, animal husbandry, pottery making, and the use of tools. These innovations constituted an agricultural revolution, the details of which are lost in pre-history. Archeological studies have established, however, that a farming community flourished as long ago as 10,000 years at the Neolithic village of Jarmo in the foothills of the Zagros Mountains in present-day Iraq. A surplus of food was produced, freeing workers from labor in the fields and permitting the support of an urban population. The full potential of the Neolithic Revolution could, however, be actualized only under different physical conditions; that is, in an environment capable of producing greater food surpluses. Such a fertile, well-watered environment

353

was found in the alluvial valleys of certain great rivers.

Between 4000–3000 B.C. a number of significant riverine city states began to emerge in the delta area of the Tigris-Euphrates system in Mesopotamia. The area, known as Sumer, witnessed the emergence of the world's first civilizations (Fig. 15.1). The Sumerian civilization, a city civilization, witnessed the formation and growth of such early riverine city states as Ur, Kish, Eridu, and eventually Babylon. The Sumerians share with the ancient Egyptians the distinction of being the earliest inventors of writing.

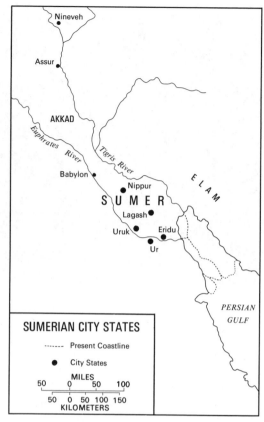

FIGURE 15.1
SUMERIAN CITY STATES.

CITIES AND CIVILIZATION

Culture suggests agriculture, but civilization suggests the city. In one aspect civilization is the habit of civility; and civility is the refinement which townsmen, who made the word, thought possible only in the *civitas* or city. For in the city are gathered, rightly or wrongly, the wealth and brains produced in the countryside; in the city invention and industry multiply comforts, luxuries and leisure; in the city traders meet, and barter goods and ideas; in that cross-fertilization of minds at the crossroads of trade intelligence is sharpened and stimulated to creative power. In the city some men are set aside from the making of material things, and produce science and philosophy, literature and art. Civilization begins in the peasant's hut, but comes to flower only in the town.

Source. Will Durant, *The Story of Civilization, Part I: Our Oriental Heritage,* New York: Simon and Schuster, Inc., 1935, p. 2. Reproduced by permission of the publisher.

In the villages of the Zagros Mountains, lands were watered by natural rains. In the Tigris-Euphrates Valley, pioneers and engineers brought water artificially to crops. The riverine environment was substantially different from that of the hill country; but the valley plains facilitated plowing, and this advantage, together with the rich alluvial soil and deposits of silt, made it easy to produce a grain surplus sufficiently large to encourage specialization and support a managerial class. Surplus food from the city's hinterland supported merchants, artisans, priests, soldiers, and others. As specialization increased among the valley people, trade, transportation, and communication were stimulated. Trade between the valley people and the hill people became increasingly important. Eventually the river cities became centers of trade and communication for the surrounding areas. By 3000 B.C. a host of innovations had been perfected; so many, in fact, that a new technology was beginning to transform the world in a fashion undreamed of a few centuries earlier. Containers stored and preserved the surplus food (the potter's wheel permitted the mass production of these vessels); the plow was used to cultivate the light alluvial soils; the loom wove cloth; the sailboat carried cargo from the Persian Gulf to the foothills of the Zagros; copper metallurgy encouraged the rise of an artisan class; mathematics and astronomy led to the calendar; and writing facilitated record-keeping for the area's merchants and for the recording by priests of law, literature, and religious beliefs. This hydraulic civilization demanded unheard of human efforts to control the

FIGURE 15.2

ABOUT 2500 B.C. THE INDUS VALLEY WITNESSED THE FLOURISHING OF TWO CITIES, MOHENJO-DARO AND HARAPPA, ALONG THE INDUS RIVER IN TODAY'S PAKISTAN. Ruins of Mohenjo-Daro are shown here.

forces of nature. Accordingly, ditches, canals, temples, palaces, and pyramids were built on a scale heretofore impossible.

Urban communities developed in the Nile Valley about 3100 B.C., or shortly after the rise of similar communities in Mesopotamia. Whether the notion of city living evolved separately in Egypt, India, and China or diffused from Mesopotamia

is still debated. All that is known is that the world's first cities took shape in Mesopotamia around 3500 B.C., and spread widely during the third and second millenniums B.C. One thousand years later (2500 B.C.), the Indus Valley witnessed the rise of two cities, Mohenjo-Daro and Harappa, along the Indus River in what is today Pakistan (Fig 15.2). After another thousand years (1500 B.C.) city life appeared along the Huang He (Huang Ho) in China. Of the five urban hearth areas, cities arose last at about the time of Christ in the New World. (Fig. 15.3).

In the Americas, most scholars believe that diffusion played an insignificant role in the rise of pre-Columbian cities and that the cities of Mesoamerica (Central America) developed independently. It should be noted that Mesoamerican cities evolved without the advantage of animal husbandry, or of the wheel, the plow, and the alluvial setting characteristic of the old world riverine urban hearths. It has been suggested that the growing of corn, a crop that can produce substantial surpluses, compensated for the lack of tools and a nonriverine environment.

The city was a rich source of innovation. Large numbers of specialists concentrated in small areas accelerated social and cultural change. New ideas and inventions flowed in and out of these early trade centers. This encouraged new speculation in religious, philosophic, and scientific matters, and also in technology, which in turn made possible the further expansion of cities.

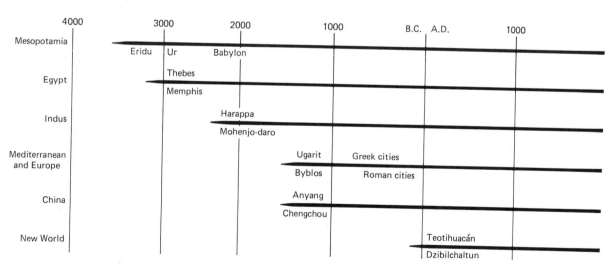

FIGURE 15.3
THE SEQUENCE OF URBAN EVOLUTION BEGAN WITH THE FIRST CITIES OF MESOPOTAMIA.

HISTORY BEGINS AT SUMER

355

Cities in The West

By 1000 B.C. the centers of urbanization and trade began to shift out of Mesopotamia and the Nile Valley into the Mediterranean Basin. Here Phoenician, Cretan, Greek, and Roman cities arose. The Roman Empire did most to diffuse city life into the rural areas of western Europe. Rome may have been the first city to approach a population of 1 million. Rome's size reflected the enormous extent and productivity of its *hinterland*; that area associated with it, both politically and economically. Roman civilization spread the idea of city life throughout the Mediterranean Basin and to the north along the Rhine and Danube rivers, into Gaul, and across the English Channel to Britain.

The Dark Ages followed the collapse of the Roman Empire. Rome and many borderland cities declined in population. Some cities disappeared as roads degenerated and commerce declined. Civil disorder discouraged trade and commerce; many towns lost their economic functions as a result. Cities survived the Middle Ages only if they were located at strategic sites, or were the seats of bishops (cathedral towns), or grew up in the shelter of the fortified chateau of a feudal lord. Surviving cities, as well as new ones, reflected the influence of a political and religious elite.

The next phase of urban growth began about 1000 A.D., when there was an impressive growth in Europe's population. This was also an era of stabilization and relative peace. Increasing amounts of land were cleared and cultivated. About 1150 A.D. the first *polders*, land reclaimed from the ocean, were diked in Flanders along Europe's northwestern coast. Europe began to "colonize" herself. Cities began to take shape after the economic renaissance of the tenth century. Extensive commercial contacts developed in the Middle East began to revitalize trade. With trade, urban life revived. Because of the Crusades and increased commercial links with Constantinople and other cities of the East, Venice became the largest city in medieval Europe. During the late Middle Ages town life spread out of the Mediterranean area to northern and western Europe. This marked a new era in the migration of civilization. By 1400, the whole of western and central Europe was covered with small towns and villages bustling with activity. Most towns of the Middle Ages were fortresses and centers of administration, located where overland routes and rivers converged. Most had fewer than 10,000 inhabitants. Figure 15.4 depicts the principal medieval towns and the trade routes connecting them. Many of these early settlements are among the largest cities in Europe today.

The market place characterized the medieval towns and the medieval town was a market for the surrounding countryside. Trade stimulated industry. Local foodstuffs and cloth formed the basis of trade in the Middle Ages.

The growth of cities had a profound effect upon medieval feudal society. Trade and industry led to an increase in the number of shopkeepers and merchants—an independent group that would emerge as an influential middle class, or bourgeoisie. A new notion of wealth emerged within this group; a notion that wealth consisted of money and commodities, and not just land. Towns became centers of intellectual life as knowledge of reading and writing spread. No longer were these abilities considered the reserve of the clergy. Medieval contributions to mapmaking and such navigation aids as the rudder and marine compass facilitated transportation. To encourage trade, the merchant class opened banks and "invented" the bill of exchange and draft, obviating the need to carry bullion on horseback throughout Europe. By the end of the medieval era the bourgeoisie began to supersede the aristocracy as the dominant social force of Europe. The bourgeoisie were to play a leading role in two great future movements: the Renaissance and the Reformation.

The Industrial Revolution of the eighteenth century had far-reaching effects on the growth and nature of cities. Technological advances in manufacturing, mining, transportation, and agriculture led to an expansion of European trade and exploration. Exploration ultimately led to colonization and the growth and expansion of cities in Africa, Asia, and the Americas. Here coastal cities that could be reached by European vessels were either established, or expanded to handle increased trade. As the breadbaskets of North America made food more plentiful, Europe experienced an unprecedented growth in its population. The factory system, based on the steam engine and other inanimate sources of energy, sprang

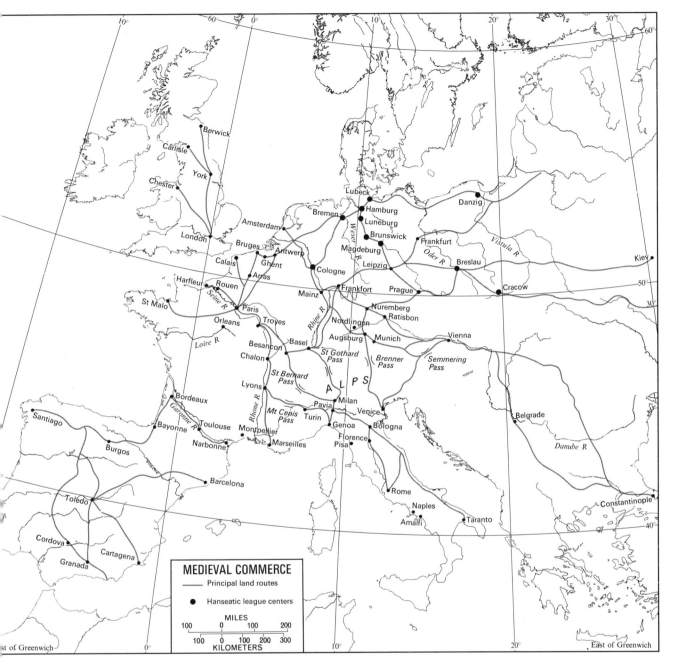

FIGURE 15.4
MEDIEVAL COMMERCE.

up throughout Europe and led to the formation of the world's first industrial cities. Many of these new cities, in Europe as well as other parts of the world, were located close to sources of bulky raw materials in order to minimize transportation costs. Accordingly, many such cities were close to coal and iron ore deposits. Manchester and Birmingham in the United Kingdom; Essen and Dusseldorf in the Ruhr Valley of Germany; Magnitogorsk in the Soviet Union; and Pittsburgh, Pennsylvania in the United States are examples. Other cities developed downstream, as outports of established cities, when ocean vessels became too large for harbors upstream. Cuxhaven for Hamburg, Germany; Kobe for Osaka, Japan; and Le Havre for Paris, France, are examples of such outports.

The economic opportunities and social amenities of industrial city life exerted a strong pull on migrants. By 1900 the United Kingdom became the first predominantly urban country (Table 15.1). As other countries industrialized, they also became urbanized. Only a few small cities dotted the Atlantic coastline of the United States in 1790; a mere five percent of the country's population lived in urban places at that time. These seaport trade centers were little more than commercial extensions of western Europe. By 1830 a few scattered small cities could be found west of the Appalachian Mountains and along the Mississippi Valley. Steamboats, canals and later, railroads, did much to open the Midwest to settlement. River and railroad towns sprang up, often at break-in-bulk points where storage and processing facilities were available. By 1860 Chicago was the nation's leading railroad center. Agricultural and industrial growth fostered increased urban growth in the early 1900s. Industrial cities in the Northeastern urban-industrialized core expanded, and mining cities sprang up as boom towns. The Great Plains were settled and railroads linked Pacific coast cities to the East. By 1920 the majority of the United States' inhabitants lived in urban areas. The automobile era (beginning in 1920) did not foster the growth of new cities so much as it permitted the expansion of existing cities. The movement of people out of crowded central cities will be discussed in the section on suburbia.

TABLE 15.1 THE TEN LARGEST CITIES (population in thousands)

1500			1700		
Peking	China	672	Constantinople	Turkey	700
Vijayanagar	S. India	500	Yedo	Japan	688
Cairo	Egypt	450	Peking	China	650
Nanking	China	285	London	Britain	550
Hangchow	China	250	Paris	France	530
Canton	China	250	Ahmedabad	Moguls	380
Constantinople	Turkey	200	Osaka	Japan	370
Gaur	Bengal	200	Isfahan	Persia	350
Tabriz	Persia	200	Kyoto	Japan	350
Paris	France	185	Cairo	Turkey	350

1900			1970		
London	Britain	6,480	Tokyo	Japan	20,450
New York	United States	4,242	New York	United States	17,252
Paris	France	3,330	Osaka	Japan	12,000
Berlin	Germany	2,707	London	Britain	10,875
Chicago	United States	1,717	Moscow	Russia	9,800
Vienna	Austria	1,698	Mexico City	Mexico	9,000
Tokyo	Japan	1,497	Paris	France	8,875
St. Petersburg	Russia	1,439	Calcutta	India	8,800
Manchester	Britain	1,435	Los Angeles	United States	8,712
Philadelphia	United States	1,418	Buenos Aires	Argentina	8,625

Source. Tertius Chandler, "The Forty Largest Cities: A Statistical Note," Historical Geography Newsletter, VII (1977), pp. 22–23.

THE URBANIZED WORLD

Today, all industrialized countries are highly urbanized. These countries include Japan, Canada, Australia, the United States, the Soviet Union, and a host of European nations. Indeed, these nations constitute the developed world. But rapid urbanization is also occurring in the world's poor, less developed countries where urban growth is outstripping the pace of industrialization and economic growth. Natural increases (the excess of births over deaths) and rural to urban migration are the principal causes of the phenomenal urban growth in the less developed world. Population growth in the Third World countries is very high and this increase affects the growth of cities: directly through births to parents who are urban dwellers, and indirectly through migration of rural folk forced off the land by population pressure. Migrants from depressed rural areas seek steady employment in the cities. Too often, they find makeshift jobs in the tertiary sector, where some of them hawk cheap goods on the street or work as domestic servants. While all large cities suffer from a host of problems, the problems of the third world are acute, especially in housing, traffic congestion, and waste disposal (Fig. 15.5). The rapid urbanization of the Third World is expected to continue. Latin America's urban population, for example, is expected to grow to 75 percent of its total population, while Africa's urban population will triple by the year 2000. Estimates are that between 1950 and 2000 the urban population of the less developed world will increase eight-fold. If these estimates hold true, two-thirds of the world's 3.1 billion urban population will be living in the Third World by the year 2000. Palliatives most often suggested are economic development of rural areas and the lowering of the birth rate. Figure 15.6 depicts world urban growth projections from 1950 to 2000.

The speed of urban growth has also increased the number of large cities. In 1950 there were 962 cities in the world each with a population of 100,000 or more inhabitants. In 1970 there were approximately 1725 such cities, and in 1975 nearly 2000. If this trend continues, there will be a projected 3600 cities with a 100,000-plus population in the year 2000. By the year 2000, according to United Nations estimates, there will be more than

FIGURE 15.5
SUB-STANDARD DWELLINGS OF RIO DE JANEIRO. Rapid urbanization is occurring in the world's poor less developed countries where urban growth is outstripping the pace of industrialization and economic growth.

400 cities in the world with more than 1 million people each. Two-thirds of these cities will be in the less developed countries (Fig. 15.7).

Figure 15.8 depicts the growth rates of the world's 20 largest urban centers. Those cities in less developed nations are growing at the fastest rate. If its present growth rate continues, Mexico City will become the world's most populous urban center by the year 2000. At present, more than 4 million of the city's 11 million people live in shantytowns. It is not uncommon for many people to share a single room, sleeping in shifts day and night. In this, Mexico City is characteristic of many cities of the Third World (Fig. 15.9).

In spite of the rapid growth of Third World cities, the less developed countries and regions are not the most urbanized. As a general rule, the more industrialized a country, the more urbanized it is. The developed regions are, as a whole, about 70 percent urban while the less developed regions are about 30 percent urban. Nonetheless, by the year 2000, nearly 40 percent of the world's total population will be living in cities.

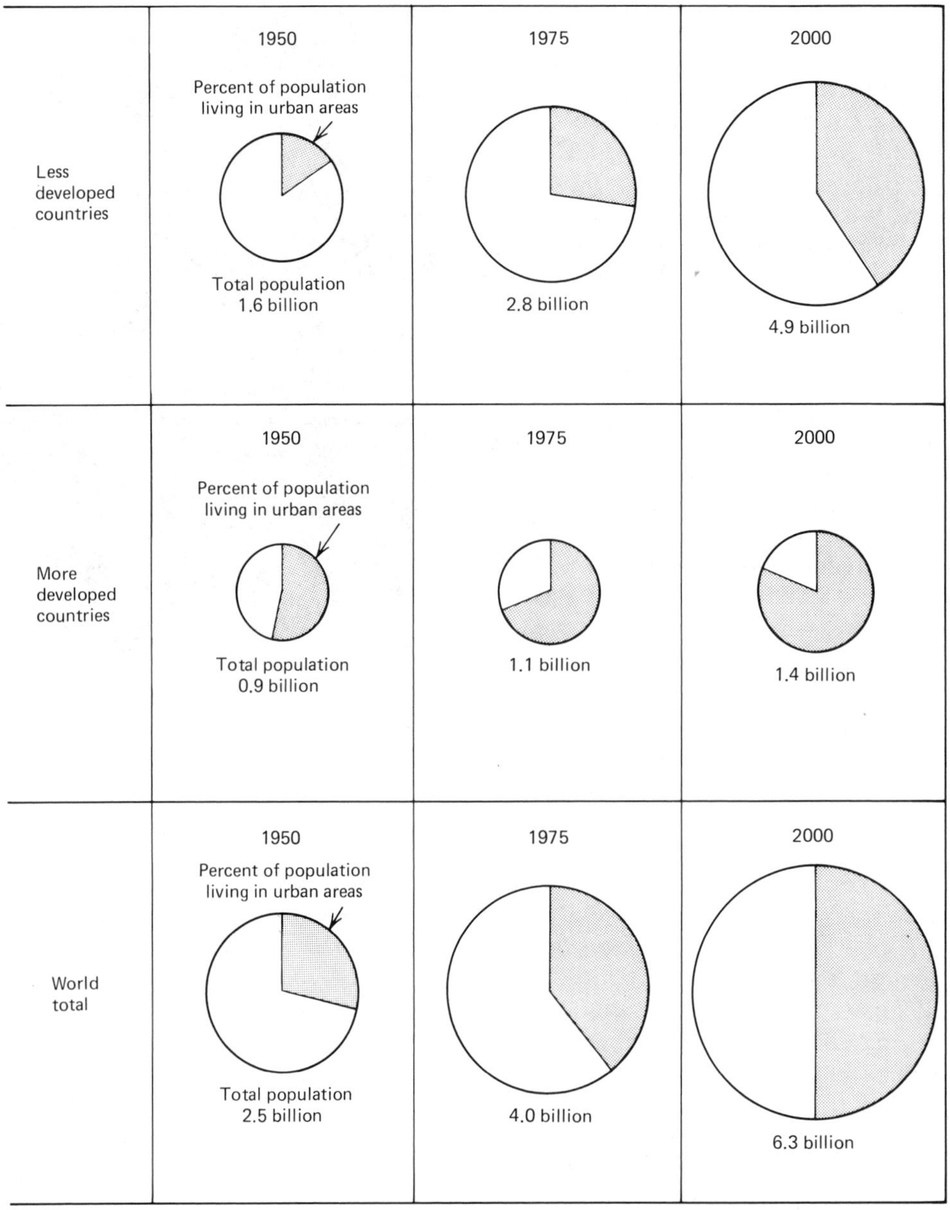

FIGURE 15.6

WORLD URBANIZATION: 1950, 1975, 2000. Source: Population Reference Bureau, Inc., Washington, D.C.

THE NATURE OF CITIES

Why do people come together to live in cities? Perhaps because the city is considered all things to all men. Economic, social, and cultural activities abound and offer numerous opportunities to those able to partake in them. The city is, however, first and foremost a center of accessibility. A city may be located at the mouth of a river that drains a productive hinterland; it may be located at the confluence of two strategic streams; or it may be at the crossroads of two different transportation arteries. The city, as a point of maximum accessibility, is not only a corporate entity with its own laws and regulations but also a collection of people and establishments engaged in highly specialized activities. As such, the city may be considered a market—a market that handles, processes, and consumes vast amounts of food, fibers, and a host of other raw materials. The city is a center of trade and transportation and a point of transshipment for goods imported by ship, plane, railroad, and truck (Fig. 15.10). The city is also a center of communication. Imagine the millions of pieces of information transferred daily via telephone, telegram, the newspaper, and the mail. The city is also the best place for businessmen to meet and exercise the art of persuasion.

The city is the center of man's cultural traditions. Here man is most often reminded of his cultural past. Independence Hall in Philadelphia, the Gateway Arch in St. Louis, the Alamo in San Antonio, and many other monuments attest to our continued interest in history. These monuments also serve as magnets for tourism and, in some instances, have stimulated office growth and renewal near them. Finally, the city is a center of anonymity and this produces tolerance—a trait that is maximized in cities and minimized in villages. The reading on *"Blue People"* suggests why. Because the city is a center of activity and new ideas and thus dependent upon interchange among people, the human experience is heightened in cities.

The Location of Urban Settlements

Site. Farmers use land as an instrument of production, but people engaged in manufacturing, commerce, and services use land only as a site. The first question a geographer asks himself when he is studying a city is, "Where is it?" Where is the actual ground upon which it stands? That immediate area is a city's *site* and consists of the terrain on which a city began and over which it spread. Site selection may have been of considerable importance when the city was originally founded. Consideration had to be given to proximity to drinking water; to freedom from floods, swamps, and malaria; and perhaps to a good defensive position. Istanbul (Constantinople), a city surrounded on three sides by water, is an example of a defensive site location. Some early sites lend character and personality to many cities and have come to characterize them: the lakefront in Chicago, the open squares in downtown Savannah, the riverfront in St. Louis, the hills of San Francisco, and the French Quarter of New Orleans are examples.

Situation. While site refers to a city's absolute location, *situation* refers to a city's relative location; that is, its location relative to the physical and human characteristics of the surrounding area. Early situational advantages were often more

---

BLUE PEOPLE

There are also changes which are not the result of new technologies but the result of the congregation of great numbers of people who, in their coming together, create new phenomena the dimensions of which we fail to understand. I will take the case of the "blue" people. If we assume that one out of every 500 people on earth is different in some way from others, let us say a "blue" person, this means that in the era of villages with normal, "green" people, there would be one "blue" person per village. This "blue" person was often called "crazy" by his fellow villagers, even though he might have been the genius of the era. If he were a genius he was isolated and more often than not was unable to thrive. In the big city of "red" people there are many "blue" people who can unite and express a new movement of thought, art, politics, anything which can be for the majority good or bad, right or wrong. In an Urban System of ten million people 20,000 "blue" people can create many new movements. A country with 200 million can easily assemble 400,000 followers of any movement, be it blue, red, or yellow. Woodstock was such a meeting of "blue" people, of which there will be many more, from liberation movements to political parties. . . . If we remember that we have, in addition to people who are "blue," others of every color of the spectrum, we will more easily understand the formations that come to life in an urban era.

Source. Constantinos A. Doxiadis, "Man Within His City," in *Teaching About Life in the City*, edited by Richard Wisniewski (Washington, D.C.: National Council for the Social Studies, 1972), p. 225.

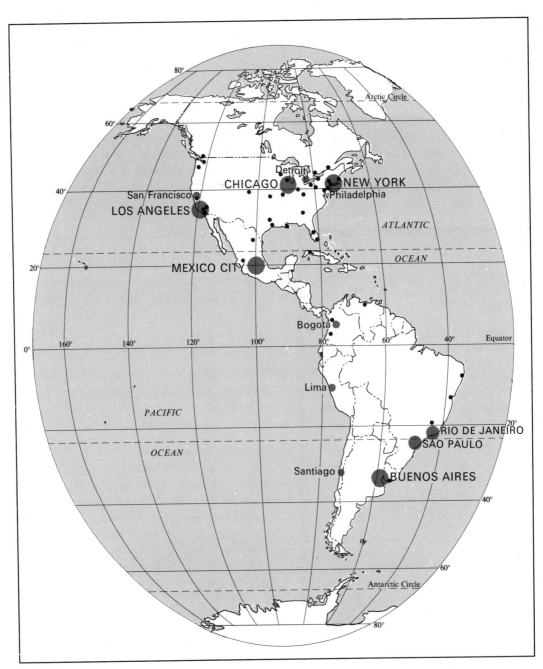

FIGURE 15.7
WORLD URBAN CENTERS.

important to urban growth than site locations. Many of today's important cities began where they could take advantage of junction points of transportation arteries, where "break-in-bulk" functions were performed. Very early in their his-tory cities such as Boston, New York, Charleston, and New Orleans became regional trade centers that occupied favorable situational locations rela-tive to their surrounding hinterlands. Many early site advantages cease to be significant today. Sim-

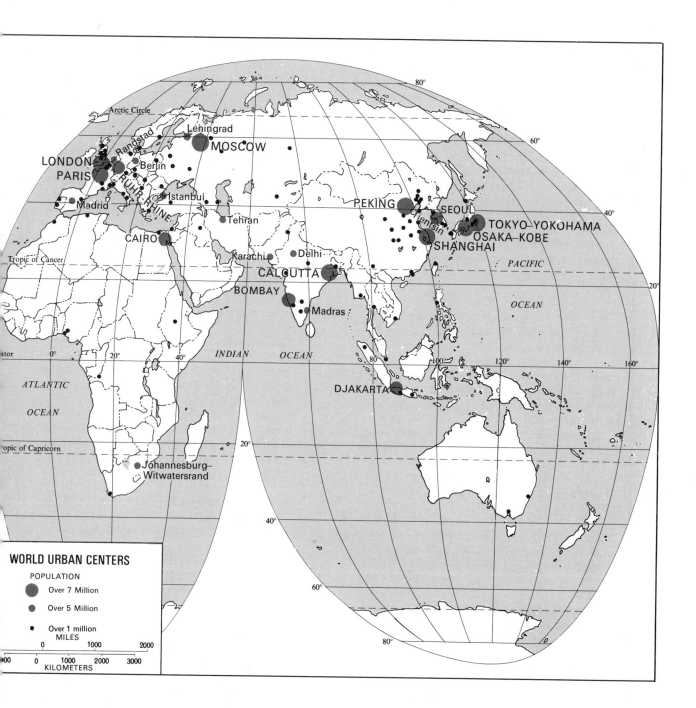

WORLD URBAN CENTERS

POPULATION

● Over 7 Million

● Over 5 Million

● Over 1 million

MILES
0   1000   2000

0   1000   2000   3000
KILOMETERS

ilarly, the advantages of a relative location can change with improvements in communications and transportation. A small town located at an important highway intersection serving travelers and local residents may find its locational advan-

tages disrupted when a large interstate cloverleaf is built ten miles down the road. The town's situation has changed. The situation of towns and cities does indeed often change, as the reading on Timbuktu demonstrates.

THE NATURE OF CITIES

TIMBUKTU

Timbuktu is a real city, despite its unreal over-tones. Why is it where it is, and why has the settlement on this site grown big? (Consult an atlas map of north Africa.) Location on the Niger River is clearly relevant, for Timbuktu lies near the margin of the Sahara Desert, and, in the absence of adequate rainfall or lakes and of ground water for wells, settlement in this area must be limited to sites where water is available. *Site* refers to the actual ground on which a city or other phenomenon rests. It is often paired or contrasted with *situation*, which refers to spatial relations with other places, principally in terms of accessibility rather than of simple distance. Situation is thus both a relative and a changeable matter. Timbuktu's situation—its relations with other places—has been much more important than the local conditions of its site in accounting for the city's growth and decline. Northward is the desert. Southward the land becomes more humid and therefore more densely populated, with a type of productive economy different from that of the desert. Settlements which lie in the zone of transition between one set of physical or economic conditions and another set may be convenient centers for the exchange of the different sorts of goods which each area produces. Timbuktu has prospered as a trade center in part for this reason. The city also has the advantage of easy access to the moister and more productive area to the south by means of the Niger River, which flows southward to the coast. Towns and cities are specialized concentrated settlements which have grown primarily because they have access to wider areas around them and thus can perform on behalf of these wider areas economic, political, or social functions which are most conveniently carried on in a central place, such as government, marketing, manufacturing, or the servicing of trade routes. The differences in the physical nature of the areas to the north and south served by Timbuktu have promoted a *complementary* (filling mutual lacks) trade relationship between them, since they produce and need to exchange different kinds of things. Regional differences do not always result in exchange. The amount of exchange depends on the effective demand in one region for the commodities produced for sale in another region and on the cost of transport between the two places.

Timbuktu is a convenient center through which the complementary exchange between the different areas around it can take place. This is not only because of its median location, but because transport carriers are likely to change there, from land to river, from desert transport to humid area transport. Timbuktu has in fact been called "the place where camel meets canoe." Places where carriers change, or *break-in-bulk* points, will have economic opportunities for the support of their populations. This will apply not only to loading and unloading, storage, or servicing of transport carriers but also to the financing and insuring of trade and to the processing or manufacturing of goods passing through. The goods must be unloaded at the break-in-bulk point: transport and loading-unloading costs for those goods are therefore minimized at that point, and a variety of routes are likely to focus there, funneling goods through the city. These are important matters for any manufacturing or service industry dependent on the assembly of raw materials or the distribution of finished goods. The trade through Timbuktu has never been very great by modern standards, because the tributary areas which the city serves are sparsely populated and unproductive (the desert), or largely organized economically on a subsistence rather than a commercial exchange basis (the moister areas to the south). Hence Timbuktu is not a very big city, but it is the biggest and almost the only city in this part of northwest Africa.

Timbuktu would be very much smaller, however, if the desert had been all which lay to the north. The desert produced almost nothing for exchange, but beyond it is the moister Mediterranean coast, which did produce something different from the areas south of Timbuktu. This coastal region also exchanged goods with humid central Africa over the most practical route, which goes through Timbuktu. Some of these goods were redistributed from north Africa by cheap sea transportation to wider markets in Europe and elsewhere, where under different physical and cultural conditions the same goods, such as ivory, gold, spices, or slaves, were not produced. These wider relations were reflected in the size and functions of Timbuktu, hundreds or even thousands of miles away. The goods which moved along the route north of Timbuktu had, however, to be of a cetain sort. Here there was no river to lessen transport costs, and it was a very long haul to the Mediterranean coast, through an area largely empty of population and devoid of opportunities for exchange. The carriers (camels) are very expensive, measured by the unit cost of transport or the cost per ton-mile (for carrying one ton one mile), mainly because their individual capacities are small. Transport costs tend to vary inversely with the capacity of the carrier, which is the principal reason why water transport is so cheap. The length of the route between Timbuktu and the Mediterranean coast increased the expense, as did the lack of trade-generating centers along the way. The latter are important factors in making a transport route profitable. Therefore this route could carry only goods which were very high in value by weight, such as gold, ivory, slaves (who provided their own transport), or salt (which is very high in value in most preindustrial economies).

With the development of mechanized sea transport and the widening of Africa's commercial relations with the rest of the world, the goods which used to move through Timbuktu to wider markets began to move instead through west African seaports. Timbuktu has become a much smaller place and more exclusively dependent on local trade and services. Change, especially technological change, is continually altering spatial relations and spatial interaction patterns. At its height, about the fifteenth century, Timbuktu was a commercial metropolis. Because of its size and its excellent access, it was also a famous cultural center for much of the large area of the western African Sudan, and it supported a well-known Moslem University. With the defection of the trade routes which fed it and their interruption by political disorders, Timbuktu's glories faded. By the end of the nineteenth century, the French found it a largely ruined town. Both political and technical change were thus responsible for Timbuktu's decline, as they affected the city's relations with other places or its pattern of spatial interaction.

Source. Rhoads Murphey, An Introduction to Geography, Chicago: Rand McNally, 1971, pp. 11–12.

Urban Functions and Specialization

All cities exist for a reason; that is, all cities have a role or a function to perform. That role is to perform services for surrounding areas, no matter how large or small the area and no matter how many smaller cities and towns are located within that area. Three functional types of cities have been identified. The first consists of *central places*, settlements that offer a host of services for small, often rural, areas. Second, *transportation cities*, such as New York and New Orleans, that offer "break-in-bulk" and other services for very large regions. Third, *specialized-function cities* dominated by one activity such as manufacturing, government, or mining, and that may serve a national or international market. Many cities may represent a combination of the three functional types.

By examining a city's labor force, geographers have been able to develop various classifications to determine the nature of a city's economic specialization, or economic function. As cities grow, their economic functions become more complex as the economy adds a broad mixture of new activities. It is not uncommon, however, to find one or two dominant economic functions. The classification scheme used here was developed by C. D. Harris in the 1940s. Although a number of shortcomings exist within the system and functional changes have occurred within some of the cities, Harris' work is recognized as a classic in the literature.

Harris classified manufacturing into two categories. In the first category, employment in manufacturing is overwhelmingly dominant; in the second, manufacturing is dominant but important secondary activities exist. Other categories are wholesale, transportation, resort-retirement, retail, diversified, mining, university, and political cities. You may want to determine if the university classification applies to the town where you study. Do any of the classifications apply to your home town?

Figure 15.11 depicts the functional classification of United States cities according to Harris' classification scheme. It is not surprising that in an industrialized country like the United States, the most prominent cities were classified as manufacturing, diversified, and retailing. Collectively these three accounted for 80 percent of the cities classified. Manufacturing alone represented 43.5 percent of the cities classified. Most of the large urban areas specialize in manufacturing, transportation, and wholesaling, while smaller cities were clas-

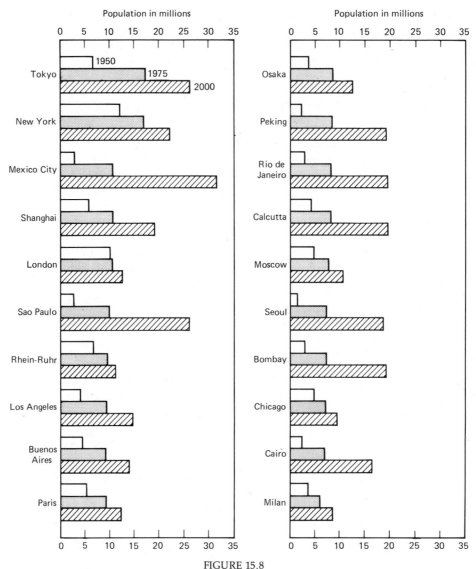

Population in millions

FIGURE 15.8
GROWTH OF WORLD URBAN CENTERS. Source: Population Reference Bureau, Inc., Washington, D.C.

sified as retailing, educational, and mining centers.

A number of generalizations may be deduced from the maps shown in Figure 15.11. The distribution of manufacturing cities coincides with the "Manufacturing Belt," located within the northeastern quadrant of the nation. In addition, two narrow extensions southward are found in the Great Valley and Piedmont areas of the Southern Appalachians. Outside of these areas, manufacturing cities are notably absent.

Diversified cities tend to be concentrated in a transitional zone between the "Manufacturing Belt" of the northeast and the band of retail cities to the west (Fig. 15.12). Retail cities also are located outside of the "Manufacturing Belt" and are found in the agricultural heartland of the nation. They are small cities and act as service centers to surrounding rural areas. Wholesaling cities are found mainly in the Midwest and South. Many of these are associated with the assembly of produce. More prominent cities within this classification are Dallas, Seattle, and San Francisco. Each of the latter is engaged in the distribution of goods over wide areas.

Transportation cities include railroad centers as well as river and ocean ports. Mining towns tend to be small; most are associated with the extraction of coal (Fig. 15.13). Educational (university) cities are often small settlements dominated by large state universities. Noteworthy examples are Gainesville, Florida; Chapel Hill, North Carolina; Boulder, Colorado; Champaign-Urbana, Illinois; Norman, Oklahoma; and Columbia, Missouri. Finally, the resort-retirement communities reflect the amenities of beach, mountain, and desert. Most of these communities are concentrated in the "Sunbelt" of the South.

The Basic-Nonbasic Concept

People living in cities are engaged in specialized activities. These activities imply that cities are centers of trade. In other words, the specialized goods and services produced by a population and not consumed by that population are exchanged for the specialized goods and services produced by other cities and regions. Harris' classification of cities was based upon total employment figures. As we shall see, however, the labor force of a city can be divided into two parts: 1) those employed in *basic* industries, or "city forming" employment that depends upon areas outside the city for its market, and 2) the *nonbasic* component, the "city serving" employment activ-

FIGURE 15.9
IF ITS PRESENT GROWTH RATE CONTINUES, MEXICO CITY WILL SOON BECOME THE WORLD'S MOST POPULOUS URBAN CENTER.

THE NATURE OF CITIES

FIGURE 15.10

THIS AERIAL VIEW SHOWS PARTS OF NEW JERSEY, LOWER MANHATTAN, AND BROOKLYN. The city is a center of trade and transportation and a point of transhipment as it handles the goods brought into it by ships, planes, railroads, and trucks.

ity that is sustained from money generated within the area where it is found.

A city does not serve just those people living within its own municipal boundaries. A city can exist only when it sells its goods and services beyond its borders. When Detroit is recognized as the automobile capital of the United States, we realize that automobiles produced within that city are sold mainly outside its borders. The automobile industry of Detroit, then, is *basic* to that city.

Nonbasic industries of Detroit produce goods and services that are to be sold within the city. Examples of nonbasic industries are television repair shops, grocery stores, laundries, taverns, clothing stores, and the like. Similarly, most of the steel produced in Pittsburgh, Pennsylvania, and Gary, Indiana, is sold outside those cities. Each city has its own nonbasic service industries. Whenever an industry of a city produces an item that is intended to be "exported" and consumed mainly

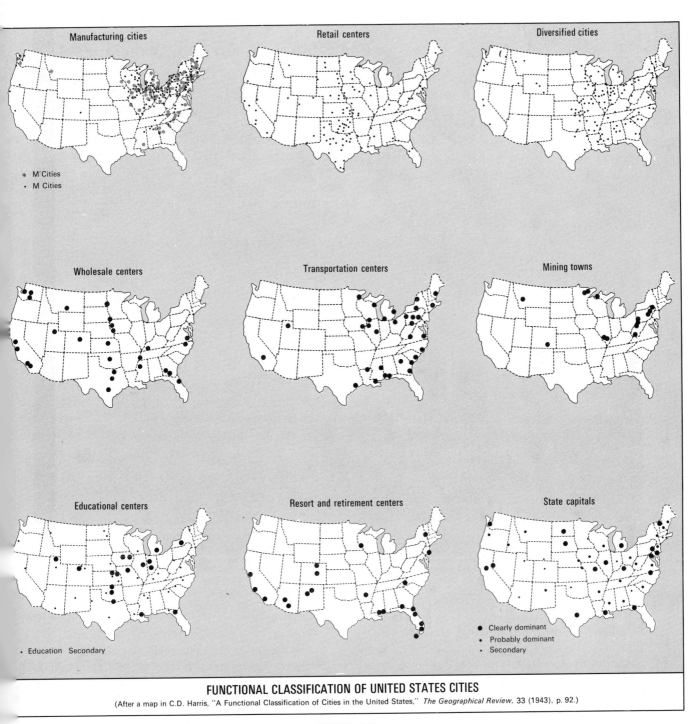

Manufacturing cities

• M'Cities
• M Cities

Retail centers

Diversified cities

Wholesale centers

Transportation centers

Mining towns

Educational centers

• Education Secondary

Resort and retirement centers

State capitals

• Clearly dominant
• Probably dominant
• Secondary

FUNCTIONAL CLASSIFICATION OF UNITED STATES CITIES

(After a map in C.D. Harris, "A Functional Classification of Cities in the United States," *The Geographical Review*, 33 (1943), p. 92.)

FIGURE 15.11
FUNCTIONAL CLASSIFICATION OF UNITED STATES CITIES.

THE NATURE OF CITIES
369

FIGURE 15.12
GATEWAY MEMORIAL ARCH, ST. LOUIS, MISSOURI. Diversified cities tend to be concentrated in a transitional zone between the Manufacturing Belt of the northeast and the band of retail cities to the west.

outside of the city, then that industry is an integral component of the city's basic function. As noted earlier, the basic industry of a city may include activities other than manufacturing. Transportation in Duluth, Minnesota; Portland, Maine; and Norfolk, Virginia, generates basic jobs just as those in manufacturing do. A service rather than a product is "exported." Similarly, the authors of this text, who live in a small (population 35,000) "college town," are components of the basic industry of this university community.

Let us draw one further analogy to demonstrate this important concept of basic and nonbasic industries. Consider a frontier mining town where

100 miners are employed in the town's only basic industry—gold mining. Assume that each of the 100 miners is married and has two children. The basic industry thus supports 400 people. But the 400 people demand services: schools and churches have to be built, grocery and clothing stores and livery stables are operated, newspapers are published, professional personnel are needed, and saloons have to cater to visiting cowboys. It has been suggested that there is an average basic/nonbasic ratio of 1:3; that is, for every miner employed in the town's basic industry, three people may be employed in a nonbasic industry. Thus, with 100 miners our community supports 300 peo-

le employed in the various nonbasic service industries listed above. Let us further assume that each of the 300 nonbasic personnel are married men with two children. That gives us 1200 people supported by the nonbasic industries. In other words, the basic mining industry (made up of 100 miners), not only supports its own 300 dependents but also economically supports the 300 nonbasic personnel and their 900 dependents for a grand total of 1600 people. (Imagine the size of a modern city that has 50,000 or 100,000 people employed in the basic "export" industries.) But let us return to our miners. Let us assume, for a moment, that the gold vein has run out. What are the consequences? The gold is gone and the 100 miners must seek work elsewhere. The basic industry is gone and nothing is left to support the numerous nonbasic service workers. They too will eventually leave and our little mining community will become a ghost town. This can and has happened to thousands of contemporary cities that have lost industries that represented a portion of the town's basic industry.

All too often we think of the benefits that new and old basic industries bestow upon our cities. While it is true that basic industries stimulate urban growth in the local economy, many tend to believe that nonbasic industries are parasitic and

FIGURE 15.13
POCAHONTAS, VIRGINIA. Most mining towns are associated with the extraction of coal. Mining towns tend to be small.

undesirable. However, the nonbasic industries may be used by the entrepreneur of a basic industry. Such services may consist of construction expertise, maintenance, truck terminals, and the like, each of which is a service to the businessman. This may mean that the businessman does not have to concern himself with these details and, as such, the nonbasic services become important and necessary and are not parasitic. Indeed, the well-developed nonbasic services of an area may be an advantage and attract basic industry to it. It is best perhaps to assume a symbiotic relationship between the two and recognize that one cannot get along without the other.

Christaller's Central Place Theory

It will be recalled that one of the three functional types of cities we have considered is central places; places that perform services for the surrounding area. Walter Christaller, a German geographer writing in the 1930s, formulated his *central*

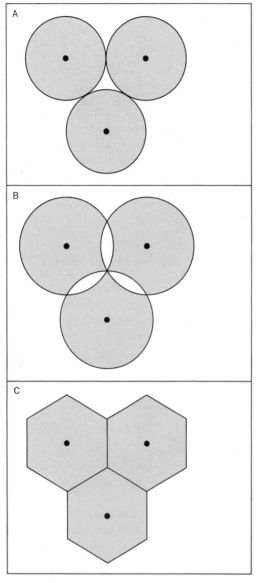

FIGURE 15.14

(A) TANGENT CIRCLES LEAVE THE SHADED AREAS UNSERVED. (B) OVERLAPPING TRADE AREAS CREATE COMPETITION. (C) HEXAGONAL TRADE AREAS LEAVE NO AREA UNSERVED AND CREATE NO COMPETITION.

same purchasing power, and 4) consumers act rationally in space in accordance to the econom principles of distance. Christaller was concerne only with central places and not with disperse areas that derived their incomes from farming mining, and monastery or military settings.

Assuming the characteristics Christaller proposed for his model, then an even distribution of urban settlements would develop on the feature less plain to service and supply goods to the surrounding *tributary area*. If consumers followed the fourth principle which states that the price of goods increases with distance from the point of production (hamlet, village, or city), then a circle would be the ideal tributary area (Fig. 15.14A). However, with this arrangement those living in the unshaded areas of the circle would not be provided with the good in question. If they are to be so provided, then the circular patterns must overlap (Fig. 15.14B). However, consumers in the shaded areas will choose the center closest to them; the result is the formation of hexagons—the most efficient form because it leaves no area unserved and creates no competition (Fig. 15.15C).

Cities, towns, villages, and hamlets all have trade areas and spheres of influence. People come to these central places to purchase goods and services; central places, acting as collecting and distribution centers, serve their respective surrounding areas. Businesses in hamlets and villages sell *low order* goods (such as bread and milk) that have a low threshold level; that is, they require fewer customers in order to be economically viable. A department store or county hospital in a town or city offer *high order* goods and services and have a high threshold level, for they need many more customers to make themselves viable.

Figure 15.15 will help further to explain the concept of a central place hierarchy. Assume the people living in the rural area surrounding a hamlet need to purchase bread, milk, or other convenience commodities. These are, as mentioned, low order commodities—items that are readily available in each of the numerous hamlets. Let us assume, however, that Farmer Jones, living within the hexagonal trade area of the hamlet, has occasion to take his family to lunch or buy some fertilizer or seed. In this case he will travel to the village—a community that not only sells the milk and bread he could have bought in the hamlet but also the lunch (in the town's only diner), fertilize

place theory that established the following principle for his theoretical model: 1) the area was a featureless plain devoid of all physiographical and cultural features, 2) a uniform transportation system permitted movement with equal effort in all possible directions, 3) the population of the rural area was evenly distributed and possessed the

and seeds that are available only in villages. Obviously, the village offers all the services of the hamlet, and more. If we review Figure 15.15, depicting Christaller's hierarchy of central places, we will see that each village's tributary area is bounded by six surrounding hamlets. The residents of the hamlets, like our visiting farmer, will also have occasion to travel to the village. Now it is Saturday and time for Mr. Jones to go to town. The town is a third order center and will have its tributary area delimited by six villages. The town offers all the services offered in villages and hamlets, and more. In town Mr. Jones visits the farm implement dealer while his wife does her weekly shopping and his daughter attends the local theater. Later he has his hair cut at the only barber shop within thirty miles while his wife shops at the local Sears and J.C. Penney stores—establishments not found in villages or hamlets. On the way home he "window shops" for a new tractor and a used car. Once a year Mr. Jones takes his family to the state fair held in the state's second largest city, a fourth order central place bounded by six towns. Here he can buy all the services offered in the town, village, and hamlet, and more. If he wanted, he could have bought bread and milk, could have gotten a hair cut, and could have purchased a used car—all the things available in the third and lower order centers. But he came to buy his daughter a wedding dress, an item found in a specialty shop and only to be purchased in a high fourth-order center. Farmer Jones

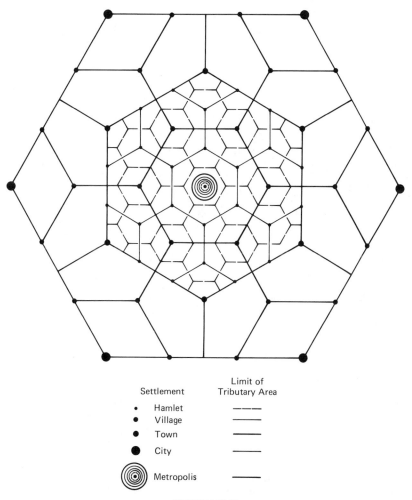

Settlement	Limit of Tributary Area
• Hamlet	— — —
• Village	——
● Town	——
● City	——
◎ Metropolis	——

FIGURE 15.15
CHRISTALLER'S HIERARCHY OF CENTRAL PLACES.

may never travel to Christaller's metropolis. The metropolis, the fifth order center, surrounded by six cities sells high order commodities—items that are readily available only in a metropolis: exquisite gems, large bank loans, works of art, and specific pieces of technical equipment. The metropolis will serve the needs of men with more business and material needs than those of Farmer Jones, although within its boundaries one will be able to buy bread and milk, get a haircut, shop at Sears, buy a used car and a wedding dress, and more.

Internal Form and Structure of Cities

Although a variety of activities take place within a city, they tend to be concentrated in specialized areas. These concentrations have produced a highly stylized and repetitive arrangement in American cities. Patterns of land use have become so predictable that students of the city have developed models to enhance their understanding of urban life. A model is a generalization, simplification, or abstraction of the real world in which superflous material is discarded in order to facilitate the investigation of a specific problem. Three models of urban form are presented in the following sections.

The Concentric Zone Model. This model was first postulated by sociologist Ernest W. Burgess in 1923. This classical descriptive model remains the most popular generalization of urban land use. Using Chicago as a prototype, Burgess hypothesized that cities expanded radially in a series of six zones (Fig. 15.16A). Zone 1, the *central business district* or CBD, is an area of maximum accessibility containing department stores, office buildings, banks, hotels, theaters, and a variety of retail shops. Zone 2, the adjacent *fringe* of the CBD, is an area of wholesaling, truck and railroad depots, and light manufacturing. Zone 3, the *zone in tran-*

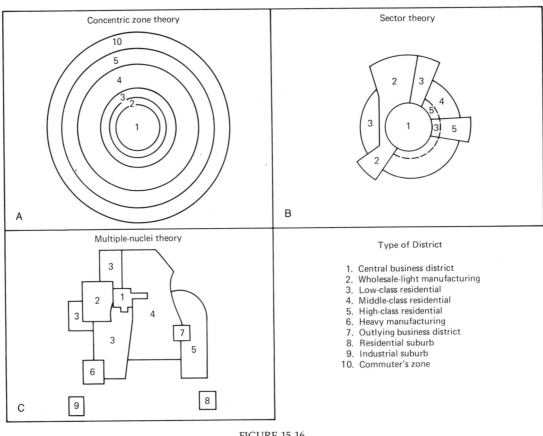

FIGURE 15.16
URBAN MODELS.

sition, is an area of residential deterioration consisting of slums and rooming houses. This was often the home of first generation immigrants. Today it houses the disadvantaged. Here anonymity is maximized and the area is a haven for skid row down-and-outers, and other groups that often are not compatible with each other. Zone 4 is the home of *independent workingmen's homes.* This zone consisted of second generation immigrants who were blue-collar industrial workers. These individuals had accumulated enough wealth to escape the transition zone and purchase a "two-flat" dwelling, often in an immigrant neighborhood that we now refer to as the streetcar suburbs of the 1890s. Zone 5, the *zone of better residences,* houses the native-born middle-class Americans who live in single-family dwellings. Zone 6, the *commuters zone,* consists of upper-class residences found in dormitory suburbs and satellite cities located along rapid-transit lines.

Although the Burgess model is founded upon the American socioeconomic scene as it existed in the nineteenth and early twentieth centuries, it serves as a useful generalization in setting the stage for a look at urban problems.

The Sector Concept. Another model of urban land use is the sector concept formulated by Homer Hoyt in 1939 (Fig. 15.16B). Using rent levels, he found residential neighborhoods did not form concentric circles. Instead, he noted that better residential areas tended to be located in one or more pie-shaped sectors that developed along transportation routes directed toward high ground and freedom from floods; toward non-industrial waterfronts; toward open country on the outskirts of the city; toward areas with developed retail and service facilities; and, in some instances, toward prestige locations in the central city. In Boston, for example, the high-rent residential area has traditionally been located on the western side of the city. Since 1936 a new area has developed on the southeastern periphery of Boston. The high-rent residential areas of Minneapolis have always been located in the southwestern sector of the city. Basically, the sector theory states that, as a city grows, high-rent areas follow wedges or sectors from the center of the city to suburban areas. Similarly, those areas of a city that develop as low-rent areas tend to retain that character as the sector extends through the process of city growth. Hoyt

studied the internal residential structure of 142 cities in the 1930s. Even today there are many urban residential patterns that support his conclusions. Low-income residential areas tend to be located along the least desirable land, away from the high-rent areas, near railroad lines and industrial areas. Hoyt rejected the concentric zone model of Burgess because the latter assumed that residential patterns would be concentrically spread throughout the city, with the lowest income and residences found near the city center while higher incomes and better residences would be located at the periphery of the city. It is felt by some that Hoyt's model is an improvement over that of Burgess.

Multiple Nuclei Model. The third model is known as the *multiple nuclei theory.* It was formulated in 1945 by two geographers, Chauncey Harris and Edward Ullman, as a modification of the concentric zone theory and sector theory. Figure 15.16C depicts this model of urban form. The authors did not give the central business district (CBD) the same weight and significance given in the previous models. They argued that land use patterns of a city developed around several nuclei. The significance of this theory is more understandable when we realize that certain specialized activities group themselves because they profit from mutual association. Interactions among establishments are called linkages. Banks and wholesale merchants are examples of such establishments. On the other hand, unlike activities are detrimental to each other—stockyards and retail functions, heavy industry and light industry, are examples of unlike activities.

Major Elements of Urban Form

The central business district (CBD) is the most recognizable component of any American city. The CBD is usually located on or near the original site of the city. It has been called the "heart" of the city for it is the focus of transportation arteries and thus the most accessible part of the city. Because all CBD's have similar characteristics, a visitor to an unfamiliar city would know immediately when he has entered the "downtown" area. Here are found contiguous retail shops and department stores whose wares may be viewed through large windows (Fig. 15.17). The CBD also has the city's

FIGURE 15.17

HOUSTON, TEXAS. The central business district contains office buildings, department stores, banks, hotels, theaters, and a variety of retail shops.

heaviest pedestrian traffic and the highest land values. Land here is sold by the front foot rather than by the acre. Tall buildings and a high intensity of land utilization reflect man's attempt to maximize the use of this expensive but highly accessible property. Generally speaking, the larger the city, the taller its CBD buildings will be.

The CBD has traditionally been an office center for financial institutions such as brokerage houses, banks, and insurance companies. City Hall and court houses attract attorneys, many of whom have their offices nearby; other offices will house physicians and dentists. A few blocks from the center of the CBD are theaters, restaurants, and hotels. A greater variety of evening activities take place in this area than in any other. Civic centers and sports stadiums may also be nearby.

CBD's may, over a number of years, change their boundaries and grow in one direction or another. The area from which a CBD is moving is called the "zone of discard." Low-grade retail shops characterize this area: pawn shops, used-clothing stores, and cheap bars. The direction in which the CBD is moving is the "zone of assimilation," a more desirable area often consisting of newer shops, showrooms, headquarter office buildings, and hotels.

Retail sales in the CBD have been declining for many years. This partly refects the growth of outlying shopping centers in suburban areas, where the majority of urban dwellers now live, and the increased use of the automobile. It is not surprising that the CBD has become a topic of controversy. Is it a vital organism that a healthy city needs? Should dying and often dilapidated CBD's be restored? While residential and industrial areas look much alike from city to city, the CBD with its historic buildings and landmarks makes each city distinctive. Yet, some CBD's are run-down and suffer from parking problems and traffic congestion. The decline of many CBD's reflects the population loss suffered by most central cities. Planned suburban residential areas, planned industrial parks, and planned shopping centers have fostered deconcentration, leaving behind a less affluent population. Still, some cities have attempted to reverse the decline in retail sales by revitalizing the downtown area and in some cases by converting the main shopping street into a mall—a concourse that encourages people to gather in front of each store. If, however, present trends continue, the CBD will become just one of several nodes of commercial activity in a city.

All cities have outlying commercial centers. The largest, known as regional centers, mirror the retail types found in the CBD. These centers are usually located at large highway intersections. One or more department stores serve as lead stores, with dozens of secondary specialized shops built around them. Unlike the pedestrian-oriented CBD, the shopping center is automobile-oriented and consists of acres of parking spaces (Fig. 15.18). Smaller *community centers* may house a bank, supermarket, drugstore, and several variety shops. The *neighborhood center* is built around a supermarket, service station, and one or two other stores. Finally the *convenience center* will consist of a laundromat, a "convenience" grocery, and a service station.

Since World War II, there has been an exodus of firms engaged in light manufacturing from the city center to the suburbs. Here large inexpensive tracts of land could be used to build single-story buildings, and land for future expansion was available. Heavy industry also has been relocating. Its locational requirements once called for flat land adjacent to water transport. But the development of highways and large, efficient trucks permitted heavy industry to leave central areas for outlying locations.

Residential districts represent the largest percentage of all urban land use. Owning a space-consuming single family dwelling has become a social imperative for millions of Americans (Fig.

FIGURE 15.18
PARAMUS, NEW JERSEY. Unlike the pedestrian-oriented CBD, the suburban shopping center is automobile-oriented and consists of acres of parking spaces.

FIGURE 15.19
HOUSING TRACT NEAR SAN JOSE, CALIFORNIA. Owning a space-consuming single family dwelling has become a social imperative for millions of Americans.

15.19). This type of housing will continue to add to urban sprawl, which in turn increases commuting time. Condominium living, popular in some parts of the country, may be considered a wiser use of residential land. But residential neighborhoods remain homogeneous, for we continue to segregate ourselves by economic status and by race. The homes of the wealthy are not located next to the homes of the poor.

CHANGING URBAN GROWTH PATTERNS

New forms of transportation are responsible for the changing patterns of city growth. In the preindustrial city, where transportation was by foot or animal power, a compact urban structure was required. The horsedrawn omnibus was the first form of public transportation in American cities. Later the omnibus was put on rails, and the horsedrawn railway car, first used in New York City in 1832, spread rapidly and was the principal form

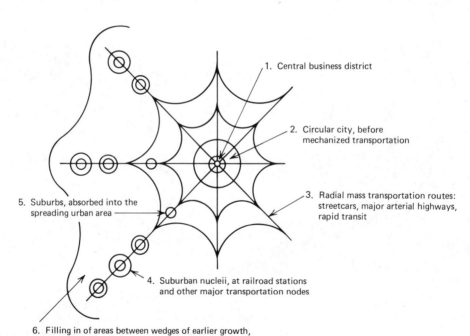

1. Central business district

2. Circular city, before mechanized transportation

3. Radial mass transportation routes: streetcars, major arterial highways, rapid transit

5. Suburbs, absorbed into the spreading urban area

4. Suburban nucleii, at railroad stations and other major transportation nodes

6. Filling in of areas between wedges of earlier growth, generally at lower densities, with access by automobile only

FIGURE 15.20
THE EVOLUTION OF URBAN DEVELOPMENT. Source: Harold M. Mayer, *The Spatial Expression of Urban Growth*. Washington, D.C.: Association of American Geographers, Resource Papers for College Geography No. 7), 1969, p. 40. Reprinted by permission.

of internal transportation in most cities by the 1860s. The horse railway caused cities to assume a star-shaped outline (Fig. 15.20). At the same time a few steam railroads operated commuter services, and a number of small suburban towns developed along railroad lines radiating from the central city. But it was with the development of the electric trolley car in the 1880s that the movement of population was accelerated toward the edge of cities. The next innovation was rapid transit. The multiunit electric train, either elevated or as a subway line, was designed to eliminate congestion in a number of the nation's largest cities. Good transportation throughout all areas of the city and to the suburbs facilitated an out-migration from the crowded central portions of many cities.

Although earlier transportation forms set the stage for peripheral spread, the automobile opened vast areas beyond the limits of mass transit systems. Most industry, now oriented to the truck, is less dependent than formerly upon railroads and waterways. Today the wedges between radial transportation arteries are rapidly being filled in. Automobile-created mobility means that compact urban structures are no longer necessary (Fig. 15.21). New residential areas are spread out and are characterized by low population and structural densities. It has been suggested that, 100 years ago, a population growth of 1000 would require

a 4-hectare (10-acre) additional area. Today, a 1000 population growth requires an additional 40 hectares (100 acres) or more. It is obvious that changes in transportation modes are related to changes in the form and structure of urban areas.

The Suburban Fringe

All aspects of the urban phenomena are showing increasing signs of decentralization. The 1970 census documented the fact that the majority of America's urban dwellers were living in the suburbs, and the data of the 1980 census are expected to show the continuation of this trend. Moreover, industry, commerce, and wholesaling are also leaving the central city while most new business is being established in the suburbs. In 1973, for the first time, employment in the suburbs exceeded that of the cities.

Suburban retailing began shortly after World War II with the appearance of small community shopping centers. Beginning about 1960, the large regional shopping center, usually on a site of 30 acres or more, set the stage for the early 1970s. By that time a new type of retail center emerged known as a superregional mall, an innovation still being developed throughout the United States. Malls may cover hundreds of hectares and include two or three department stores and several hundred smaller shops. Thousands of parking spaces on millions of square feet of land accommodate tens of thousands of shoppers every day. The superregional center performs most of the functions traditionally associated with the CBD; it has been called the "outer city." Here one can shop, have lunch, enjoy a movie, see a play at a dinner theater, have a car repaired, attend political rallies, and transact business at one of the local banks. It is not surprising that more than half of all retail sales now take place in the nation's shopping centers.

When most suburbanites commuted to jobs in the central city, their home areas were called "bedroom communities." This is no longer true, however. Since 1973, most metropolitan jobs are located in the suburbs. This work pattern reflects one other new urban phenomenon—the suburbanization of industry and of office-based employment. New freeways, interstate highways, and circumferential expressways developed in the

FIGURE 15.21
HAYWARD, CALIFORNIA. Automobile-created mobility means that compact urban structures are no longer necessary.

FIGURE 15.22

TARRYTOWN, NEW YORK. Industrial and office parks tend to be located in attractive settings adjacent to interstate and circumferential highways.

late 1960s and early 1970s made all points of the metropolis equally accessible. The increasing use of trucks freed many industries from rail and water terminals located in the central city. Many of the new suburban industries are located in industrial parks, most of which have been developed since 1960. These parks tend to be located in attractive pastoral settings adjacent to interstate and circumferential highways (Fig. 15.22). Attractively landscaped frontage, with fountains, ponds, and sculptures, give a positive image of the firm to the public. Office-based employment, traditionally associated with the CBD, has also been moving to the suburbs, where most of it is found in planned office parks, a somewhat newer phenomenon than industrial parks.

It should be pointed out that the ability of businesses to choose freely any number of urban or suburban sites mirrors the many changes taking place in today's economy. Increasingly, more people are employed in the tertiary sector of the nation's economy. Distribution, warehousing, information storage, and the communication industries represent today a larger percentage of the labor force than ever before.

The combination of superregional shopping centers and industrial and office parks (usually located at the intersection of two interstate highways), themselves surrounded by residential areas, suggests the development of suburban mini-cities. The suburbs are urbanizing and becoming increasingly independent of the central city. It has been suggested that we are moving toward an urban civilization without great cities. The central city in the United States, having lost some of its economic functions, now suffers from a declining tax base. At the same time, out-migration of the more affluent continues, leaving behind the disadvantaged segments of the metropolitan population.

Since most new jobs being created are located in the suburbs, blue-collar inner city residents (mainly non-white) now find they must commute longer distances. This is known as reverse commuting. Reverse commuting presents many problems to lower-income city residents. They own fewer automobiles, and mass transit in suburban areas is notoriously inadequate. Moreover, the high price of suburban homes makes it impossible for low-income people to move close to work. Although white-collar suburban residents once commuted the greatest distance, they now have the opportunity to move closer to work. Naturally, this reinforces the suburban trend, a trend whose speed, however, might be slowed by the ever-increasing price of gasoline.

Megalopolis and the Urban Future

Megalopolis is a word coined by French geographer Jean Gottman to refer to the large urbanized area of the northeastern seaboard of the United States. Here are found a group of closely-spaced cities stretching from Boston to Washington, D.C. The word *megalopolis* comes from two Greek words: *mega,* meaning "great," and *polis,* meaning "city-state." Although the term has been expanded and is now used to refer to any area of extensive urban coalescence, Gottman chose to use the word to describe the northeastern seaboard, because no other section of the country has such a large, dense population spread over such a large area— 800 kilometers (500 miles) by 80 240 kilometers (50-150 miles). The area is dominated by five cities—Boston, New York, Philadelphia, Baltimore, and Washington, D.C.—four of these five are important seaports. The area remains the commercial and financial center of the nation and has nearly all types of industry. Here the world's highest standards of living are found. Moreover, the area contains only 2 percent of the nation's land but 20 percent of its population.

While the northeastern seaboard is the largest and most complex megalopolis in the nation, other agglomerations are developing. Among these are Southern California, Chicago-Milwaukee, Pittsburgh-Cleveland, Southeastern Michigan, and the Bay Area-Central California. It is expected that some of these developing megalopolises, especially along the Great Lakes, will eventually merge into each other.

In 1920 more than 50 percent of the United States population lived in urban areas. Today the figure is about 75 percent. Urbanism, however, is unevenly distributed. The northeastern seaboard, the Great Lakes States, and the Pacific Coast are the regions with our largest metropolises. The urban areas within these and other regions will continue to grow as they expand into the suburban fringe and beyond, eventually coalescing with neighboring urban centers. In less than 100 years, however, super-metropolitan areas will be found in many new areas: southern Texas, eastern Florida, the Gulf coastal "Old South," the Puget Sound lowland, central Colorado, and others. But the area of greatest urban growth will undoubtedly take place in the "Sunbelt" of the South along the Gulf and Atlantic coasts.

Declining Metropolitan Population

Beginning in the 1970s, the United States population statistics revealed an abrupt reversal of long established trends to urbanization. The longstanding trend toward metropolitan growth and nonmetropolitan decline are now the reverse of each other as more Americans are moving away from urban areas than are moving to them. Whether this is a temporary or long-term trend is unclear. The fact remains, however, that one-sixth of the nation's metropolitan areas (not just central cities) are losing population.

An equally important trend is the growth of nonmetropolitan areas, and even of areas outside the influence of large cities. The growth of population in small cities and towns, and rural nonmetropolitan areas remote from urban centers, has been referred to as America's "rural renaissance."

Americans are a very mobile people, and although employment opportunities have traditionally influenced the direction of migration, other considerations seem to be gaining in determining what constitutes an attractive place in which to live. Jobs are no longer the sole attraction as people weigh the real and perceived costs of metropolitan life with those of nonmetropolitan areas. Moreover, rising affluence, retirement benefits, and the availability of public assistance have produced a footloose population that can live in amenity-rich areas such as Colorado, the Ozarks, and northern New England, as well as in the traditional retirement settlements of Florida and the Southwest.

Many metropolitan areas are experiencing "shrinking pains" as they face the problems of underused schools, abandoned housing, and outmoded public facilities. Selective outmigration, especially of the young and mobile, tends to leave behind the elderly and low-income citizens, a situation that strains the local tax base. While many small towns and rural areas welcome growth, others may resent arriving migrants and the problems they cause. Local officials may feel frustrated at their inability to cope with growing congestion, sprawl, and support costs.

Just as people within urban communities interact, so too do the nations of the world community. International cooperation, or the lack thereof, is an important consideration within the field of political geography, our next topic of discussion.

KEY TERMS AND CONCEPTS

hinterland

site

situation

central place

basic industries

nonbasic industries

central place theory

tributary area

concentric zone model

central business district

transition zone

sector theory

multiple nuclei theory

regional shopping centers

suburban fringe

megalopolis

rural renaissance

DISCUSSION QUESTIONS

1. Discuss the historical development of urban life. Where were the earliest centers of urbanization?

2. What effect did the Industrial Revolution have upon the location and growth of cities?

3. Discuss world urban growth projections for the year 2000.

4. What is meant by the statement that "A city is a center of accessibility"?

5. Discuss the notions of site and situation as related to the location of urban settlements.

6. Discuss the functional classification of United States cities according to Harris' classification scheme.

7. Discuss the basic-nonbasic concept of urban function as related to your home community.

8. Describe the three principal models of urban form.

9. What aspects of urban life are showing signs of decentralization?

10. What recent changes are altering long-established trends toward metropolitan growth?

REFERENCES FOR FURTHER STUDY

Bourne, L., and J. Simmons, eds., *Systems of Cities: Readings on Structure, Growth, and Policy,* Oxford University Press, New York, 1978.

Gottman, J., *Megalopolis,* The M.I.T. Press, Cambridge, Mass., 1961.

Mayer, H., and C. Kohn, eds., *Readings in Urban Geography,* University of Chicago Press, Chicago, 1959.

Mumford, L., *The City in History,* Harcourt Brace Jovanovich, New York, 1961.

Murphy, R., *The American City,* McGraw-Hill Book Co., New York, 1966.

Yeates, M., and B. Garner, *The North American City,* Harper & Row, New York, 1971.

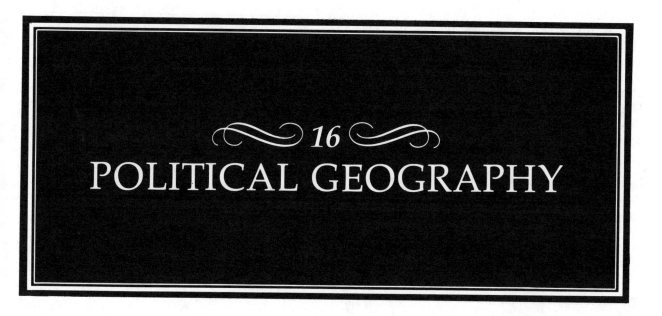

16
POLITICAL GEOGRAPHY

M AN IS A SOCIAL BEING. HE LIVES IN GROUPS, AND EVEN at the most simple stage, the groups exhibit characteristics of political organization. These characteristics may be tacit and informal but they exist nevertheless in the forms of internal organization and external interaction. Inevitably, there is an interplay between geography and politics; this field of investigation is known as political geography.

Political geography provides a basis for evaluating the realities of world politics. It furnishes a basis of understanding of politics for the individual citizen, for the citizen is affected by the political decisions of leaders. The citizen has a stake, therefore, in ensuring that policies pursued in his name are wise. Political geography provides a beginning for objective analysis of such policies.

THE NATION AND THE STATE

In political geography some everyday terms are assigned specific meanings. Three important terms are *state*, *nation*, and *nation-state*. A state is a sovereign, independent political unit. As such it must have *territory*, *population*, and *organization*, the so-called *elements of the state*. Thus, France is a state. Today there are more than 150 states in the world.

A nation, on the other hand, is made up of a group of people held together by a common emo-tional bond which, whatever its antecedents, gives them a feeling of being unique; so unique that they feel they must have their own particular state. A nation-state is an ideal situation in which a state and a nation coincide absolutely and completely. Thus, a nation-state would be made up of a territory inhabited by a population all of whom subscribe to the national ideas. The government would enjoy the unanimous support of all citizens, and no discontent would be found in the land. Obviously, such a happy circumstance is beyond reach in practice. The nation-state thus remains a goal towards which mankind constantly strives.

Territory

Size. A state must have territory over which it is sovereign. States vary enormously in area. It is necessary, however, that sovereignty, or control, be exercised over *some* area. Without territory, a state cannot exist.

States are found in a great variety of sizes. The largest, the Soviet Union, covers 22 million square kilometers (8.6 million square miles), an area almost as large as the entire continent of North America. Other large states include Canada, China, the United States, Brazil, and Australia, all of which exceed seven million square kilometers (2.9 million square miles) in extent. The traditional powers of western Europe are much smaller;

France, the largest, is similar in size to Texas, while West Germany is approximately the size of Colorado. A group of *micro-states* makes up the smallest members of the international community. Of these, Vatican City, with an area of 44 hectares (106 acres), is the smallest state. But in spite of this, Vatican City is a sovereign, independent, self-governing state.

Shape. States also differ in shape. It is desirable that a state be *compact* (Fig. 16.1). Thus, a "model" state would be circular in form with the *core area* located at the center of the circle. Such a configuration would make for ease of circulation and administration and also reduce disruptive stress due to regional variations.

FIGURE 16.1
COMPACT STATE.

CORE AREAS

The core area of a country is generally that portion of the country which is densely inhabited and contains a major concentration of the population. A core area usually contains the administrative capital, most of the large cities, a large part of the industrial infrastructure, the most productive farming region, and the densest transport network. The core area constitutes the nucleus, or heartland, around which the state has crystallized. A strong core area makes for a strong state, while any instability or inadequacy in the core area will tend to undermine cohesion and vitality. Competing core areas can be particularly troublesome in a state.

An *elongated* state can experience difficulties in terms of defense, administration, and national unity. As examples, we will examine Italy and Norway. Both are elongated, but differ in their human landscapes. Italy has two core areas. One core centers on Rome and represents Mezzogiorno, the impoverished southern half of Italy. Although the south is primarily a rural and agricultural region, farming is unproductive due to lack of moisture and thin soils. There is little industry and the area offers the classic pattern of underdevelopment and outmigration. A rival core area, representing modern Italy, centers on Milan in the north, an area of greater economic development than the south (Fig. 16.2). This second

core area is industrialized, prosperous, and more egalitarian. There is more rainfall in the north and farming is relatively more productive. Northern Italy is also cosmopolitan; it is linked to and enjoys much circulation with adjoining regions of western and central Europe. These two core areas, the north and the south, Milan and Rome, exhibit strong regional feelings mixed with a degree of mutual antagonism. Continual disagreements between the regions have induced some degree of instability into Italian politics. Italy has had more than 40 governments since World War II, and the existence of two rival core areas within this elongated country provides at least a partial explanation.

Norway is elongated, but has only one core area. This core is formed by Oslo and the surrounding lowlands. Aside from Oslo, the settlement pattern of Norway consists of a thin scattering of towns and villages along the coast. None of these agglomerations are of sufficient size or influence to act as a rival to Oslo in terms of national leadership. Thus, for Norway, elongation does not result in political stress.

A state can be *fragmented*. Indonesia, extending across thousands of islands, is an extreme example of fragmentation (Fig. 16.3). The Philippines, Japan, and Italy are also fragmented. The United States is fragmented inasmuch as Alaska and Hawaii are integral parts of the country. Fragmentation can induce political stress, although it is the

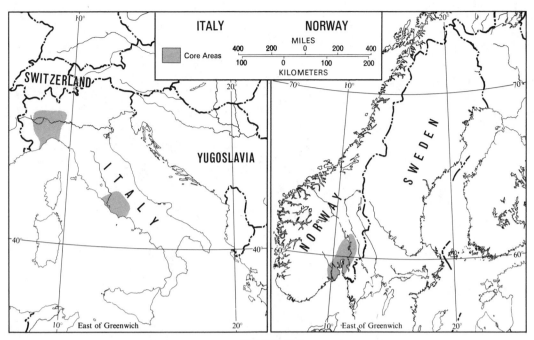

FIGURE 16.2
ELONGATED STATES.

human landscape rather than boundaries drawn on a map that is important. Alaska and Hawaii do not act as rival core areas and thus create no disruptive stress within the nation. By way of contrast, consider Pakistan. Pakistan was created in the late 1940s when self-government was achieved in the Indian subcontinent. Pakistan was to serve as a homeland for the Moslems. Its territories were laid out to include those regions where Moslems were dominant. These territories were established as West Pakistan and East Pakistan, separated by about 1400 kilometers (900 miles) of India (Fig. 16.4). The eastern sector, containing most of the people, is a low-lying tropical area. The bulk of the dense rural population in East Pakistan is made up of Bengali-speaking rice farmers. West Pakistan is a dry area where irrigated wheat is the dominant crop. The capital, Karachi, was in the Urdu-speaking western sector, and this area dominated the politics of the country. Ignoring their religious affinities, the East Pakistanis came to resent this situation, and to feel that they were being exploited by an alien group who were a different kind of people. This led to rebellion, warfare, and eventually, in 1971, to the establishment of the independent state of Bangladesh, formerly East Pakistan.

A state may be *prorupt.* A proruption is any section of national territory that juts out so as to form a narrow extension, or corridor. The usual reason for creating a proruption is to establish connection with navigable water (Fig. 16.5). Thus, the Leticia Corridor extends southward from the main body of Colombia to the Amazon River. The Caprivi Strip connected German Southwest Africa with the Zambezi River. The Polish Corridor provided access to the Baltic Sea. The Afghan Strip, on the other hand, was created so as to form a barrier between Imperial Russia and British India.

A state may be *perforated.* Italy is perforated in that Vatican City, an independent state, exists within and is completely surrounded by Italian territory. San Marino is also surrounded by Italy and thus constitutes another perforation, or *enclave,* within the Italian state. The city of West Berlin constitutes a perforation within East Germany. (Fig. 16.6).

Population. A state must have population. The populations of different states vary enormously. China is the largest country, with about 975 million people. Vatican City, the smallest, has only about 1000 residents. Other states range between these two extremes. India, with 676 million peo-

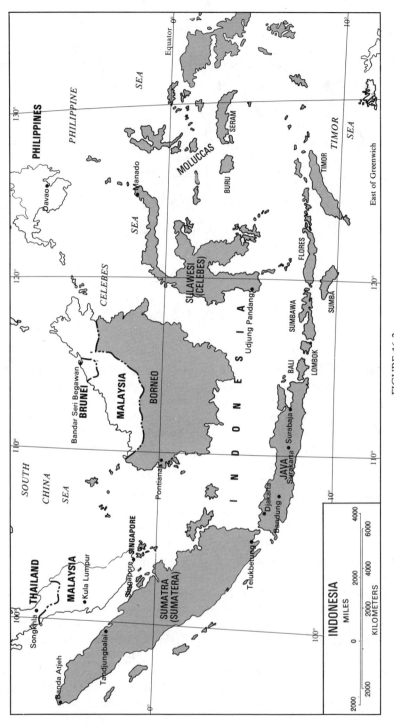

FIGURE 16.3
FRAGMENTED STATE.

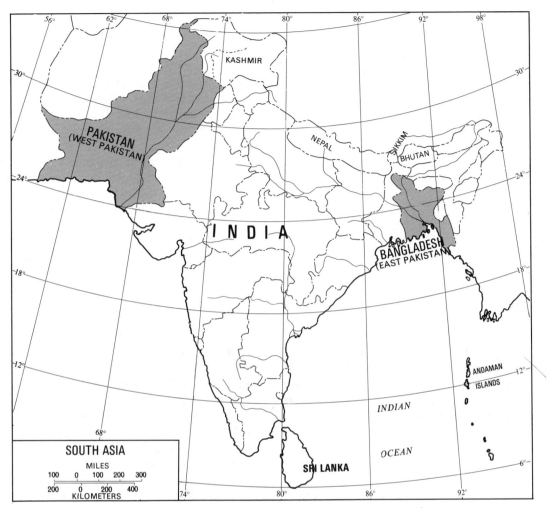

FIGURE 16.4
FRAGMENTED STATE (WEST PAKISTAN AND EAST PAKISTAN).

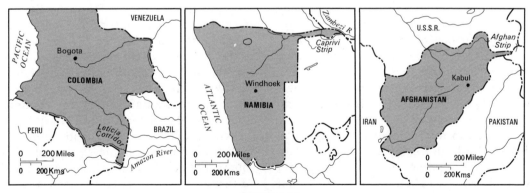

FIGURE 16.5
PRORUPT STATES.

THE NATION AND THE STATE
389

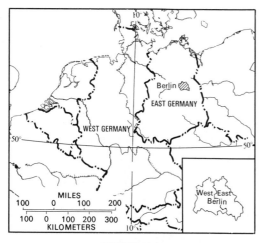

FIGURE 16.6
PERFORATED STATE.

ple, is second largest; the Soviet Union with 266 million people, is third; and the United States, with 223 million people, is fourth.

A large population may be considered a national resource in that it comprises a labor force that can work productively on farms and in factories. If necessary, a large population can also serve as a source of military manpower. However, too large a population can be counterproductive if it exceeds the carrying capacity of the resource base; high man/land ratios are one reason for low living standards in the Far East.

Populations can differ in a qualitative sense. They can differ in levels of health, an area in which marked contrasts exist between the developed and the less developed countries. In the developed countries, infants are born under the care of physicians in hospitals. Children receive proper food, medical check-ups, and dental care. They are inoculated against a host of diseases, are required to go to school, are educated to the limits of their interests and capacities. As adults, they have had every advantage in terms of training and health care to enable them to compete effectively in the world. At the same time, their energies add to the productive potential of their country.

Many children in the impoverished regions of the Third World do not enjoy these advantages. They do not receive proper medical care, proper food, or sufficient education. Accordingly, their vitality suffers as well as their flexibility and skills. They fail to reach their full potential and, since people make up the body of any state, the state suffers accordingly.

Organization. A state must be organized. Disorganized groups occupying an area do not constitute a state. Organization means some form of governmental structure. This government must be in more or less effective control in order for a state to exist.

There are several forms that governmental organization can assume. One method of viewing this question is to categorize states as *unitary* or *federal*. A unitary state is one in which power is centralized at the national level. A federation is a looser structure with more local autonomy. As an example, the United States is a federation, and according to the Constitution, all powers not specifically assigned to the federal government in Washington, D.C., are retained by the member states. Thus, for Americans, the daily pattern of events tends to be dominated by local law. People in Chicago, for instance, marry and divorce in accordance with the laws of Illinois. For most criminal offenses they are brought to justice in Illinois courts. Their children attend schools that are run by a local school board. France, on the other hand, is highly centralized. In France, marriage, divorce, criminal justice, and the schools are all dominated by national law. France is a unitary state.

Of the states in the world, all but several dozen are unitary. Because the federal system is more flexible at the local level, it lends itself particularly to large, diverse states where it can conveniently accommodate regional differences. On the other hand, the more rigid unitary structure is better suited to a small, compact, homogenous state. Most of the traditional European States are unitary, whereas a list of federal states would include Canada, the United States, the Soviet Union, Australia, Brazil, and Argentina, all of which are large.

Federal states can differ in organization because of the possibility of wide variance in how a given constitution is administered. For example, the Soviet Union has a constitution similar in many respects to the American constitution, and incorporates similar guarantees of civil liberty. The manner of administration, however, is completely different, and the Soviet citizen who takes his "Bill of Rights" too literally may find himself discussing his attitudes with the KGB.

Furthermore, federal states can differ in the degree of power retained at the local level. Brazil and the Soviet Union, although federal by constitution, are both centralized in a de facto sense—Brazil under a military regime, the Soviet Union under the Politburo. In Canada a large measure of power is held by the provincial governments. In fact, so much power is held by the Canadian provinces as to introduce instability into the national structure. Therefore, Canada might well be called a *confederation* where the power of the central government is severely limited and the member states are sovereign. This same issue, incidentally, underlay the disagreements leading up to the Civil War in the United States. Because the North won the war, the country is today a federation, and the Confederate States of America are part of history.

Another aspect of political organization has to do with authoritarianism versus democracy. This fundamental question deals with the competence of man to govern himself and with the proper relationship between the individual and his government. The authoritarian, or elitist, view is that the masses, for their own good, should be governed from above; the citizen exists to serve the state, and individualism is a luxury that society cannot afford. The democratic, or Jeffersonian, ideal, however, is egalitarian and places a high value on the individual. Governments exist to serve the people; if governments fail in this respect, the people have the right to vote them from office. Thus, in a democracy, power comes from below.

No final answer has been found to this continuing debate. What is certain, however, is that most people of the world live under authoritarian regimes, while democracy flourishes only in limited regions of the world. Authoritarian governments justify their existence on the grounds of supposed efficiency, or because their populations are "politically immature;" whatever the rationale, at least three-quarters of the population of the world are denied any effective voice in running their own affairs.

CENTRIFUGAL AND CENTRIPETAL FORCES

A state can be subject to disruptive forces which induce stress in the body politic and also to unifying forces which tend to bind it together. According to a nomenclature borrowed from physics, these forces have been designated as *centrifugal* (outward, destructive) and *centripetal* (inward, cohesive). This terminology finds common use in political geography today. In order for a state to survive, the centripetal forces must be at least as strong as the centrifugal forces.

Centrifugal forces can stem from ethnic variances or differences in language and religion. It should be emphasized, however, that cultural diversity does not always induce stress. In Belgium, for example, tension exists between the Waloons and the Flemish, who speak different languages. No such problem exists in Switzerland, however, which has four official languages. Similarly, religious differences pose a problem in Northern Ireland but not in the United States.

Profound centrifugal forces are at work within the troubled African nation of the Sudan (Fig. 16.7). The Sudan, Africa's largest country, sprawls across two distinct regions: the dry north, peopled by Arabic-speaking Moslems, and the humid forests of the south, inhabited by Christian and animistic Negroid tribes. Khartoum, the capital, lies in the north. Thus, the government is dominated

FIGURE 16.7
THE SUDAN.

by northerners, and the center of political and economic activity is concentrated there. This has led to economic disparities between the better developed north and poorer south. Moreover, the government has attempted to impose the Arabic language and the Islamic faith on the south. Political friction manifested by violence and bloodshed has existed between the two sections since the Sudan became an independent country in 1956. Attempts to maintain peace have resulted in some local autonomy; the southern section now has its own elected assembly to deal with matters of internal concern.

Although centrifugal forces are frequently easy to identify and understand, centripetal forces, inasmuch as they tend to be intangibles, can be harder to identify. What holds a state together? Is it some spiritual factor? Geographer Jean Gottman argued it was a "strong belief based on some religious creed, some social viewpoint, or some pattern of political memories, and often a combination of all three." He expressed the sum total of these intangibles in the term *iconography* when he stated:

> *Thus regionalism has what might be called an* iconography *as its foundation: Each community has found for itself or was given an icon, a symbol slightly different from those cherished by its neighbors. For centuries the icon was cared for, adorned with whatever riches and jewels the community could supply. In many cases such an amount of labor and capital was invested that what started as a belief, or as the cult of or even the memory of a military feat, grew into a considerable economic investment around which the interests of an economic region united.* *

This symbolism or imagery can manifest itself in various ways as, for instance, respect for the flag, belief in a heroic legend, or adherence to a philosophical view. As an example, consider the following:

> *We hold these truths self-evident, that all men are created equal, that they are endowed by their Creator with certain inalienable rights, that among these are Life, Liberty, and the pursuit of Happiness. . . . That to secure these rights, Governments are instituted among Men, deriving their just powers from the consent of the governed.*

This statement, incorporated in the Declaration of Independence, captures the spirit and essence of

America. It is an ideal to which Americans subscribe. As such it is an abstraction. It remains, nevertheless, an important part of the American "myth" and an adhesive that holds the nation together.

NATIONAL POWER

Power can be defined as the ability to influence events. Political geography deals with two kinds of power. *Internal power* concerns the ability of a government to influence events within its own territory. Internal power rests on the popularity of a regime and/or the ministrations of an internal security force. Internal power is nongeographical; a small, weak state can be a tight little dictatorship. *External power*, on the other hand, is our more conventional concept of the notion of power; it is a measure of a country's ability to influence events beyond its own borders. External power is geographical in nature and is an integral part of the elements of the state: territory, population, and organization. A powerful state, therefore, should have a territorial base that is large and well endowed with resources. The population should be numerous, healthy, intelligent, and skilled. This population should be organized efficiently, with the energies of the people devoted to goals which enjoy their wholehearted support. Any inadequacy in these elements will diminish a country's power base.

External power can be classified as economic power, military power, or power over opinion. These aspects of power are interrelated and difficult to separate. Power has a material side based on economic productivity. Thus, the farms, mines, and factories of a country provide the basis for gross national product, which can be accepted as a rough index of economic power. Economic power can be brought to bear in the international sphere through the purchase of strategic materials from abroad, by making these materials available to allies while denying them to enemies, by the achievement of *autarky* (economic self-sufficiency), and by the creation of armaments. War is the weapon of last resort in settling international disputes. Any evaluation of military strength requires caution because more is involved than just soldiers and equipment. The French, for instance, had more tanks and more troops on the Western

---

* Jean Gottman, *A Geography of Europe* (New York: Holt, Rhinehart and Winston, Inc., 1969), p. 76.

Front in 1940 than the Germans, but the Germans overwhelmed the French in less than a month. The reason may lie in morale, organization, training, and leadership. These factors, including belief in the cause, are of vital importance. The less tangible power over opinion may be used to persuade small dissident groups to support and cooperate with policies adopted by the state. This power attempts to make the population cohesive by appealing to the minority to join the majority in achieving the ends pursued by the state. Appeals are often made in broad terms, such as the call to "make the world safe for democracy."

Frontiers and Boundaries

In the geographic sense, a *frontier* is a disorganized zone that lies beyond the ecumene. A *boundary*, on the other hand, is a plane that separates one state from another. A boundary is a creation of the state and serves as a geographical point to which the authority of the state is exercised (Fig. 16.8).

The functions of a boundary are defensive and fiscal. The defensive function of a boundary is traditional. In earlier times, boundaries were often delimited with stone walls and fortifications. Static defenses, however, have been rendered obsolete by modern warfare, but a boundary still represents a trip-wire, the crossing of which by hostile forces signals a de facto state of war.

The fiscal function of a boundary has to do with control over the passage of goods and people. It is at the boundary that a host of officials stand ready to inspect cargoes, collect tariffs, and exercise passport control.

FIGURE 16.8

THE BERLIN WALL IS A HIGHLY VISIBLE DEMARCATED BOUNDARY. Boundaries act as points to which the authority of the state is exercised. Invariably signs serve instructive purposes.

THE NATION AND THE STATE

Three steps can be involved in the establishment of a boundary. These are, in sequence: *definition, delimitation,* and *demarcation.* First, a boundary is defined by means of a sentence or paragraph in a treaty. This written agreement may be specific and unequivocal (e.g. the boundary follows the 49th parallel), or it may be vague and subject to various interpretations (e.g. the boundary follows the crest of the Andes). The next possible step is delimitation, which refers to the precise interpretation of the language of the treaty for the purpose of establishing exactly where the

---

THE FRONTIER HYPOTHESIS

Historians Frederick Jackson Turner and Walter Prescott Webb have held that democracy is characteristic of those societies that have free access to an open frontier. The reasoning is as follows: In a frontier situation, the central authority is remote and governmental institutions are poorly developed. Citizens, therefore, cannot look to the government to solve their everyday problems. They are thrown, rather, on their own resources; they are forced to make their own decisions, to be individualistic. Individualism, in turn, requires loose governmental structures; that is, democracy. Futhermore, feedback from the frontier fosters democratization throughout the entire body politic.

There is some evidence to support this view. Democracy flourished in both Greece and Rome when the frontiers of the classical world lay open before them. During the Middle Ages, when Europe went on the defensive and, from its point of view, the frontier closed, democracy declined. And following the Age of Exploration, the countries that have been most democratic have been those that have been most actively involved with various frontiers, as for instance the United Kingdom, the United States, Canada, Australia, and France.

---

boundary should lie. And a third possible step, demarcation, refers to marking a boundary by some visible means, such as pillars or a fence. Few boundaries are demarked in anything more than a perfunctory sense. The Berlin Wall, on the other hand, is an extreme example of demarcation.

Boundaries are subjective. By this is meant that boundaries are established and placed as the result of a decision arrived at by participating groups. Any such decision, therefore, constitutes an evaluation of presumably relevant criteria and is therefore irretrievably subjective. There is no such thing as an objective boundary. It follows that the concept of a *natural boundary*, which enjoyed considerable vogue at one time, runs counter to logic. So-called natural boundaries, such as mountains, lakes, or streams are really *physiographic boundaries*. Thus, the boundary separating France and Germany follows the Rhine in part because a decision has been made to that effect. The boundary formed by the Rhine is properly designated as physiographic and to characterize it as natural is misleading (Fig. 16.9).

Mountains serve more effectively as boundaries than do rivers. Since the purpose of a boundary is to separate peoples, a crestline serves well in this sense, as steep uplands tend to be thinly populated and rugged terrain inhibits circulation. Rivers, on the other hand, attract settlement and they also serve as arteries of traffic. Thus, rivers bring people together. It follows that a river basin has a geographical unity that is disruped by a boundary running through it. Moreover, questions arise as to who should control traffic on the river. Finally, rivers shift their beds, leaving territory on one side or the other. All in all, rivers make bad boundaries.

Some boundaries can be classified as *geometric.* Most geometric boundaries are straight lines (a few are arcuate) and tend to conform to parallels or meridians, although a few are oblique (Fig. 16.10). Generally speaking, geometric boundaries are established in areas that are thinly populated and/or poorly mapped. Frequently they have been laid down by people with little or no knowledge of the area. Much of the Sahara, for example, was delimited by French statesmen in Paris who simply drew lines on a map in accordance with their own sense of convenience. Geometric boundaries are generally established before an area is occupied. (Fig. 16.11).

Boundaries can be classified according to their genetic sequence vis-a-vis the evolution of the cultural landscape. Thus, a boundary can be *antecedent,* meaning that it was established before the area was settled to any significant extent. Most

FIGURE 16.9

THE RIO GRANDE IS A PHYSIOGRAPHIC BOUNDARY THAT MARKS THE JURISDICTIONAL LIMITS OF THE UNITED STATES AND MEXICO.

boundaries in Antarctica can be accepted as antecedent. Antecedent boundaries tend to be geometric. Other boundaries can be laid down during, or after, settlement. These are *subsequent boundaries*; they usually separate cultural groups. Thus, subsequent boundaries tend to conform to the cultural landscape.

Another type of boundary is *superimposed*. A superimposed boundary is laid down after the settlement pattern has reached full development, but it ignores the cultural landscape. As it happens, a superimposed boundary is almost always established as a truce line at the end of a war and is maintained by one or more strong alien powers. The superimposed boundary that separates East Germany and West Germany was fixed in place by the Soviets and the western Allies. Similarly, the boundary separating North Korea and South Korea was established not so much by the local people as by the Chinese and the Americans.

A boundary that has been abandoned will leave behind evidence of its previous existence. Such evidence inscribed on the cultural landscape forms what is known as a *relict* boundary. It can find expression in the pattern of place names, in architectural differences, in ethnic zonations of language or religion, or in differences in the pattern of land use. Traveling through southern Louisiana, one enters a different cultural realm, evidence of early French occupation of the lower Missis-

FIGURE 16.10
GEOMETRIC BOUNDARIES.

sippi Valley. The entire area is part of the United States, but a discernable relict boundary still encircles much of Louisiana.

International boundaries continue to evolve in response to changes in the international political environment. The boundaries separating the member-states of the European Community, for instance, have lost a considerable part of their functional importance. Other boundaries in the Third World, however, have "hardened" in response to the enthusiastic nationalism of some of the newly emergent states. And the "iron curtain" that separates eastern and western Europe can be viewed as an ideological boundary as much as an international boundary.

OWNERSHIP OF THE OCEANS

Our planet earth is a blue planet—blue because approximately 70 percent of its surface is covered by the ocean. This water mass has been used by man for a variety of purposes, but until recently no nation could claim ownership of this area. Modern technology available for exploitation of ocean resources has now advanced to a stage when it is inevitable that nations will seek areas of the ocean for their own use.

In historical times the seas have been used principally by man for navigation and trade and for warfare and fishing. Salt has been extracted from seawater and seaweeds have been used for food and fertilizer. These uses barely scratched the surface of the ocean's resources, and from the time of the Greek city states the notion of free access to the seas was widespread among trading nations. This is not to say that this doctrine was not challenged. Medieval Venice claimed jurisdiction over the Adriatic, and Denmark and Norway claimed ownership of much of the North Atlantic in the fourteenth century. Finally, in 1493 the ownership of the oceans was divided by Pope

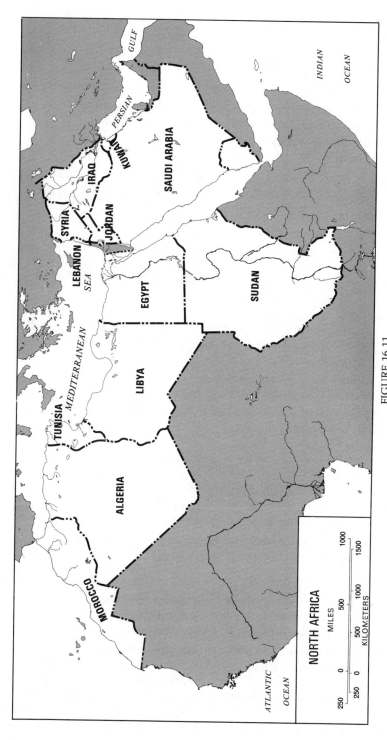

FIGURE 16.11
GEOMETRIC BOUNDARIES IN NORTH AFRICA.

Alexander VI between Spain and Portugal. The control these nations exercised over the seas depended upon their naval prowess, and this was contested by other nations, most notably by the Netherlands and Britain.

The doctrine of freedom of the seas was expressed again by Hugo Grotius in 1605 in a thesis entitled *Mare Liberum*, and this concept prevailed until 1945. Jurisdiction over the ocean by a nation was limited to its territorial sea, a narrow band of water that normally extended three nautical miles from the coast. The territorial sea was defensible by coastal batteries and effective control could be exercised over it. Commerce with overseas territories dictated that the use of the oceans in an orderly fashion was to the benefit of all.

In modern times, similar concerns—military security, freedom of navigation for merchant vessels, fishing, mineral resources, energy, scientific research, and the quality of the environment—have caused nations to reassess their attitudes towards the ocean. Commerce, particularly the transport of oil, has grown with the industrialization of nations to the extent that 90 percent of all international trade is carried by sea. Ocean trade quadrupled between 1950 and 1975 and is expected to quadruple again by the year 2000. Oil and gas reserves in the sea floor represent 40 percent of the estimated global petroleum resources (Fig. 16.12). Fisheries, too, have increased from a catch of 16 million tons in 1950 to 70 million tons in 1975 and, with careful management, could increase again before the end of this century. The deep ocean floor is important as a source of minerals. Manganese nodules, about the size of a potato and rich in manganese, cobalt, and nickel, are found in many areas of the deep ocean. The ocean itself contains about 165,000 tons of minerals in solution for each cubic mile of sea water, and has been described as the world's largest continuous body of ore.

In order to exploit resources within the ocean, there has to be some form of law or jurisdiction. Nations have already competed for control of fishery resources with gun boats; the "cod war" of the 1970s between Iceland and the United Kingdom is an example. Disputes over the oil-rich regions around the Paracel and Sprately islands in the South China Sea have led to confrontations between China and Vietnam. Similar political and legal problems persuaded the League of Nations

FIGURE 16.12

REPRESENTING NEARLY 40 PERCENT OF THE ESTIMATED GLOBAL PETROLEUM RESOURCES, OFFSHORE OIL AND GAS RESERVES ARE BEING EXPLOITED AT INCREASING RATES.

in 1930, and the United Nations in 1958, 1960, and 1972 to convene diplomatic conferences in an attempt to evolve a new law of the sea.

The third United Nations Conference on the Law of the Sea (UNCLOS), convened in 1972, is expected to complete its work in late 1980. Some 152 nations are taking part. Agreement has already been reached on several issues. UNCLOS delegates have agreed that the oceans "are the common heritage of mankind," and as such all countries are entitled to share in the benefits of ocean exploitation. It has been agreed that each coastal state shall have a territorial sea of 19 kilometers (12 miles) and an economic zone extending 300 kilometers (188 miles) beyond the territorial sea. Within this economic zone, states will manage the living resources (e.g., fish) and share any surplus with other states, giving preference to land-locked states in their region. Non-living resources (e.g., petroleum and minerals) in the economic zone will be the property of the coastal state, with the proviso that resource sharing with developing countries will be encouraged. Agreement has been reached on navigation rights, passage of vessels through straits, the regime for scintific marine research, and the principles for the transfer of technology to developing states. UN-CLOS has also agreed that the deep ocean floor will come under the jurisdiction of an International Sea Bed Authority, although the arrangements for mining manganese nodules and other minerals by states and/or the Authority have not yet been fully agreed upon. There is agreement, however, that a special international court will be created to hear disputes between nations, but no agreement has been reached on methods of drawing sea boundaries between states. Agreement has also to be reached on jurisdiction over the continental shelf and continental margin beyond the economic zone. The oceans, once free for all forms of activities regardless of nationality, are thus passing into national and international spheres of jurisdiction.

COLONIALISM

The drive to colonize would seem to be a trait of mankind. National energies spill over national boundaries, and colonies are acquired. Colonies thus constitute a new frontier into which surplus population and investment capital can be channelled.

The drive to colonize can manifest itself in a variety of ways. Nationalistic zeal can inspire men to go forth bearing the benefits of "law and justice" to those so unfortunate as to be without—dominating people for their own good, in other words. In South Africa, Cecil Rhodes worked from the basic premise that British institutions were superior and could not fail to be of benefit wherever implanted.

In terms of economic motivation, it is tempting to think that a colony can be developed as an exclusive source of raw materials and as a market, and that advantages will accrue from this arrangement. This may be the case at times, but conditions change and profit margins can be altered by changing circumstances. Consider the British experience of India. In the early nineteenth century, British trading companies encouraged the growth of cotton in India, shipped the cotton back to England for processing, and then exported cotton piece goods to India. The Indian market was protected and local competing manufacturers were discouraged. This was an ideal monopolistic sit-

uation, and British interests profited. The monopoly, however, was based on technology, and as the techniques spread, the British lost their advantage. Competing mills were built in India, and at the same time administrative costs, both military and civil, escalated. Gradually, from the viewpoint of the accountant, India became a burden on the British economy. In fact, it can be argued that Britain's prominence as a world power depended not so much on the exploitation of her empire as on her leadership as a manufacturing nation. The empire, in other words, was a luxury Britain could afford because of the power of her domestic economy. Britain would have been better off making purely commercial arrangements with overseas areas rather that by seeking to conquer and rule them. (This latter view is in keeping with American attitudes, which are basically anticolonial.)

Nations also acquire colonies for the purpose of military advantage. British bases in the home islands and at Gibraltar, Suez, Aden, Singapore, Cape Town, Australia, and the Falkland Islands enabled the British Navy to control traffic on the oceans. Thus, in time of crisis, the United Kingdom was able to draw upon worldwide sources of supply for defense of the homeland. At the same time, Britain was able to blockade the European continent, cutting off the flow of supplies on a line stretching from Gibraltar to Murmansk. Overseas bases, thus, played a vital role in the defense of the British homeland.

Germany, on the other hand, experienced no military benefits from its experiments in empire building. In the years leading up to World War I, Germany held bases in Africa and in the Pacific. When war broke out in 1914, Germany, not being a naval power, was quickly cut off from its overseas territories, which fell into the hands of the Allies. Because of lack of control over the means of accessiblity, the German colonies made no contribution to the war effort of the Fatherland. From the viewpoint of either economic or military considerations, the net contribution by any given colony can be positive or negative, depending on the circumstances of the situation.

Colonization attends the flowering of flourishing civilizations. In the Western world, several major phases have been observed. The first of these accompanied the rise of the Mediterranean powers of the Classical Period. Prominent colo-

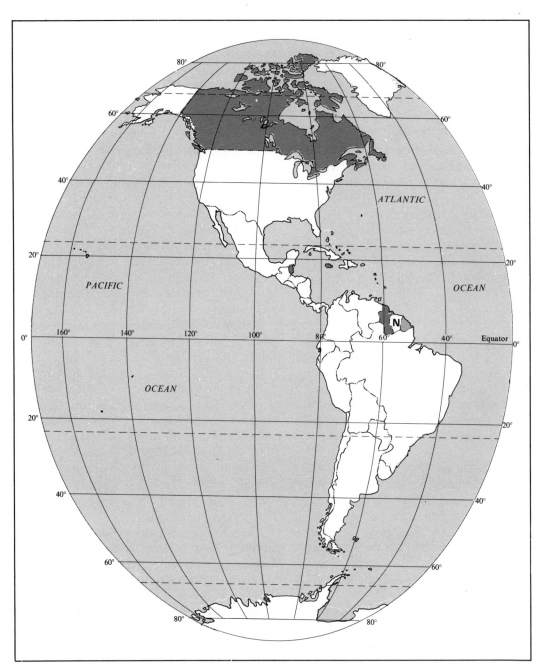

FIGURE 16.13
AREAS OF EUROPEAN COLONIALIZATION.

nizing powers of that time were the Phoenicians, the Greeks, and the Romans. After the collapse of Rome, dynamic leadership fell into the hands of the Islamic peoples, who became aggressive colonizers while Europe slept in the Middle Ages. The Renaissance and the Age of Exploration, however, inspired a resurgence of European civiliza-tion that embraced the world, establishing colonies in every corner of the globe. Beginning in the early sixteenth century, Europeans have at one time or another colonized and dominated North America, South America, Africa, Oceania, and most of Southern Asia (Fig. 16.13). This extraordinary outburst of energy spent itself as colonial

POLITICAL GEOGRAPHY
400

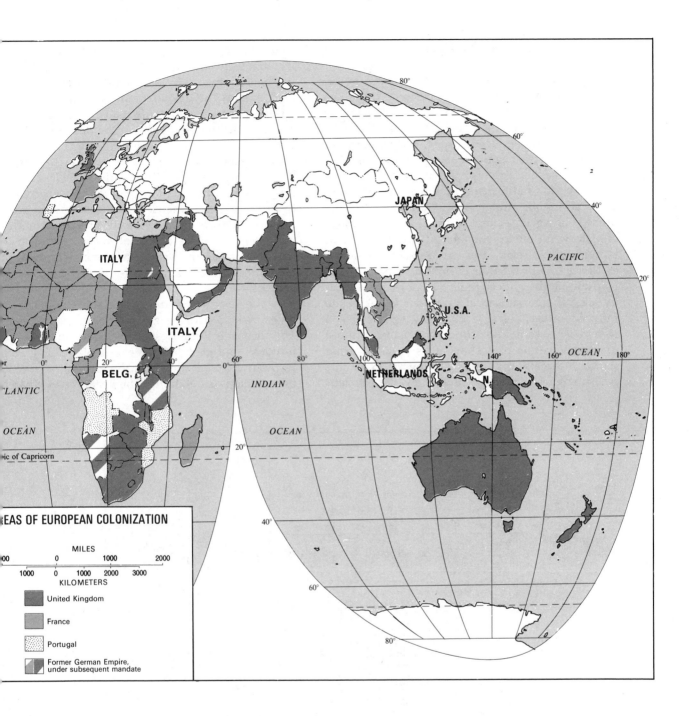

AREAS OF EUROPEAN COLONIZATION

MILES
00   0   1000   2000

1000   0   1000   2000   3000
KILOMETERS

United Kingdom

France

Portugal

Former German Empire,
under subsequent mandate

peoples throughout the world reacted against control by outsiders; in one colony after another, nationalistic sentiments have become irresistible. Sometimes by peaceful means, and sometimes with violence, colonials have thrown off the bonds of political subservience. Thus, America in 1776, Brazil in 1822, India in 1947, and Zaire in 1960 have stepped forth as free, independent, self-governing states. Man prefers to govern himself, and the four centuries of European dominance of world politics is at an end. Although the lines of political control have been sundered, a legacy remains. European culture has left its mark on the culture of societies across the globe.

SUPRANATIONALISM

The political evolution of the Western world can be divided into the following periods: the kingdoms of the ancient world, the city-states of Greece, Rome (both the Republic and the Empire), the fiefdoms of the Middle Ages, and modern nationalism which began to emerge with the Renaissance and the Age of Exploration.

The world is still in the nationalistic phase of political organization. We would seem to be poised, however, at one of the great watersheds of history, for the functional utility of nationalism is being questioned, and we are surrounded by evidence of a drift towards supranationalism. Let us examine the underlying cause of this situation by asking a fundamental question: What is the purpose of a state? It is generally agreed that a state has certain vital functions. Among these functions are the protection of its citizenry from military threats and the fostering of economic well-being. The modern state, however, is defi-

cient in both respects, because it cannot protect its citizens from nuclear attack. Political evolution has been outpaced by the technology of weapons. And, just as the development of gunpowder and artillery put an end to feudalism by rendering stone castles indefensible, so the development of the hydrogen bomb and the intercontinental balistic missile have cast a shadow over the traditional defensive capabilities of the modern nation. Similarly, modern nationalism is counterproductive in an economic sense because it works against the principles of the economies of scale. Nationalism tends to restrict size of market and also to inhibit circulation. Neither tendency is helpful in maximizing the efficient use of the world's resources.

In the twentieth century these problems have given rise to a drive towards supranationalism. This drive has manifested itself on the political, military, and economic levels. A pioneering step in political integration was the formation in 1919 of the League of Nations. The League was created

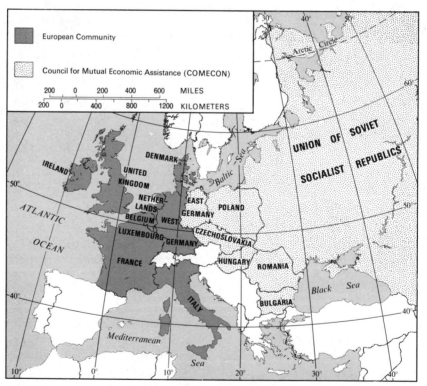

FIGURE 16.14
EUROPEAN COMMUNITY AND COMECON.

FIGURE 16.15
THE UNITED NATIONS, LOCATED IN NEW YORK CITY, IS THE WORLD'S SUPREME SUPRANATIONAL ORGANIZATION.

ern world has coalesced into the North Atlantic Treaty Organization (NATO), which finds its counterpart in the Soviet-bloc nations of eastern Europe who make up the Warsaw Pact. Other alliances unite various countries in a tangle of arrangements that shift from year to year. The United States, for example, maintains mutual security alliances with more than 40 countries. What is important in all this is that the individual nation no longer feels able to defend itself. For this reason, it seeks refuge in a structure of supranationalism.

In the economic sphere mankind has also found nationalism to be counterproductive. Economic integration therefore proceeded apace. Since World War II the European Community has come into existence and today comprises a thriving family of nations. In like manner, the birth of the Council for Mutual Economic Assistance (Comecon) signals the partial unification of the economies of the centrally-planned states of eastern Europe (Fig. 16.14). Similar arrangements are in force in Latin America, the Caribbean, and Africa. All these alliances denote a desire to escape the confines of nationalism and to improve efficiency by operating without hindrance in a supranational environment.

The supreme supranational structure remains the United Nations (Fig. 16.15). The UN follows in the footsteps of the League of Nations and has the same motive—the prevention of war. The United Nations is not perfect, the machinery contains defects. The concept of the United Nations should not be discarded without a sober evaluation of the consequences, which might well be suicidal for mankind.

At the political level, there is more international agreement today than a few decades ago. The desire for peace, agreement, and harmony are stronger than ever before. People, however, are beginning to realize that the ecological balance between mankind and the earth is tenuous and delicate and that the need for harmony in the environmental realm is as important as it is in the political realm. Thus, our final chapter examines the relationships between people and the earth and explores the issue of progressive environmental deterioration, and the evidence that contemporary mankind lives in increasing disharmony with nature.

as a forum to settle international disputes without violence. World War I had just ended, and a feeling existed that the human family could never again afford death and destruction on so vast a scale.

The League of Nations failed. The United States refused to join, and the League was fatally undermined in the 1930s by the militarists of Germany, Italy, and Japan. Nevertheless, the League of Nations was a milestone in political evolution in that a legislative body was permanently established to consider such problems as might arise in the international community.

World War II began in 1939. In the six years that followed unprecedented violence was unleashed across Europe and the Far East. In the closing days of the war the atomic bomb was used. Since then mankind, appalled by the destructive potential of modern weapons, has sought mechanisms to prevent war. Individual nations have sought security by grouping into defensive blocs. Thus, the West-

KEY TERMS AND CONCEPTS

nation

state

nation-state

elements of the state

micro-state

federal state

unitary state

confederation

centrifugal forces

centripetal forces

iconography

internal power

external power

frontier

boundary

League of Nations

United Nations

DISCUSSION QUESTIONS

1. Distinguish between a nation, a state, and a nation-state.
2. Distinguish between a unitary state, a federation, and a confederation. What are the advantages and disadvantages of each.
3. Discuss the notion of core areas. What role do they play in the development of centrifugal and centripetal forces?
4. "External power must of necessity have a geographical foundation." Do you agree with this statement? If so, why?
5. Of what significance is the shape of a state?
6. What motivations lead to the acquisition of colonies?
7. Describe the role of the United Nations in the world today.
8. What factors underlie the creation of trading blocs in the modern world?
9. Autarky means economic self-sufficiency. In what ways is autarky desirable and can it be achieved?
10. What major political problems lie before the world community today? How can they be resolved?

REFERENCES FOR FURTHER STUDY

Bergman, E. F., *Modern Political Geography*, Brown Company, Dubuque, Iowa, 1975.

Cohen, S. B., *Geography and Politics in a World Divided*, 2nd Ed., Oxford University Press, New York, 1973.

deBlij, H., *Systematic Political Geography*, 2nd Edition, John Wiley, New York, 1973.

Pounds, N. J. G., *Political Geography*, 2nd Ed., McGraw-Hill Book Co., New York, 1972.

Sprout, H., and M. Sprout, *Toward a Politics of the Planet Earth*, Van Nostrand, New York, 1971.

Ward, B., and R. Dubos, *Only One Earth*, Norton, New York, 1972.

17

MANKIND IN DISHARMONY WITH NATURE

AS THE TENDER OF EDEN, ADAM WAS GIVEN THE CHARGE to: "be fruitful and multiply, and replenish the Earth, and subdue it . . ." Indeed, the human family has achieved these tasks. No other organism in earth history has attained such dominion over the plant and animal kingdoms, and none have had mankind's capacity for altering the landscape.

Our influence upon the environments we inhabit has been both positive and negative. In some areas erosion has been drastically reduced; sterile soils have been made technologically fertile; and plant breeding has led to improved quality and higher yielding grains and forests. Fulfilling Isaiah's prophecy, even the "deserts have been made to blossom as the rose." We would be remiss, however, if only our achievements were lauded. Mankind's greatest cities serve as sources of atmospheric pollutants. We inject our streams with wastes and then decry their unsightly and unuseable qualities. And, out of ignorance, or perhaps indifference, sites hazardous to inhabitation are occupied. Mankind's negative impacts upon the environment represent the "growing pains" of an immature planet population. Just as infants and adolescents may demand rights without accepting responsibility, the peopling of earth has historically been associated with the pillaging of global resources and with minimum concern for maintaining the dynamic processes that offer resource viability. Yet, in recent years a new trend has emerged. Conservationists and environmentalists have been vocal about abuses of land, air, and water resources. As a result, public concern about environmental deterioration has been awakened and concerted action taken to reduce its threat.

ENVIRONMENT POLLUTION

The word pollution may mean different things to different people. It can be applied to a concentration of either noxious gases or particulate matter in the atmosphere, to the presence of pathogenic bacteria in drinking water, or to the nutrient enrichment of a stream or lake. However, the common ingredient in all polluted environments is the presence of undesirable matter. Because all of the earth's substances can be considered resources, a pollutant is, in reality, "a resource out of place." To cite a few examples: thermal pollution is nothing more than energy in the form of heat; an oil spill on the high seas represents the dispersion of valuable combustible fuels; sewage is rich with plant nutrients; and although mercury ingested by animals is dangerous to health, it is an important industrial element. Each of the foregoing substances has resource qualities, but in their displacement and concentration in alien environments they become pollutants.

Pollution is usually considered to be a phenomenon associated with humankind. Yet, under natural conditions erupting volcanoes spew ash particles into the air; lakes are slowly filled-in with sediments and organic matter; and streams are supplied with sediments that make their waters turbid. It is possible to argue that no environment, even one that is pristine, has pure water or air, and that it is unwise and environmentally disruptive to produce such conditions. There are tolerable levels of pollution within which we can all exist comfortably. However, associated with modern society are many new forms of pollution which with their accelerating intensities of environmental disruptions are capable of destroying the resources upon which mankind depends. These forms of pollution are our concern.

Atmospheric Pollution

We all recognize that air is vital to our existence. But, how important is air? As a rough estimate, humans can survive for five weeks without food and five days without water, but only for five minutes without air. On a daily basis, we require approximately one kilogram (2.2 pounds) of food and two kilograms (4.4 pounds) of water, but about 14 kilograms (30.8 pounds) of air. But the quality of air our respiratory system demands is just as important as the quantity. When heavily laden with foreign matter the atmosphere becomes an agent of destruction. The beauty of landscapes becomes obscured; property is damaged; and plants, animals, and people can be injured or killed.

Air pollution hides the beauties of nature. When air is dry and pure large objects, such as mountains, can be seen for more than 160 kilometers (100 miles). As dust, smoke, and industrial particulate matter is mixed with clear air, atmospheric transparency decreases and visibility is reduced. In effect, the sky is dirty and the mountain peaks are seen not nearly as magnificently as before. In many industrial regions average visibility is now commonly less than 10 miles. As described by meteorologist Louis J. Batten, "polluted air acts like a translucent screen pulled down by an unhappy God."

Of greater immediate importance is the impact of air pollution on the health of plants, animals, and people. Noxious gases can enter through the pores (stomata) of plants and damage their inter-

nal structure. Among those gases most toxic are sulfur dioxide, hydrogen flouride, and ozone. Others that can also be harmful when highly concentrated are chlorine, nitrogen dioxide, and hydrogen chloride. A classic example of the impact of toxic gas upon plantlife occurred near Ducktown, Tennessee, where two copper smelters with short smokestacks emitted sulfur dioxide. This chemical was carried downwind, killed plants, and chemically altered the soil to the extent that now, 60 years later, the area is still mostly barren (Fig. 17.1). Damage of this nature is visible evidence of contaminating gases. Much more difficult to assess is the effect of air pollution upon the productivity of plants that do not exhibit external deterioration. For example, what is the yield of fruit from a tree "breathing" contaminated air in comparison to one "breathing" clean air? Although this is a difficult value to determine, one estimate places the damage to plant life and growth in the United States in the hundreds of millions of dollars each year.*

People and animals suffer from air pollution in much the same ways. Their eyes and respiratory systems are most vulnerable to attacks by foreign substances in the air, and living in a dirty atmosphere causes them ailments that can result in death. Eyes are sensitive targets for atmospheric poisons. A common complaint about smog is eye irritation that causes tears to flow. Fortunately, however, this involuntary tearing permits the eyeball to be bathed with liquid. Without this relief the incidence of blindness would be much greater than it is today.

The respiratory system can be seriously affected by contaminated air. Inhaling floods the nose, throat, and lungs with oxygen as well as with undesirable gases and particulate matter. Exhaling removes a portion of the tiniest pollutants. But some of the intermediate sized particles can remain lodged in the lungs and may initiate an ugly chain of events that subtly destroys the efficiency of the respiratory system. Certain gases have the same effect. In some instances, emphysema, a deterioration of cellular material, is produced or aggravated by gases. In other cases, the lungs may be coated with particles that interfere with their main function of transporting oxygen from the

---

* C. Stafford Brandt, "Effects of Air Pollution on Plants," *Air Pollution*, New York: Academic Press, 1962, pp. 255–281.

FIGURE 17.1
LANDSCAPE DESTROYED BY SMELTING FUMES NEAR DUCKTOWN, TENNESSEE.

atmosphere to the bloodstream. When the normal flow of air in and out of the lungs is disrupted, the heart must pump more rapidly and consequently is strained. If the heart is already weak from other afflictions, the added strain may be too much. Here is where air pollution is most deadly: when it attacks those who have the least strength—the old and the very young. If it strikes one who is both aged and sick, the likelihood of survival is limited.

Most of the time the effects of air pollution on human health are slow and invisible. Particles slowly accumulate on the linings of the lungs like moss spreading over a forgotten wall. By the time the problem is discovered, the infection is normally so entrenched that it can never be completely cured. Other times its death sickle is swift and concentrations of poisons may accumulate above lethal levels quickly, as is demonstrated in the following case study of the London Smog of 1952:

The third of December 1952 was a delightful winter day in London, England. The weatherman reported that a cold front had passed in the night, and by noon the temperature reached 5.5° C (42° F). The relative humidity was about 70 percent. The wind blew pleasantly from the north, and the fluffy cumulus clouds for which the English climate . . . is famous dotted the sky. All in all, it was a beautiful day.

On December 4 a high pressure system moved from the northwest and was centered a few hundred miles to the west of London. The winds had turned slightly and were coming from the north-northwest, but they were slower than they had been. Several layers of clouds almost obscured the sky. A higher deck of cloud, at about 10,000 feet, could be glimpsed through breaks in the lower expanse of uniform, dark gray stratus clouds, that shut out the sun as well as the sky. At noon the temperature was (3.3° C) 38° F, and the relative humidity was a moist 82 percent.

The smell of smoke was in the air. From thousands of chim-

neys the unburnt remains of coal—the gases, the soot, and the specks of ash—floated silently into the atmosphere. Large particles fell to the rooftops, the streets, on hats and coats. The smaller smoke particles drifted with the air. When playing children ran in or out of houses, gusts of air carried the particles and gases indoors.

On December 5th, the high pressure center had moved almost over London. The winds were very light. Patches of fog reduced visibility, making it difficult to get about. The odor of smoke was becoming stronger. The winds were too weak to carry away the chimney's outpourings. In the lowest few thousand feet of the atmosphere both smoke and moisture accumulated. People began to complain to their neighbors. Cab drivers muttered about the fog.

The next day conditions worsened. Dense fog blotted out the sky entirely. The city was under the western end of the high pressure cell. Visibility was measured in tens of feet. All airplane flights were canceled, and only the most experienced driver ventured on the road with his automobile. Pedestrians groped their way along the pavements.

As the air hung virtually stagnant over the city, the smoking stoves, furnaces, and fireplaces fed it with poison. The fog droplets captured some of the smoke gases and particles. It was not a clean fog any more. No longer tiny droplets of clean water, it was composed of a mixture of smoke and fog, a mixture we call smog. The smog bathed the city in its own debris, attacking all living things. People felt it in their eyes. Tears streaked down faces. Every breath meant a lungful of polluted air. Wherever groups of people congregated, coughing could be heard.

But the weather in London on December 7 and 8 was no better. The smog was terrible. The old and sick, who just a few days earlier had been enjoying what was then a balmy breeze from the north, were suffering badly from the foul air, finding it hard to breathe. Even some of the young were barely enduring; the ones with respiratory diseases found it hard to get oxygen into their lungs. To asthmatics, the smog was torture. Patients crowded London hospitals, casualties of the smog. Many did not survive.

On December 9 there was a slight improvement in the weather. The fog was still present, but the wind was blowing slowly and fairly steadily from the south. Some clean air was mixing with the smog and diluting it. The next day a cold front passed over England. Brisk, west winds brought in air from the North Atlantic. Londoners, filling their lungs again with fresh, clean air, heaved a collective sigh of relief. In retrospect the five days seemed a nightmare.

In the period of smog about four thousand people died, directly or indirectly victims of its effects. Most of the dead had been weakened already by age or lung troubles, and the week of pollution had been too great an extra strain. Who knows how long they would have lived had not nature conspired to bring about five consecutive days of heavily polluted air?

In addition to the fatalities there were uncounted thousands whose illnesses were aggravated severely or who developed respiratory ailments for the first time. Finally, among the sufferers there were the families of the sick and dead, the survivors whose lives were changed by their losses. By any standard, certainly, it was a catastrophe, mass homicide by poison, with the weather an accessory before the fact.

The great London smog came about because moist, foggy air over the city stagnated while huge quantities of smoke were spewed into it. The atmosphere over London became a dump for the finely divided waste matter that rose from smokestacks and chimneys. The use of coal certainly was an important contributor to the problem, but it was not the sole cause.*

London is not unique in having air pollution problems. Almost every major city of the world has to contend with the effects of dirty air containing irritating substances. Yet, it is wrong to think of large urban complexes as being the only source of foul air. Isolated smelters and manufacturing plants in rural areas, mining and quarrying operations, and forest and grass fires all contribute to air pollution. Nonetheless, as depicted in Figure 17.2, the generators of energy and the exhausts of vehicular traffic are the dominant suppliers of atmospheric contaminants; and these agents are concentrated in urban areas.

Water Pollution

People are basically "water-beings." The average male human body is about 67 percent water by weight; the female body is approximately 52 percent water by weight. Yet relatively little water is needed to maintain our systems: about 2 kg (4.4 pounds) per day. Thus, it would seem that the world's population need not be concerned about a water shortage. This is not the case, however. As societies advance both culturally and technologically their demands for water increase. To enjoy the benefits of a flushing toilet, an automatic dishwasher, or even the paper you write upon, large quantities of water must be available. Large municipal systems within the United States currently supply water at an average daily rate of 600 liters (157 gallons) per capita. This amount meets domestic, commercial, and industrial needs. Thus, a population of 500,000 requires about 100 million

---

* Louis J. Battan, *The Unclean Sky* (Garden City: Doubleday, 1966), pp. 1–4. Reprinted by permission of the author.

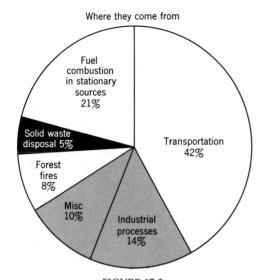

Where they come from

FIGURE 17.2
SOURCES OF ATMOSPHERIC CONTAMINANTS.

cubic meters of water annually (equivalent to 70 million gallons a day). Our nation has 35 municipal systems capable of supplying this amount or more. The largest municipal systems, in New York and Chicago, have demands that are 15 to 20 times this rate. Undoubtedly our per capita dependence upon water will increase in the years ahead, while the quantity of our water resources will remain unchanged.

In many ways the pollution problems associated with water are similar to those of the atmosphere. While both are essential to life, air and water may contain large stores of pollutants, cause health hazards, and reduce the beauty of the landscape. Also, like air pollution, water contamination in the United States has principally resulted from three ongoing processes: population growth, metropolitan expansion, and technological progress.

In those agrarian societies where population densities are low and people are widely dispersed, water pollution problems are few. Most wastes are capable of being broken down as part of nature's orderly recycling mechanisms. As people increase in numbers and crowd closely together, however, their waste products and those of their industries are concentrated in a relatively small area, and the threats of water pollution and hazards increase dramatically. In the United States, for example, more than two-thirds of the population live in urbanized areas that occupy about 7 percent of the

land. Although all waste pollution problems cannot be attributed solely to these densely settled areas, such areas are, nevertheless, major contributors to water contamination.

One of the basic factors contributing to water pollution is that very little of the water used in municipal systems is actually consumed. Large quantities of water are withdrawn from streams, used, and then quickly returned. This normally results in little loss in the stream's volume and negligibly affects its flow. The used water, however, brings to the receiving stream a wide variety of pollutants that reduce its quality, including:

1. *Biological contaminants.* These are living organisms; germs of all types, including those with the potential of causing outbreaks and epidemics of human diseases. This is a continuing public health concern. The source is mainly house-hold waste; more specifically, human sewage. With dilution and time, harmful germs ultimately die off in the stream. Chlorination of sewage effluents reduces the number of living organisms.

2. *Organic contaminants.* These are composed of unstable material that utilizes dissolved oxygen in the stream in the natural process of stabilization. The dissolved oxygen supports the type of bacteria needed to consume organic matter. Thus, streams have the ability to cleanse themselves of organic pollution, provided such loadings are not too heavy. Below points of such stream pollution the oxygen in the stream is used up, generally much faster than it is replaced. Within limits, and with time, the stream recovers from the shock of such pollution. On the other hand, if overloaded, the dissolved oxygen in the stream is depleted; fish die, the aquatic balance is upset, and the stream becomes literally a sewer and a cesspool. In a sense, it becomes useless water. The source of organic pollution is largely cities and industries.

3. *Inorganic pollution.* These are acids, alkalies, and salts from mining operations, oil fields, and a host of industrial and commercial pursuits. Such pollution is persistent, and where excessive, degrades stream usage.

A river carrying a load of municipal wastes will usually contain pesticides, fertilizers, and silt particles derived from farmlands through which it flows. In flushing these contaminants toward the sea, such a stream may pass by and meet the water needs of a half dozen or more municipal systems. Each of these systems adds contaminants to the stream and uses water that has the input of all upstream polluters.

Drinking water has traditionally been chlorinated and filtered to render it safe. Although chlorination does help:

> *. . . there is growing evidence that high content of organic matter in water can somehow protect viruses from the effects of chlorine. Infectious hepatitis is spreading at an alarming rate in the United States, and a major suspect for the route of transmission is the "toilet-to-mouth pipeline" of many water systems not made safe by chlorination. Indeed, the safety of purifying water with chlorine has been questioned by geneticists because certain chlorine compounds, which are sometimes formed by the chlorination process, can cause mutations that may lead to heriditary defects. But considering the high level of dangerous germs in many of our water supplies, we probably will have to accept any risks involved in chlorination.*
>
> *Water pollution from sewage provides one of the classic examples of diseconomies of scale accompanying population growth. If a few people per mile live along a large river, their sewage may be dumped directly into the river and natural purification will occur. But if the population increases, the waste-degrading ability of the river becomes overstrained, and either the sewage or the intake water must be treated if the river water is to be safe for drinking. Should the population along the river increase further, more and more elaborate and expensive treatment will be required to keep the water safe for human use and to maintain desirable fishes and shellfishes in the river. In general, the more people there are living in a watershed, the higher the per capita costs of avoiding water pollution will be.**

Radiation Pollution

Humankind has evolved to its present state within a radioactive environment. We are bombarded by cosmic rays from outer space, by emissions from radioactive elements within the earth's crust, and circulate—along with other organisms—radioactive substances such as potassium–40 through our bodies. *Radioactivity* is a naturally occurring process wherein atoms of certain substances undergo spontaneous nuclear disintegration and emit both high-speed particles and penetrating electromagnetic rays.

Natural sources of terrestrial radioactivity normally generate low levels of radiation. Within the last 50 years, however, major technological advances in the use of radioactive resources have been made. As a result, the number of people subject to intensive exposure levels of radiation has increased, a situation having serious implications for our planet's population. Medical analysis of genetic defects, stillbirths, and cancer attribute a large proportion of all three to the radiation environment in which we live. Any additional exposure to radioactivity will magnify health problems already in existence.

Radiation pollution exists when levels of radioactivity exceed the planetary norm. A major review of the effects of radiation pollution estimates that the average person in the United States is exposed to 182 millirems of radiation. Of this total, 102 millirems come from natural sources, 73 millirems from medical exposures, 4 millirems from global fallout, and a small fraction of 1 millirem from the generation of electricity by nuclear power.[*] The medical and dental professions are responsible for the bulk of man-made radiation exposure. According to the director of Health Physics for the Oak Ridge National Laboratory, excessive medical radiation exposures are possibly causing between 3,000 to 30,000 unwarranted deaths in the United States each year (Fig. 17.3).[†] Next in importance is radioactive fallout from nuclear weapons. Above-ground testing of nuclear weapons releases radioactive particles. Having long life-spans, some particles can be circulated by the atmosphere about the planet for numbers of years. From a global perspective this increase in radiation exposure is relatively minor, averaging only 1 to 2 percent of the total radiation environment. Such averages, however, hide more

---

* Paul R. Ehrlich, Anne H. Ehrlich, and John P. Holdren. *Human Ecology: Problems and Solutions*, San Francisco: W. H. Freeman and Co., 1973, p. 127.

* *Millirem* is one-thousandth of a *rem*, a measure of radiation absorbed by the entire human body. It is measured as the dosage of radioactivity absorbed by living tissue in amounts of 100 ergs of energy per gram of tissue. Data on millirem exposure is from: National Academy of Sciences National Research Council, Advisory Committee on the Biological Effects of Ionizing Radiation, *The Effects on Populations of Exposure to Low Levels of Radiation* (1972).

† Paul R. Ehrlich, Anne H. Ehrlich, and John P. Holdren. *Human Ecology: Problems and Solutions*, San Francisco: W. H. Freeman and Co., 1973, p. 141.

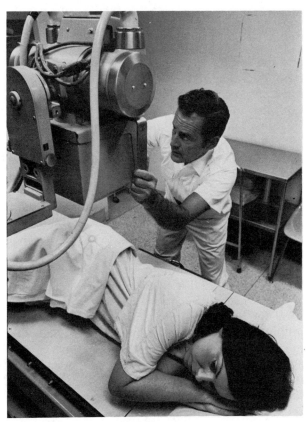

FIGURE 17.3
A MAJOR SOURCE OF MEDICAL RADIATION: THE X-RAY MACHINE.

than they reveal. The testing of nuclear weapons takes place at given sites. Thus, radioactive elements may be highly concentrated in certain areas, but appear relatively minor when sampled from random locations around the world. In addition, the long life-span of many radioactive elements means that with each above-ground nuclear test, more elements accumulate in the environment.

A third potential source of radiation pollution could eventually arise from the generation of electric power by nuclear reactors. This is possible in several ways:

"from mining and processing the fuel, from the operation of the power plant itself, from the transportation and reprocessing of spent fuel elements, and from storage of the long-lived radioactive wastes. In addition to these 'routine' processes, there is the possibility that an accident at a nuclear plant will release much larger quantities of radioactivity—

potentially as much as the fallout from hundreds of Hiroshima-sized fission bombs."*

It is not likely that such an accident would occur. Nevertheless, the potential for increased exposure to radioactive materials is a realistic concern.

Increased radiation exposure poses a serious threat to earth's organisms. The risks, however, must be weighed against the benefits. Hundreds of thousands of people are relieved of suffering, pain, and death because of X-ray diagnosis; the research leading to the development of nuclear weaponry has had technological ramifications aimed at "peace-oriented" utilization of atomic resources; and the generation of electric power by nuclear reactors, as we approach an age wherein fossil fuels are becoming scarce, has a basis of validity. No earth environment is pollution free; humans live comfortably within specific limits of pollution. Monitoring the level of pollution, including radiation, and living within limits that are crucial to the survival of organisms is a challenging task that faces this and future generations.

Noise Pollution

"Turn down that stereo, you're waking-up the dead!" This statement, or one like it, can be heard within a large number of households throughout the United States. It typifies the emergence of a new type of pollution, *noise pollution*. The problem of noise pollution recently became a public issue when it was discovered that many teenagers suffered permanent hearing loss because of long exposure to amplified rock music and when supersonic aircraft with their "sonic" booms began disturbing people and causing structural damage in buildings. These are not, however, the only forms of noise pollution. The roar of power lawnmowers, chain saws, motor boats, and the noise of a truck's exhaust system all reduce aural tranquility.

Noise is normally measured in decibels, a scale of sound intensity devised as follows: zero decibels represents the threshold of human hearing. A tenfold increase in sound intensity adds 10 units on the decibel scale, a 100-fold increase adds

* Testimony of Dr. Karl Z. Morgan, Director of Health Physics, Oak Ridge National Laboratory, in *Environmental Effects of Producing Electric Power*, Part 2, Vol. I., U.S. Government Printing Office, Washington, D.C., 1970, p. 1257.

20 (Table 17.1). Evidence indicates that intense noise of even short duration can cause a temporary loss in hearing. And it is believed that noise exceeding the 70-decibel range can cause irreversible damage to the nervous system, including permanent loss of hearing. Noise also leads to stress and probably to stress-related diseases such as hypertension and peptic ulcers. In industry, high levels of noise have been directly related to poor performance, high accident rates, and absenteeism.

Noise pollution is a growing menace to our society's health and well-being. Yet, it is a problem that is more readily solvable with technology than most pollution problems. In the United States, the Environmental Protection Agency has been assigned the task of identifying, monitoring, and establishing threshold limits for noise pollution. Although regulations have been effective in limiting population exposure to excessive noise by traffic, aircraft, and product design, a considerable burden of responsibility for noise reduction eventually rests with the average citizen.

ENVIRONMENTAL DETERIORATION

Environmental deterioration is defined as the physical and aesthetic degradation, or loss, of earth's cultural and natural resources. Although all forms of pollution contribute to environment deterioration, the term is reserved for other ways in which resources are either diminished in quality or have lost their utility. These include: urban

---

TABLE 17.1 NOISE LEVELS

Noise Level	Decibels
Threshold of hearing	0
Normal breathing	10
Whispering	30
Conversation	60
Food blender	80
Heavy automobile traffic	100
Jet aircraft taking off	120

---

blight, reduction of the land's productive capacity, and extinction of earth organisms.

The city has become the focus of human activity and the symbol of human achievement. Yet, it is here that deterioration of cultural resources is most obvious. *Urban blight,* the loss of quality of both life and landscapes in cities, is pronounced in slums and ghettos, and along neon lit, paper plastered thoroughfares (Fig. 17.4). The vista of ghetto-dwellers often consists of aged buildings suffering from lack of maintainance. Boarded-up windows, cracks in walls and ceilings, falling plaster, and poor ventilation and heating systems, are distinguishing features of their habitats. For slum inhabitants a concern about wildlife does not focus on our nation's fish and game animals; rather, their attention is directed toward "ghetto ecology," the wildlife in their homes—rats, mice, and cockroaches. Such environments are less than ideal for human beings, and provide a "life for millions in concrete and asphalt prisons filled with too many people and too much smog, trash, noise, and violence."* It has been reported that each day the average New Yorker breathes in the toxic equivalent of two packs of smoked cigarettes because the city's atmosphere contains thousands of tons of airborne pollutants. Evidence also indicates that traditional cultural patterns break down in cities and that urban crime rates are five times as high as in rural areas. The physical manifestations of urban blight can be largely countered with technology and more creative design of houses and neighborhoods. Unfortunately, elimination of urban deterioration requires a massive expenditure of capital, something that the citizens and governments of cities are lacking.

Land productivity has been reduced by humankind in many ways and in varied global environments. Lack of conservation practices have led to serious soil erosion, causing agricultural pursuits to be abandoned (Fig. 17.5). Over-grazing and the extension of grain cultivation into moisture deficient regions deteriorates biotic and edaphic resources, enhancing desertification. About 2000 years ago, western India was covered by vegetation. Poor cultivation practices, lumbering, and

---

* Robert Detweiler et al. *Environmental Decay in its Historical Context,* Glenview, Ill.: Scott, Foresman and Company, 1973, p. 87.

FIGURE 17.4
URBAN BLIGHT IN THE SOUTH BRONX.

overgrazing combined with natural climatic change are responsible for the replacement of original forests with the arid landscape of the Thar Desert.* Similar processes are operating in the Sahara where the desert is advancing southward along a broad front at the rate of several kilometers per year. Wholesale deforestation had destroyed ecosystems and led to disruption of watersheds, to the erosion or loss of soil fertility, and to siltation and flooding of stream valleys. Mining has resulted in scarred landscapes, acid tailings, and polluted water.

* M. Kassas. "Desertification versus potential for Recovery in Circum-Saharan Territories," *Arid Lands in Transition*; Washington, D.C.: *American Association for the Advancement of Science*, 1970; and B. R. Seshacher. "Problems of Environment in India," *Proceedings of Joint Colloquium on International Environmental Science*, Report 63-562, U.S. Government Printing Office, Washington, D.C., 1971.

In the course of occupying land and modifying environments, humankind has changed the structure of the biotic populations it encounters. Species of wild animals have been reduced in numbers or disappeared. One species that has vanished from North America is the passenger pigeon. Another is the great grizzly bear of California. Bison that once numbered in the millions suffered near extinction. The few surviving herds are found in carefully guarded parks and refuges. Among animals now listed as endangered species are the mountain caribou, marten, wolverine, fisher, various waterfowl, the whooping crane, trumpeter swan, wild turkey, and California condor. In addition to the problem of species extinction, the introduction of exotic (nonnative) wildlife into new environments has also had serious repercussions. Rats and mongooses brought into the United

FIGURE 17.5
CONSTRUCTION OF THIS SUBDIVISION, NEAR OMAHA, NEBRASKA, WITHOUT ADEQUATE CONTROLS PRESENTED THIS ERODED APPEARANCE.

States have threatened native birds and mammals; walking catfish and other exotics compete with or prey upon native fishes, and some 50 species of native freshwater snails are threatened or have been eliminated by snails introduced from abroad. Disease brought in by parrots and myna birds caused the loss of over 11 million chickens in California in 1973. And imported monkeys have infected humans with turberculosis and hepatitis. Other birds, snails, and primates may carry human pathogens, and small turtles bought in pet stores are estimated to cause 40,000 cases of salmonella poisoning per year. Bites from captive or escaped wild species—from lizards to lions—are an increasing problem. The Director of the U.S. Fish and Wildlife Service said recently that current information shows that injury caused by imported wildlife is more widespread and serious than previously believed.*

HAZARD ZONE OCCUPANCE

When people have excessive waste products their environment normally experiences degradation. In a sense, the problem of pollution is a

* Council on Environmental Quality. *Environmental Quality*, U.S. Government Printing Office, Washington, D.C., 1974.

universal phenomenon. Even very remote regions of the earth experience its impact, as is illustrated in Figure 17.6. Yet, there are other cases of mankind living in disharmony with nature that are more local in character. Associated with environmentally sensitive areas, these "hazard zones" are subject to geologic and atmospheric catastrophies that endanger the structures people build, and even human lives. Hazard zones are of varied types. Probably the most extensive are regions of active volcanoes and earthquake activity. Volcanoes have been known to bury cities and farmland under blankets of ash, and earthquakes have devastated cultural landscapes (Fig. 17.7). Japan's earthquake in 1923 killed more than 140,000 citizens and destroyed almost 600,000 buildings. Of all Japanese cities, Tokyo suffered the greatest physical damage. China probably has the longest historical record of earthquake activity. More than 10,000 Chinese quakes have been recorded in the last 3,000 years, 530 being classified as disasters. The worst quake occurred in the province of Kansu in 1556, when 800,000 people were killed.

Regardless of the number of times volcanoes bury the landscape or earthquakes destroy people and property, human tenacity is directed toward

FIGURE 17.6
POLLUTION IN ANZA-BAREGO STATE PARK, CALIFORNIA.

FIGURE 17.7
THIS VIEW SHOWS THE EXTENT OF THE DAMAGE CAUSED BY THE EARTHQUAKE IN THE RETAIL DISTRICT OF TOKYO THAT STRUCK THE CITY ON SEPTEMBER 1, 1923.

reinhabiting hazard zones. Today, Tokyo is rebuilt and more crowded than any other time in its history; the same may be said of Kansu Province. Mankind has similarly flaunted other catastrophic forces of nature by occupying shorelines subject to hurricane force winds, floodplains of rivers, and unstable slopes that are prone to mass movements. Because of "Hurricane Betsy," 150,000 people had to be evacuated from the delta area of Louisiana in 1965. The 1963 Vaiont Dam Flood in northern Italy resulted in the devastating loss of almost 3000 lives and inestimable damage. Landslides, mudflows, avalanches, and a host of other localized hazardous sites have all taken toll of people and property.

Why are hazard zones occupied? Sometimes they represent the only land available for settlement. At other times esthetic reasons, the sheer tenacity of people, ignorance of the inherent dangers, or economic pressures cause humankind to live where it does. In many ways the settlement of hazard zones can be compared to the problems of pollution. Both can be accommodated by their respective environments if caution is excercised and modifications of the environment do not exceed nature's limits.

EPILOGUE

Understanding the geography of an area requires a composite knowledge of the physical and cultural elements that provide it with character. This regional "personality" is built upon an evolution of forms and processes, some operative since the beginning of time. Cosmic forces, for example, fashioned the earth's internal structure some 5 billion years ago. Since then, physical processes acting upon and near the surface have provided our planet with a rich diversity of landforms, soils, and atmospheric and biotic components.

The genus *Homo* has had a brief history on the globe; a mere 2 million years. *Homo sapiens sapiens*, modern man, has had an even briefer historical record (about 50,000 years). Yet this species, alone of all the biotic world, has learned to manipulate planetary resources. Earth is the home of humankind and few areas have escaped the impact of our presence. People fashion new landscapes reflective of their cultural heritage. Cultures, or ways of life, have evolved in response to human needs. And human needs are, in part, related to the environments that people inhabit. Thus, the viability of cultural systems is intimately dependent upon systems operative within the physical environment. However, maintaining functional physical systems has, historically, been of secondary concern to mankind as it inhabited the globe. Presently, our planet's people face population and environmental crisis in areas where resources have been overtaxed and the environment misused. The challenge of the future, and the fate of humankind, largely rests with learning how to counter these abuses and to live in harmony with nature.

KEY TERMS AND CONCEPTS

pollution
atmospheric pollution
water pollution
biological
   contaminants
organic contaminants
inorganic pollution
radiation pollution
radioactivity

noise pollution
environmental
   deterioration
urban blight
reduction of land
   productivity
endangered species
hazard zone occupance

DISCUSSION QUESTIONS

1. Make a list of examples from your local area that illustrates "man in disharmony with nature."
2. What is pollution? Give examples of pollutants that could be considered usable resources.

3. What deleterious effects can atmospheric pollution have on the environment?
4. Discuss the three societal factors that have resulted in increased water pollution in the United States.
5. How do water pollution problems of an agrarian society of low population density compare with those of an industrial society with high population densities?
6. Describe the pollutants that reduce the quality of a stream used for municipal purposes.
7. Describe the pollutants that reduce the quality of a stream that drains agricultural land.
8. What significance does radiation pollution have for humankind
9. Identify forms of noise pollution to which you have been exposed. Suggest ways in which their harmful effects could be reduced.
10. Make a list of ways in which environmental deterioration has affected the area in which you live. To what can you attribute the cause of each case of environmental deterioration? What possible solutions exist for the problems you have identified?
11. What is a hazard zone? Why do people live in hazard zones?
12. What steps can be taken to reduce the danger associated with occupance of hazard zones?

REFERENCES FOR FURTHER STUDY

Battan, Louis J., *The Unclean Sky*, Doubleday, Garden City, N.Y., 1966.

Brodine, Virginia, *Air Pollution*, Harcourt Brace Jovanovich, New York, 1971.

Ehrlich, Paul R., A. H. Ehrlich, and John R. Holdren, *Human Ecology: Problems and Solutions*, W. H. Freeman and Co., San Francisco, 1973.

Hynes, H. B. N., *The Biology of Polluted Waters*, Liverpool University Press, Liverpool, 1963.

U.S. Department of the Interior. *Man, An Endangered Species*, Conservation Yearbook No. 4, U.S. Government Printing Office, Washington, D.C., 1968.

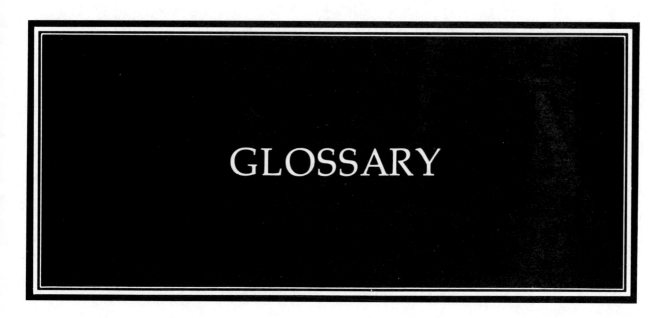

GLOSSARY

Absolute humidity The weight of moisture in the air per unit volume of air.

Accessibility Ease of access to a resource or location.

Acculturation The process of intercultural borrowing and adoption marked by the transmission of traits among culture groups.

Adiabatic heating and cooling Change of temperature in a gas due to compression or expansion and taking place without gain or loss of heat from the outside.

Adiabatic temperature lapse rate Rate at which air cools or heats when it is lifted or descends in altitude. If the air is unsaturated, the rate is $10C°/1,000m$ ($5.5F°/1,000$ ft). If the air is saturated, the rate is $6C°/1,000m$ ($3.2F°/1,000$ ft).

Advection fog Fog produced by condensation of a moist air layer moving over a cold land or water surface.

Agglomerating tendency A tendency to cluster.

Air mass Large body of air within which the vertical gradients of temperature and moisture are relatively uniform.

Alluvial fan Gently sloping fan-shaped accumulation of course alluvium formed below the point of emergence of a channel from a narrow canyon or gorge. Most commonly found in arid to semiarid environments.

Alluvium Stream-laid sediment deposit found in a stream channel.

Alto clouds Clouds of the middle height range, 2 to 6 km (6,500 to 20,000 ft).

Alumina Aluminum oxide, Al_2O_3, which occurs as an intermediate step in refining of bauxite into aluminum.

Anadromous Fish that spawn in fresh water, then make their way down-stream to the sea, where they spend most of their lives.

Anthracite Hard coal that contains only a small amount of volatile matter. Its almost smokeless flame and long-burning characteristics made it a popular household fuel.

Anticline Uparched folds in rock strata wherein there are downward dipping limbs.

Aquaculture Raising fish in captivity.

Aquifer Rock mass or layer of sediment that readily transmits and holds ground water.

Arête Sharp, knifelike divide or crest formed between two cirques by the erosive action of mountain glaciers.

Arithmetic density The number of people per square kilometer (or square mile) of an area; obtained by dividing the number of inhabitants by the area they occupy.

Artesian flow Spontaneous rise of water in a well or a fracture in the earth's crust that lifts above the level of the surrounding water table.

Arid Dry; lacking any significant amount of moisture.

Autarky Economic self-sufficiency.

Available soil water The difference between a soil's field capacity and wilting point water storage values.

Balance of payments The summation of all monetary transfers into or out of a country.

Balance of trade Exports minus imports.

Barrel (bbl) Liquid measure of oil, usually crude oil, and equal to 42 gallons, or about 306 pounds.

Bases (base cations) Certain cations present in the soil solution that are important plant nutrients; the most significant cations are: calcium, magnesium, potassium, and sodium.

Basic industry Economic activity, usually manufacturing or commercial, that depends for its market upon a region outside of the immediate area.

Batholith Massive body (perhaps part of a solidified magma chamber) of igneous rock that can extend over hundreds of square kilometers.

Bauxite The ore of aluminum.

Bed load of streams Heavy matter that is rolled or pushed along the stream channel bottom.

Beneficiation Any process applied to an ore for the purpose of enrichment, and used in particular to raise the metallic content of taconite.

Biota Life forms.

Birth rate Number of live births per 1,000 population in a given year.

Bituminous coal Soft coal; coal that is high in carbonaceous and volatile matter.

Blowouts (deflation hollows) Broad, shallow depressions formed by the action of wind removing loose surface particles.

Boreal forest Expansive needleleaf evergreen forest of the earth's subarctic regions.

Boundary Plane established by a state at the limit of its territory.

Broadleaf Leaf form that is wide in relation to its length, and is thin and comparatively large.

Calcic horizon Subsurface soil horizon of carbonate enrichment.

Capillary water Water that clings to solid surfaces by means of the force of capillary or surface tension.

Capital intensive Economic process in which capital costs comprise a relatively large item of expense.

Central Business District The main retail business region of a town or city.

Central place Cities or towns that provide goods and services for surrounding tributary areas. Central places attain an ''order'' according to the range of goods and services they offer.

Central place theory A theory designed by Water Christaller (1933) to explain the number, size, and distribution of market centers and urban areas on the basis of certain selected criteria.

Centrifugal forces In political geography, disruptive forces that create stress in a state.

Centripetal forces In political geography, binding forces that hold a state together.

Chemical weathering A chemical change that takes place in rock minerals through exposure to atmospheric conditions in the presence of water; a process of decomposition.

Cinder cone volcano Steep-sided conical-shaped volcano built of coarse ejecta.

Cirque An amphitheater, or bowl-shaped depression formed by mountain glaciers.

Cirrus clouds High clouds formed of ice and shaped into delicate white filaments, streaks, or narrow bands. They occur at above 6 km (20,000 ft).

Climate Generalized statement of weather conditions for a given location that accounts for time.

Climax vegetation Stable community of plants and animals reached at the end of a series of plant succession stages.

Climograph Graph upon which climatic variables are plotted.

Clouds Dense concentrations of suspended water or ice particles.

Coal Solid, combustible, organic material formed by the decomposition of plant material without free access to air. It is about 75 percent carbon.

Comfort energy Any form of energy whose end use is the heating and cooling of buildings and homes.

Comparative advantage Any advantage peculiar to a region, based on some characteristic of the environment, such as climate, soils, accessibility, or labor supply.

Concentric zone model One of several city structure models used in theories to explain patterns of urban land development.

Condensation A change in the state of matter from gas to liquid.

Cone of depression Area in which the water table is depressed, usually because of excess pumping of groundwater.

Confederation Form of political structure, looser than a federation, in which the central authority is severely limited, and a preponderence of power is retained by participating members.

Coniferous Cone bearing.

Continent Large body of land that stands above sea level.

Continuous process Process wherein iron ore is transformed directly into steel without interruptions.

Conurbation Contiguous urbanized area, including several neighboring, and formerly separate, towns.

Convection Process by which heat is transferred by moving matter.

Cottage industry Small scale manufacturing in or near the home.

Convective clouds Clouds formed from the lifting of warm air above heavier surrounding air.

Coriolis effect Effect of the earth's rotation tending to turn freely moving objects toward the right in the

Northern Hemisphere and to the left in the Southern Hemisphere.

Culture hearth Area in which a culture develops and from which culture traits spread.

Cultural landscape Natural landscape as transformed by mankind, acting through culture, to create a man-made landscape.

Cumuliform clouds Clouds of globular shape, often with extended vertical development.

Cumulonimbus clouds Large, dense cumuliform clouds yielding precipitation.

Death rate Number of deaths per 1,000 population in a given year.

Debris slide Small areas of detached, fast moving unconsolidated crustal debris and soil.

Deciduous Tree or shrub that sheds its leaves on a seasonal basis.

Deflation Process by which loose surface particles may be lifted or rolled along to be removed from an area by the action of wind.

Deflation hollows *See Blowouts.*

Delta Low-lying wedge of land formed of alluvium and projecting into the sea.

Demographic transition The historical shift of birth and death rates from high to low levels in a population. The decline of mortality usually precedes the decline in fertility, resulting in rapid population growth during the transition period.

Dendritic stream pattern Branching treelike arrangement of streams that converge into a single channel outlet that are formed on surfaces comprised of relatively homogeneous materials.

Density currents Deep ocean circulations that are governed by gravity and moved by density differences.

Dependency ratio The ratio of the economically dependent part of the population to the productive part; arbitarily defined as the ratio of the elderly (those 65 years and older) plus the young (those under 15 years of age) to the population in the "working ages" (those 15 to 64 years of age).

Desert pavement (reg) Desert surface blown free of fine mineral particles by the wind. Reg is a layer of coarse pebbles and gravel.

Dew point Temperature at which the air is fully saturated and below which condensation will occur.

Diastrophism Reorganization of solid rock materials by forces within the earth.

Diffusion Dispersal or spread of ideas and innovations.

Dike Platelike layer of igneous rock formed by intrusive igneous activity, often found in a near vertical position and typically cutting across the strata of older rock formations.

Dissolved stream load Mineral matter in a soluble state carried along by a stream.

Diurnal Daily.

Doldrums Belt of calm and variable winds located in the vicinity of the equator.

Drainage basin Total surface area occupied by a drainage system and bounded by a drainage divide.

Drainage system Branched network of stream channels and their adjacent land, bounded by a drainage divide. These stream channels converge to a single channel outlet.

Drizzle Form of precipitation comprised of water droplets less than 0.5 mm (0.02 in) in diameter.

Drumlin Lens-shaped hill of glacial till formed by plastering of till beneath moving, debris-laden glacial ice.

Dune Sandy deposit formed when rapidly moving air, laden with sand particles, encounters an obstruction that causes the wind to lose velocity and deposit its load.

Earthflow Moderately rapid downhill flow of masses of water-saturated soil or crustal debris.

Earthquake Trembling or shaking of the ground resulting from a shock wave associated with movements of the earth's crust.

Economic distance Distance figured on basis of a cost gradient.

Economic nationalism A feeling that a country should maintain control over its own resources and means of production.

Economic reach Distance a market can reach out to in order to attract goods or customers. Also known as "radius of pull".

Economics of scale The principle that industry is most efficiently organized in large units.

Ecumene Permanently inhabited portion of the earth.

Elements of the state Territory, population, and organization.

Endangered species Life forms facing extinction.

Energy A quantity having the dimensions of a force times a distance which is conserved in all interactions within a closed system. Energy exists in many forms and can be converted from one form to another. Common units are Calories, joules, BTUs, and kilowatt-hours.

Energy intensive An economic process in which energy costs comprise a relatively large item of expense.

Environmental perception The concept that people of different cultures perceive, interpret, and use similar natural environments differently.

Epeirogeny A term meaning "continent making". It refers to crustal movement—chiefly vertical—affecting large portions of a continent.

Ephemeral plants Small desert plants that complete a life cycle very rapidly following desert rainfall.

Epiphytes (air plants) Plants that live above the ground out of contact with the soil. They receive their nutrient needs from the atmosphere.

Equatorial trough Low-pressure trough centered more or less over the equator and lying between the two belts of trade winds.

Equinox Days on which the illumination period equals the darkness period, approximately March 21 and September 23.

Erg A sandy desert.

Esker A narrow, often sinuous ridge of coarse gravel and boulders deposited in the bed of a meltwater stream that was enclosed within a tunnel in a stagnant ice sheet.

Evaporation A change in the state of matter from liquid to gas.

Evapotranspiration The combined water loss to the atmosphere by evaporation from the soil and transpiration by plants.

Evergreen A tree or shrub that holds most of its green leaves throughout the year.

Export coefficient Exports of a country as a percentage of Gross National Product.

External power The ability of a government to influence events outside the boundaries of its own country.

Extratropical cyclone A wave cyclone tracking eastward along the polar front.

Extrusive vulcanism A form of vulcanism wherein fluid rock moves from beneath the surface to points above the surface.

Factory ship A relatively large "mother ship" from which smaller fishing vessels work. Many factory ships have facilities for canning or freezing the catch.

Fault A sharp break in rock with displacement of the block on one side with respect to the adjacent block.

Federation A form of political organization that features a considerable measure of local autonomy.

Field capacity The maximum capacity of soil to hold water against the pull of gravity.

Fiord A narrow, deep ocean embayment that occupies a glacial trough.

Fission The splitting of heavy nuclei into two lighter nuclei with the release of large amounts of energy and one or more neutrons. The emitted neutrons can be absorbed by and initiate fission in other, similar, nearby nuclei, thus producing a chain reaction.

Floodplain An area of low, flat ground that is present on one or both sides of a stream channel and subject to flooding about once annually.

Fog A cloud layer in contact with a land or sea surface.

Folds Wavelike patterns in rock, resulting from crustal compression.

Fossil fuels Any naturally occurring fuel such as coal, crude oil, or natural gas, formed from the fossil remains of organic materials.

Free trade A doctrine that artificial barriers to trade, such as tariffs, should be reduced or eliminated.

Friction of distance Resistance to circulation, generally applied to the movement of economic goods, and figured in terms of costs.

Front Surface of contact between two unlike air masses.

Frontal clouds Clouds formed when relatively warm and/or moist air is forced to rise over denser, cooler air masses.

Frontier A disorganized sparsely settled zone that separates states.

Functional (nodal) region An area organized on the basis of linkages and circulation.

Fusion The formation of a heavy nucleus from lighter nuclei, such as hydrogen isotopes, with an attendant release of energy. The mass of the new, heavier nucleus is less than the combined masses of the original nuclei; the lost mass appears as energy.

Geography A field of study concerned with the interrelationships that give character (or personality) to area.

Geostrophic wind Frictionless air in which pressure gradient force and the Coriolis effect are in balance. Winds tend to flow along the isobars.

Geothermal energy The heat energy in the earth's crust whose source is the earth's interior. When this energy occurs as steam, it can be used directly in steam turbines.

Geyser A fountain of heated groundwater and steam that normally spouts at irregular intervals.

Glacial moraine A till deposit formed at the terminus of a glacier. Types of moraines include: end, terminal, recessional, and ground moraine.

Glacial plucking A removal of masses of bedrock from beneath a glacier as the ice moves forward.

Glacial trough A deep, steep-sided valley of a U-shaped cross-section formed by mountain glacier erosion.

Glaciers A large natural accumulation of land ice that shows evidence of present or past flow.

Gravitational (free) soil water Surplus precipitation that percolates through the solum to lower levels under the influence of gravity.

The Green Revolution A term applied to efforts to increase food production in the Third World by introducing high-technology methods (high-yield plants, fertilizers, and pesticides).

Gross National Product The total value of goods and services produced in a country in a given year.

Ground (terrestrial) radiation Long-wave energy emitted from the earth's surface.

Groundwater Subsurface water that occupies a saturated zone of loose earth material beneath the crustal surface.

Growing season The length of time between the last frost of spring and the first frost of fall.

Growth rate The rate at which a population is increasing (or decreasing) in a given year due to natural increase and net migration, expressed as a percentage of the base population.

Hail A form of precipitation consisting of pellets or spheres of ice with a concentric layered structure.

Halophyte A plant tolerant of relatively high quantities of mineral salts in the soil.

Hanging valley A stream valley that has been truncated by erosion so as to appear stranded above the main valley into which it formerly flowed.

Hardwood A broad-leaved, generally deciduous tree, as distinguished from a coniferous tree.

Heavy industry The kind of industry that is frequently organized in large units, and requires the assembly of massive amounts of bulky materials, which are generally processed by use of large heavy equipment. *Example:* a steel mill.

Hinterland The area surrounding a town or city which has some association with the urban center.

Hominid Manlike in physical structure.

Homo The genus of mammals consisting of mankind, usually considered as belonging to the order Primates which contains also the monkeys, apes, and lemurs.

Homo erectus (upright man) The forerunner species of *Homo sapiens.*

Homo habilus The most ancient representative of the genus *Homo.*

Homo sapiens (wise man) The species of the genus *Homo* from which modern man evolved.

Homo sapiens sapiens Modern man.

Horse latitudes *see* Subtropical High Pressure System.

Humid climate A moist climate in which surplus precipitation occurs in at least one season of the year.

Humidity The amount of water vapor present in the air.

Humus The more or less stable fraction of soil organic matter that remains after the major portion of plant and animal residues has decomposed.

Hurricane *See Tropical Cyclone.*

Hydraulic action The excavation of unconsolidated materials (gravel, sand, silt, and clay) by the force of flowing water exerting an impact and drag effect upon the bed and banks of the stream channel.

Hydrologic cycle A model of moisture movement into and out of the atmosphere.

Hygroscopic water Soil water that is tightly bound to mineral matter within the soil. It is unavailable for plant use.

Iceberg A mass of ice floating in the ocean and derived from the terminal end of a glacier.

Iconography The spirit or traditions of a group tending to hold it together.

Igneous rocks Rocks that have solidified from a high-temperature molten state.

Inanimate power Power from inanimate sources such as fossil fuels, or water power, solar power, nuclear power.

Industrial Revolution The transformation of industry, pioneered in Great Britain about 1750, that accompanied the use of inanimate power and led to the centralization of manufacturing. A large increase in productivity was brought about.

Insolation The portion of solar radiation that the earth intercepts.

Interchangeable parts Standardized parts that permit mutual substitution in a machine.

Interfluve Upland area between streams.

Internal power The power of a government to control events within its own country.

Intrusive vulcanism A form of vulcanism wherein magma is forced into rock formations that lie beneath the surface of the earth.

Investment climate All the factors (physical, economic, political, social) of an environment that might affect an investment decision.

Ironstone Plinthite that has irreversibly hardened.

Island A small body of land that stands above sea level.

Isobar A line on a map connecting points of equal barometric pressure.

Isohyet A line on a map connecting points of equal precipitation.

Isotherm A line on a map connecting points of equal temperature.

Isothermal Equal temperature or no temperature change.

Joint A fracture in a body of rock wherein there is no displacement of the rock on either side of the fracture.

Jungle A type of tropical forest, consisting mainly of low, dense plants.

Labor intensive operation An economic process in

which labor costs comprise a relatively large item of expense.

Laccolith A lens-shaped body of igneous rock formed by intrusive igneous activity.

Lamprey A slender eellike fish widely distributed in both fresh and salt water. The lamprey has a jawless sucking mouth and is parasitic on other fish. It has caused serious depletion of valuable food fish in the Great Lakes.

Land breeze Local wind along coasts that blow from land to water during the night.

Landslide *See Rock slide.*

Land Use Theory A theory proposed by Johann von Thünen that attempts to explain the concentric pattern of land use that will tend to come into existence around a market place.

Latent heat Energy that is absorbed and held in storage when water is converted from a liquid to a gas, or from a solid into a gas.

Latitude Angular distance north or south of the equator.

Lava Magma that has emerged onto the earth's surface.

Lava flow plateau A volcanic tableland formed where fluid magma reaches the surface along extensive fractures in the crust.

League of Nations An international agency established in 1919 with headquarters in Geneva, Switzerland. Its purpose was to provide a mechanism for the peaceful settlement of international disputes.

Liana Thick, woody vines.

Light industry The kind of industry that is frequently organized in small units, and produces a small, light product. *Example:* manufacture of pharmaceuticals.

Lignite A brown low-grade coal of recent geological origin. It contains large amounts of volatile matter and its use is limited to heating and steam raising.

Lingua franca An auxiliary hybrid "trade" language used as a commercial tongue. Swahili and pidgin English are examples.

Location factor Anything that influences the location of an economic activity.

Loess Wind deposited sediments made up chiefly of silt.

Long wave radiation Energy of long wavelength form emitted from bodies of relatively low temperature.

Longitude Angular distance east or west of the Prime Meridian.

Magma chamber A source region at considerable depth below the surface of the earth from which fluid rock moves in the processes of vulcanism.

Mass production Production of goods in quantity, usually by machinery.

Mass wasting The spontaneous downward movement of soil and crustal debris under the influence of gravity.

Megalopolis A concentration of urbanized settlement formed by a coalescence of several metropolitan areas. Most often used to refer to the urbanized northeastern seaboard of the United States from Boston to Washington, D.C.

Mesopause Upper limit of the mesosphere, approximately 80 km (48 mi) above the earth's surface.

Mesosphere The outer layer of air in the lower atmosphere. It extends from 50 km (30 mi) to 80 km (48 mi) above the earth's surface. Temperature normally decreases as altitude increases within the mesosphere.

Metamorphic rock Rock that has been either physically or chemically altered in the solid state by the action of heat, pressure, shearing stress, or infusion of elements, all taking place at a depth substantially beneath the surface.

Micro-state A very small independent state, such as Vatican City.

Millibar A unit of atmospheric pressure equal to one thousandth of a bar. A bar is a force of one million dynes per square centimeter.

Monsoon winds Winds that seasonally reverse their direction of flow.

Mudflow A very fluid and rapid downhill flow of masses of saturated soil and crustal debris. Mudflows are common in arid and semiarid climates.

Multinational Corporation (MNC) A company, generally large, that is international in scope.

Multiple Nuclei Theory One of several city structure models used in theories to explain patterns of urban land development.

Nation A group of people held together by a common loyalty or emotional bond that causes them to feel unique.

Nation-state A state inhabited by a population all of whom subscribe completely to the national ideals.

Natural gas A naturally occurring mixture of hydrocarbons found in porous geologic formations under the earth's surface, often in association with petroleum. It is almost pure methane, CH_4, but also contains small amounts of various more complex hydrocarbons.

Natural increase The surplus (or deficit) of births over deaths in a population in a given time period.

Needleleaf A leaf form that is very narrow in relation to its length.

Noise pollution Excessive noise measured on the decibel scale, that affect the human nervous system.

Nonbasic industry An industry or service activity that is sustained from money generated within the area where it is found.

Nonecumene The uninhabited or temporarily inhabited area of the earth.

Non-tariff barriers to trade Any of a number of devices, such as quotas or inspection requirements, that impede the international traffic of goods.

Normal fault A fault, generally steeply inclined, along which the hanging-wall block has moved relatively downward.

Normal temperature lapse rate The long-term average decrease in temperature as altitude increases. It is 6.5C°/1,000m (3.5F°/1,000 ft).

Nuclear energy The energy released during reactions between atomic nuclei.

Oil shale A sedimentary rock containing a solid organic material called kerogen. When oil shale is heated at high temperatures, the oil is driven out and can be recovered.

Nutritional density The number of people per unit of arable land.

OPEC Organization of Petroleum Exporting Countries. An organization of countries that aims at developing common oil-marketing policies.

Opportunity costs Profits available from competing alternatives

Orogeny A major geologic episode in which land masses are deformed by folding and faulting, and resulting in mountain formation.

Orographic clouds Clouds formed when air is forced to ascend the slopes of a mountain range.

Outwash plain A flat, gently sloping plain formed by the deposition of sand and gravel by the meltwater streams in front of the margin of an ice sheet.

Overland flow Broad sheets of water that are eventually carried to a stream from interfluve areas.

Oxic horizon A subsurface soil feature of tropical soils that consists of a mixture of oxides of iron and/or aluminum, quartz sands, and clay.

Ozone Gas molecules made up of three oxygen atoms.

Ozone layer A region of the atmosphere, 15 km (9 mi) to 55 km (35 mi) in altitude, wherein ozone is concentrated.

Paddy A flooded field as used in rice farming in southern and eastern Asia.

Pelletization The process of treating an ore so as to form it into pellets. Used in particular in processing taconite.

Peneplain A land surface of slight relief and low elevation.

Photosynthesis A process whereby plants utilize solar energy to produce chemical energy in the form of carbohydrates.

Physical weathering The breakup of massive rock into small particles (a process of disintegration) through the action of physical forces acting at or near the earth's surface.

Physiological density The number of people per unit of arable land.

Pioneer industry The first implantation of modern industry in a traditional society.

Plane of the ecliptic A theoretical plane in space which coincides with the earth's orbit around the sun.

Plankton Small marine organisms that float freely in the water. Phytoplankton contain chlorophyll and can engage in photosynthesis. Zooplankton are animals.

Plant communities (plant associations) A group of plants that coexist in a fashion that appears to be mutually beneficial to one another.

Plant succession A predictable evolutionary sequence in the occupance of a site by plants.

Plate tectonic theory A means to explain the processes that move the earth's crust about and change the relative positions of land masses.

Plateau An upland area, more or less flat and horizontal.

Plinthite An iron-rich mixture of clay with quartz that commonly occurs as dark red mottles within the soil.

Polar easterlies System of easterly surface winds located at high latitudes.

Polar front Front lying between cold polar air masses and warm tropical air masses.

Polar high A persistent center of high atmospheric pressure over cold surfaces at high latitudes.

Polders A tract of low land reclaimed from the sea by the use of dikes and dams.

Pressure gradient force Change of atmospheric pressure along a horizontal line.

Prevailing westerlies Surface winds blowing from a generally westerly direction in the midlatitudes, but varying greatly in intensity and direction.

Primate A mammal with five digits bearing flat nails on feet, or hands adapted for grasping. Man and monkeys are primates.

Production line A line of machinery and/or workers arranged to turn out goods in quantity.

Productivity Output per man-hour.

Quota A quota specifies some limited amount of a good that can be exported or imported.

Radial stream patterns A pattern of stream channels that are formed where an elevated structure, such as a volcano or dome, exists. The streams assume a pattern similar to the spokes of a bicycle wheel, radiating out from the central most uplifted or highest elevated landscape position.

Radiation fog Fog produced by radiational cooling of the air in contact with the earth's surface.

Radiation pollution Occurs when levels of radioactivity exceed the planetary norm.

Radioactivity A naturally occurring process wherein atoms of certain substances undergo spontaneous nuclear disintegration and emit both high-speed particles and penetrating electromagnetic rays.

Rain A form of precipitation comprised of water droplets more than 0.5m (0.02 in) in diameter.

Rainshadow Area of dry climate to the lee of a mountain barrier, produced as a result of adiabatic warming of descending air.

Ramapithecus An extinct primate believed to be the evolutionary ancestor of mankind.

Reg *See Desert pavement.*

Region An area defined and delimited on the basis of criteria that provide it with homogeneity.

Regional method An analytic approach toward determining the character of area that requires an assumption that some part of the earth contains homogeneous landscape elements, and then proceeds with an examination of the area's component factors to explain the interrelationships providing for the region's identity.

Relative humidity Ratio of water vapor present in the air to the maximum quantity it could hold at the same temperature.

Reverse fault A fault, generally steeply inclined, along which the hanging-wall block has moved relatively upward.

Rock slide (landslide) The rapid sliding of large masses of bedrock from steep mountain slopes, from high cliffs, and from unstable slopes of stream banks and shorelines.

Runoff A term encompassing the physical movement of water, under the influence of gravity, toward lower elevations, and ultimately the sea.

Saltation A leaping, impacting, and rebounding motion of materials moved by the wind or water.

Sawah Paddy rice farming.

Sclerophyllous leaves Leaves of plants that are Sclerophylls. They are hard, thick, and leathery.

Sea breeze Local wind along coasts that blow from sea to land during the daylight hours.

Seasonal aridity A climatic characteristic of a region wherein a given season normally receives scant amounts of precipitation.

Sector theory One of several city structure models used in theories to explain patterns of urban land development.

Sedimentary rock Rock formed from the accumulated sediments derived from preexisting rocks.

Selva The tropical rainforest vegetation.

Semideciduous Plants that shed their leaves at intervals not in phase with a season.

Shield volcano A domelike accumulation of lava flows

Short wave radiation Energy of short wavelength form emitted from very hot radiating surfaces, such as the sun.

Sill An intrusive igneous rock in the form of a plate and created where magma was forced into a natural parting in the bedrock.

Site The immediate area of a specific place.

Situation The regional or relative position of a place with respect to other places or regions.

Sleet A form of precipitation consisting of ice pellets, which may be frozen raindrops.

Slurry A solid suspended in a liquid so it can be pumped through a pipeline. Examples are iron ore in water, or coal in oil.

Snow A form of precipitation consisting of ice particles.

Softwood A coniferous (cone-bearing) tree.

Soil A natural layer at the outermost part of the earth's crust containing living matter and supporting or capable of supporting plant life.

Soil creep The extremely slow downhill movement of soil and crustal debris as a result of continued agitation and disturbance of the surface particulate matter by activities such as frost action, temperature change, or the wetting and drying cycles of the soil.

Soil profile A vertical representation of the soil from the surface downward through the solum.

Soil reaction The degree of acidity or alkalinity of a soil, usually expressed as a pH value.

Soil structure The manner in which soil particles are clustered together to form aggregates or lumps.

Soil texture The proportion of mineral particles of given sizes that comprise the soil.

Solar constant 2 Calories/cm²/minute, an amount of energy reaching the earth's atmosphere when it is at an average distance from the sun.

Solar energy The electromagnetic radiation transmitted by the sun. The earth receives about 4,200 trillion kilowatt-hours of solar energy per day.

Solar system A celestial family of which the earth is a member. It is composed of a star (the sun), and nine major planets plus smaller bodies.

Solum (L. *solum*) Soil.

Specific humidity The weight of moisture in the air per unit weight of air.

Spodic horizon A subsurface soil horizon in which aluminum, iron, and organic matter have accumulated.

Stable air Air that resists lifting.

State A sovereign, independent political unit having territory, population, and organization.

Strata Layers.

Stratified drift A glacial deposit of materials sorted by glacial melt waters prior to their being deposited.

Stratiform clouds Clouds of layered, blanket like forms.

Strato clouds Cloud types of the low-height family, below 2 km (6,500 ft), formed into dense, dark gray layers.

Stratopause The upper reaches of the stratosphere, approximately 50 km (30 mi) from the earth's surface.

Stratosphere A strata of air that surrounds the troposphere and extends to an altitude of 50 km (30 mi). Its lower portion contains an isothermal layer wherein temperature remains constant with increased altitude. Its upper reaches exhibit a gradual warming trend as altitude increases.

Strato volcano A volcano constructed of alternate layers of lava flows and coarse blocky volcanic ejecta.

Stream A form of channelized flow of any size, including small brooks and large rivers.

Stream abrasion The grinding action of streams as they strike suspended rock fragments against their channel walls.

Stream corrosion The process by which soluble rock materials in stream valleys are dissolved by their streams.

Strike-slip fault A fault on which displacement has been horizontal.

Strip mining Surface, or open-pit, mining.

Sublimation A change in the state of matter from gas to solid or from solid to gas.

Subsistence farmer One who produces food only for himself and his family. Does not engage in market activities.

Subtropical high pressure systems (Horse latitudes) Areas of persistent high atmospheric pressure, trending east-west and centered about 30°N and S.

Summer solstice The date when the sun is highest in altitude. For the Northern Hemisphere, this date is about June 21.

Super-ship A large commodity carrier, generally accepted as exceeding 200,000 tons deadweight.

Suspended stream load Materials carried by a stream in floatation.

Sustained yield Permanent yield.

Swidden Slash-and-burn agriculture. A type of subsistence farming practiced primarily in the humid tropics.

Syncline A troughlike fold in rock strata wherein the limbs of the fold are oriented upward.

Systematic method An analytical approach toward determining the character of area that involves examining the areal variation of individual landscape elements (whether cultural, economic, or physical).

Taconite A hard flint-like rock that serves as an ore of iron. The metallic content generally runs about 22 percent.

Taiga *See Boreal forest.*

Tariff A tax imposed on goods entering or leaving a country.

Tar sand A sandy geologic deposit in which oil is found. The oil binds the sand together.

Temperature range Difference between the maximum and minimum temperatures.

Terrestrial radiation *See Ground radiation.*

Thermal gradient In earth or water, the rate at which temperature increases or decreases with depth below the surface.

Thermals Updrafts of heated air.

Thunderstorm An intense, local convectional storm yielding heavy precipitation along with lightning and thunder.

Tides Periodic fluctuations in the elevation of the sea's surface.

Till A glacial deposit of an unsorted mixture of clays, sand, cobbles, and boulders.

Tornado The smallest and most violent cyclonic storm having intense winds and extremely low air pressure.

Trade winds Surface winds of the low latitudes.

Traditional society A preindustrial society.

Transhumance The seasonal migration of livestock. Can be either vertical or horizontal.

Transition zone A general term used to describe an area whose characteristics represent a variety of gradiations from one functional area to another.

Transpiration The loss of water to the atmosphere from the leaf pores of plants.

Tree farm Where trees are raised as a crop.

Trellis stream pattern A system of stream channels formed in surface materials made up of parallel bands of rock that contrast in their resistance to erosion. The primary stream occupies a valley eroded into soft rock, while tributary streams flow down adjacent ridges, entering the primary channel at nearly right angles.

Tropical cyclone (hurricane) An intense traveling cyclone of tropical and subtropical latitudes, accompanied by high winds and heavy rainfall.

Tropopause Upper limit of the tropopause, averaging 20 km (12 mi) in altitude at the equator and 10 km (6 mi) at the poles.

Troposphere Layer of atmosphere in contact with the earth's surface, wherein temperatures normally decrease with increased altitude.

Truck farm A farm that specializes in growing vegetables for some urban market. The produce generally goes to market by truck. Also known as *market gardening*.

Uniform (formal) region An area characterized by one or more criteria, attributes that are uniform throughout the region's extent.

Unitary state A form of political organization in which power is concentrated in the hands of a central government.

United Nations An international agency established in 1945. Headquartered in New York, the main purpose of the U.N. is the prevention of war, but it undertakes many other good works through its agencies within the Secretariat.

Unstable air Air that spontaneously rises because it is warmer than surrounding air.

Upwelling The rising of water from deep layers of the ocean to the surface.

Urban blight The loss of quality of both life and landscapes in urban settings.

Vegetation A mosaic of plant assemblages in which the individual plant is the basic structural unit.

Ventifacts Wind-blasted cobbles that have faceted surfaces joined in sharp edges.

Volcanic ash Finely divided igneous rock blown under gas pressure from a volcano.

Vulcanism The redistribution of fluid rock.

Volcano A conical, circular structure built of accumulations of lava and volcanic ash.

Water table The upper limit of the saturated zone of groundwater. It has a three-dimensional surface with a shape resembling the topography within which it is found, yet of less relief.

Weather The physical state of the atmosphere at a given place and time.

Weathering The total of all processes acting at or near the earth's surface to cause physical disruption and chemical decomposition of rock.

Wilting point The quantity of water stored in the soil, at less than which the foliage of plants not adapted to dry conditions will wilt.

Wind The horizontal movement of air.

Wind shadow An area behind an obstacle wherein wind speed is reduced.

Winter solstice The date when the sun is lowest in altitude. In the Northern Hemisphere this date is approximately December 22.

Woodland A form of forest in which the trees are widely spaced, the canopy cover being only 25 to 60 percent.

Xerophyte A plant adapted to a dry environment.

PHOTO CREDITS

CHAPTER 1

Opener: NASA. Figure 1.5: NASA. Figure 1.10: Melvin H. Burke/U.S. Forest Service. Figure 1.12 (A) ASCS, USDA, (B) Fritz Henle/Photo Researchers. Figure 1.13 and Figure 1.14: Nebraska State Historical Society. Figure 1.15: USDA.

CHAPTER 2

Opener: NASA. Figure 2.37: Air Force photograph. Figure 2.48: ESSA photo.

CHAPTER 3

Opener: NOAA. Figure 3.11: Georg Gerster/Rapho-Photo Researchers.

CHAPTER 4

Opener: Carl Frank/Photo Researchers. Figure 4.1: (A) David W. Thornton/National Audubon Society-Photo Researchers (B) Leland J. Prater/U.S. Forest Service. Figure 4.4: M. Freeman/Bruce Coleman. Figure 4.6: Karl Weidman/National Audubon Society-Photo Researchers. Figure 4.8: G. R. Roberts. Figure 4.9: Lawrence Pringle/National Audubon Society-Photo Researchers. Figure 4.10: National Film Board Photothèque, photo by Judith Currelly. Figure 4.12: USDA. Figure 4.14: Toni Angermayer/Photo Researchers. Figure 4.15: Leland J. Prater/U.S. Forest Service. Figure 4.18: Jen and Des Bartlett/Photo Researchers.

CHAPTER 5

Opener: Grant Heilman. Figures 5.9 through 5.17: From Marbut Memorial Slides, Courtesy the Soil Science Society of America.

CHAPTER 6

Opener: Alberta Government Photograph, Department of Industry and Development. Figure 6.1: (A) U.S. Forest Service (B) Union Pacific Railroad photo (C) Swiss National Tourist Office (D) Santa Fe Railway, Public Relations Department. Photo by Frank Meitz. Figure 6.4: Hugo Boehme/Rapho-Photo Researchers. Figure 6.6: Georg Gerster/Rapho-Photo Researchers. Figure 6.13: PFC Donald Whitbeck/U.S. Army Photograph. Figure 6.23: J. R. Stacy/U.S. Geological Survey. Figure 6.24: Woody Higdon. Figure 6.28 © Tom Hollyman 1977/Photo Researchers. Figure 6.43: U.S. Geological Survey.

CHAPTER 7

Opener: © Keith Gunnar 1972/Photo Researchers. Figure 7.3: Leland J. Prater/U.S. Forest Service. Figure 7.5: The Swiss National Tourist Office. Figure 7.6: F. E. Matthes/U.S. Geological Survey. Figure 7.12: G. Blouin/National Film Board Photothèque. Figure 7.19: © Pierre Berger 1971/Photo Researchers. Figure 7.21: National Archives. Figure 7.23: Parker Hamilton/U.S. Geological Survey.

CHAPTER 8

Opener: © Ray Ellis/Photo Researchers. Figure 8.2: Charles C. Curtis/U.S. Navy Photograph. Figure 8.10: United Nations. Figure 8.11: WHO. Figure 8.12: Charles Rotkin/PFI. Figure 8.13: United Nations/FAO.

CHAPTER 9

Opener: All from United Nations with the exception of (top left) Norwegian National Travel Office and (bottom right) Hawaii Visitors Bureau. Figure 9.1: (top) Paul Almasy, (bottom) David Plowden/Photo Researchers. Figure 9.3: Marburg-Art Reference Bureau. Figure 9.4: Mexican National Tourist Council. Figure 9.5: American Museum of National History. Figure 9.8: Bernard Pierre Wolff/Photo Researchers. Figure 9.9: Andrew Rakoczy/Photo Researchers. Figure 9.10: Paul J. C. Friedlander. Figure 9.12: J. Allan Cash/Rapho-Photo Researchers. Figure 9.13: Courtesy Museum of Fine Arts, Boston. Frederick L. Jack Fund.

CHAPTER 10

Opener: Georg Gerster/Rapho-Photo Researchers. Figure 10.2: S. Trevor/Bruce Coleman. Figure 10.3: H. Hull/FAO. Figure 10.4: Paul Almasy. Figure 10.6: United Nations. Figure 10.10: Robert C. Bjork/USDA. Figure 10.12: Alberta Government Services. Figure 10.13: Joe Munroe/Photo Researchers. Figure 10.14: U.S. Forest Service.

CHAPTER 11

Opener: Don Green/Kennecott Copper Corporation. Figure 11.2: Ted Spiegel/Black Star. Figure 11.4: Joe Munroe/Photo Researchers. Figure 11.5: National Film Board, Canada. Figure 11.7: U.S. Forest Service/USDA. Figure 11.8: USDA-Soil Conservation Service.

CHAPTER 12

Opener: Georg Gerster/Rapho-Photo Researchers. Figure 12.11: E. E. Hertzog/Bureau of Reclamation. Figure 12.12: Bill Pierce/Contact Press. Figure 12.14: Paul Almasy.

CHAPTER 13

Opener: Paolo Koch/Rapho-Photo Researchers. Figure 13.2: Jon Wrice Schults/Photo Researchers. Figure 13.3: Michael C. Hayman/Photo Researchers. Figure 13.4: Courtesy Bethlehem Steel. Figure 13.5: Courtesy Burlington Industries, Inc. Figure 13.7: Georg Gerster/Rapho-Photo Researchers. Figure 13.9: German Information Center. Figure 13.12: Paolo Koch/Rapho-Photo Researchers.

CHAPTER 14

Opener: Courtesy Chevron/Standard Oil Company of California. Figure 14.4: Will McIntyre/Photo Researchers. Figure 14.5: Fritz Henle/Photo Researchers. Figure 14.6: Courtesy Association of American Railroads. Figure 14.9: H. Armstrong Roberts. Figure 14.11: Courtesy Cities Service Oil Company, Tulsa.

CHAPTER 15

Opener: Everett C. Johnson/de Wys. Figure 15.2: Frances Mortimer/Rapho-Photo Researchers. Figure 15.5: John Littlewood/United Nations. Figure 15.9: René Burri/Magnum. Figure 15.10: The Port of New York Authority. Figure 15.12: Josephus Daniels/Photo Researchers. Figure 15.13: David Plowden/Photo Researchers. Figure 15.17: Courtesy Houston Chamber of Commerce. Figure 15.18: Thomas Airviews. Figure 15.19: Joe Munroe/Photo Researchers. Figure 15.21: Joe Rychetnik/Photo Researchers. Figure 15.22: Courtesy Technicon Instruments Corporation.

CHAPTER 16

Opener: William Karei/Sygma. Figure 16.8: A. L. Goldman/Rapho-Photo Researchers. Figure 16.9: Grant Heilman. Figure 16.12: Courtesy Gulf Oil Corporation, photo by Lois M. Weissflog. Figure 16.15: United Nations.

CHAPTER 17

Opener: Ray Ellis/Rapho-Photo Researchers. Figure 17.1: Kenneth Murray/Nancy Palmer. Figure 17.3 and 17.4: Ed Lettau/Photo Researchers. Figure 17.5: USDA/Soil Conservation Service. Figure 17.6: Gene Daniels/EPA-Documerica. Figure 17.7: U.P.I.

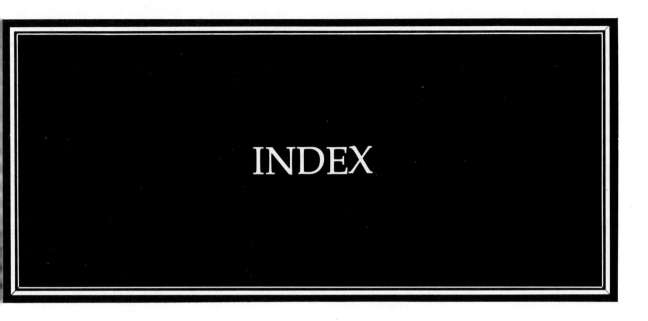

INDEX

Intrusive vulcanism, 143, 147, 423
Iron ore, 286-287
Ironstone, 132, 423
Islam, 236-239
Island, 5, 423
Isobar, 34-35, 423
Isohyet, 49, 423
Isotherm, 31, 423
Isothermal, 22, 31, 423

Jainism, 240
Joint, 149, 423
Judaism, 235-238
Jungle, 97, 423

Kaaba, 238
Karma, 240
Karst, 167-168
Koran, 238
Kshatriyas, 239

Labor, 310-311
Labor intensive operation, 308, 423-424
Laccolith, 147, 424
Lamprey, 276, 424
Land breeze, 39, 424
Landform, 139-193
Landslide, 155, 424
Land use theory, 266-267, 424
Language, 229-232
Latent heat, 25-26, 41, 424
Laterite, 132
Laterization, 132
Latitude, 29, 424
Lava, 144, 424
Lava flow plateau, 144, 145-146, 424
League of Nations, 402-403, 424
Liana, 97, 424
Life expectancy, 207
Light industry, 308, 424
Lignite, 295, 424
Lingua franca, 232, 424
Locational inertia, 313
Location factor, 308, 424
Loess, 190-192, 424
Longitude, 29, 424
Long-lots, 12
Long shore current, 185
Long wave radiation, 22, 424

Magma, 139, 142-143
Magma chamber, 143, 424
Mahayana, 241
Manufacturing, 307-329
Manufacturing belt, 366
Marine west coast climate, 70, 74-76
Market, 310
Market gardening, 246-247, 265-266
Mass production, 308, 424
Mass wasting, 154-156, 424
Meander, 160

Mediterranean climate, 70, 73-74
Megalopolis, 267, 381, 424
Mesopause, 22, 24, 424
Mesosphere, 22, 24, 424
Metamorphic rock, 139, 142-143, 424
Micro-state, 386, 424
Millibar, 35-37, 424
Milpa, 251
Mining, 282-287
Mixed farming, 246-247, 255-259
Mollisols, 128-129
Monotheistic religions, 223, 235-239
Monsoon wind, 39, 424
Mountain breeze, 39
Mudflow, 155, 424
Multilingualism, 230-232
Multinational corporation, 308, 424
Multiple nuclei theory, 374-375, 424

Nation, 385, 424
National power, 392-393
National-State, 385, 424
Natural boundary, 394
Natural gas, 293-294, 297-298, 424
Natural increase, 207, 424
Needleleaf, 92, 424
Needleleaf forest, 93, 100-103, 424
Nirvana, 241
Nodal region, 2
Noise pollution, 413-414, 424
Nonbasic industry, 367-371, 424
Nonecumene, 195-197, 425
Non-tariff barriers to trade, 337, 425
Normal fault, 149-150, 425
Normal temperature lapse rate, 22-23, 425
Nuclear energy, 295, 299-300, 425
Nutritional density, 199-200, 425

Occluded front, 52-53
Ocean currents, 179-181
Oceans, politics of, 396-399
Ocean waves, 183-184
Oil, 293-294, 295-297
Oil shale, 298, 425
OPEC, 425
Opportunity costs, 278, 425
Ore, 283
Orogeny, 154, 425
Orographic clouds, 44-45, 425
Orographic effect, 44
Outwash plain, 177-178, 425
Overland flow, 157, 425
Oxic horizon, 132, 425
Oxisols, 132
Ozone, 22, 24, 425
Ozone layer, 22, 24, 425

Paddy, 251-252, 425
Pangaea, 6-7
Parasite, 97
Pedalfer, 121

SOURCES

CHAPTER 1, Page 3, From *Our Environment: An Introduction to Physical Geography* by D. K. Fellows. Copyright 1980 by John Wiley and Sons, Inc. Reprinted by permission of the publisher. Page 7, From *Physical Geology*, 2nd ed., by R. F. Flint and B. J. Skinner. Copyright 1977 by John Wiley and Sons, Inc. Reprinted by permission of the publisher.

CHAPTER 2, Page 23, From *Our Environment: An Introduction to Physical Geography*, 2nd ed., by D. K. Fellows. Copyright 1980 by John Wiley and Sons, Inc. Reprinted by permission of the publisher. Page 25, From *Physical Geography*, 3rd ed., by M. P. McIntyre. Copyright 1980 by John Wiley and Sons, Inc. Reprinted by permission of the publisher. Page 29, From *Our Environment: An Introduction to Physical Geography* by D. K. Fellows. Copyright 1980 by John Wiley and Sons, Inc. Reprinted by permission of the publisher. Page 34, From *World Weather and Climate* by Riley and Spolton. Copyright 1974 by Cambridge University Press. Reprinted by permission of the publisher. Page 36, From *Earth, Space, and Time* by J. G. Navarra. Copyright 1980 by John Wiley and Sons, Inc. Reprinted by permission of the publisher.

CHAPTER 3, Page 77, From *Elements of Geography*, 5th ed., by G. T. Trewartha, et al. Copyright 1967 by McGraw-Hill Book Co. Used with permission of McGraw-Hill Book Co.

CHAPTER 4, Pages 93, 98, 104, 108, From *Vegetation of the Earth* by Heinrich Walter. Copyright 1973 by Springer-Verlag, Inc. Reprinted by permission of the publisher. Page 102, From *Man's Physical World* by Joseph E. Van Riper. Copyright 1962 by McGraw-Hill Book Co. Used with permission of McGraw-Hill Book Co.

CHAPTER 5, Page 126, From *Fundamentals of Soil Science* by Foth and Turk. Copyright 1972 by John Wiley and Sons, Inc. Reprinted by permission of the publisher. Page 135, From *Geology of Soils* by Charles B. Hunt. W. H. Freeman and Company. Copyright © 1972.

CHAPTER 6, Page 146, From *Physical Geology*, 2nd ed., by R. F. Flint and B. J. Skinner. Copyright 1977 by John Wiley and Sons, Inc. Reprinted by permission of the publisher. Page 151, From Leet, Judson, Kauffman, *Physical Geology*, 5th ed., © 1978, p. 51. Reprinted by permission of Prentice-Hall, Inc., Englewood Cliffs, New Jersey. Page 153, From *Physical Geology*, 2nd ed., by R. F. Flint and B. J. Skinner. Copyright 1977 by John Wiley and Sons, Inc. Reprinted by permission of the publisher. Page 150, From *Earth Science*, 2nd ed., by Richard J. Ordway © 1972 by Litton Educational Publishing, Inc. Reprinted by permission of D. Van Nostrand Company. Page 159, From Leet, Judson, Kauffman, *Physical Geology*, 5th ed., © 1978, p. 260. Reprinted by permission of Prentice-Hall, Inc., Englewood Cliffs, New Jersey. Page 162, From *Planet Earth* by A. N. Strahler. Copyright 1972 by Harper and Row, Inc. Reprinted by permission of the publisher.

CHAPTER 7, Page 172, From *The Earth: A Topical Geography* by Harm J. de Blij. Copyright 1980 by John Wiley and Sons, Inc. Reprinted by permission of the publisher. Page 184, From *Physical Geology*, 2nd ed., by R. F. Flint and B. J. Skinner. Copyright 1977 by John Wiley and Sons, Inc. Reprinted by permission of the publisher. Page 191, From *Physical Geology*, 2nd ed., by R. F. Flint and B. J. Skinner. Copyright 1977 by John Wiley and Sons, Inc. Reprinted by permission of the publisher.

CHAPTER 8, Pages 199, 200, 201, 360, From publications of Population Reference Bureau, Inc., 1754 N. Street, N.W., Washington, D.C. 20036. Reprinted courtesy of Population Reference Bureau, Inc.

CHAPTER 9, Pages 219, 220 From *The Study of Man* by Ralph Libton. Copyright 1936 by Prentice-Hall, Inc. Reprinted by permission of Prentice-Hall, Inc., Englewood Cliffs, New Jersey.

CHAPTER 10, Page 252, From *Plants, Food, and People* by M. J. Chrispeels and D. Sadava. Copyright 1977 by W. H. Freeman and Co. Reprinted by permission of the publisher.

CHAPTER 12, Page 291, From *Energy in the Perspective of Geography* by Nathaniel B. Guyol. Copyright 1971 by Prentice-Hall, Inc. Reprinted by permission of Prentice-Hall, Inc., Englewood Cliffs, New Jersey. Page 292, From *Energy Outlook, 1978 –1990* by Exxon Corp. Copyright 1978 by Exxon Corporation. Reproduced by permission of Exxon Corporation. From *World Energy*